ADVANCES

Computational
Intelligence

Theory & Applications

SERIES IN INTELLIGENT CONTROL AND INTELLIGENT AUTOMATION

Editor-in-Charge: Fei-Yue Wang
(*University of Arizona*)

Series in Intelligent Control and Intelligent Automation **Vol. 5**

ADVANCES

Computational Intelligence

Theory & Applications

Fei-Yue Wang

The University of Arizona, USA

Derong Liu

University of Illinois at Chicago, USA

 **World Scientific**

NEW JERSEY • LONDON • SINGAPORE • BEIJING • SHANGHAI • HONG KONG • TAIPEI • CHENNAI

Published by

World Scientific Publishing Co. Pte. Ltd.

5 Toh Tuck Link, Singapore 596224

USA office: 27 Warren Street, Suite 401-402, Hackensack, NJ 07601

UK office: 57 Shelton Street, Covent Garden, London WC2H 9HE

British Library Cataloguing-in-Publication Data
A catalogue record for this book is available from the British Library.

Series in Intelligent Control and Intelligent Automation — Vol. 5
ADVANCES IN COMPUTATIONAL INTELLIGENCE
Theory and Applications

ISBN-13 978-981-256-734-5
ISBN-10 981-256-734-8

Printed in Singapore

Preface

After a half century tumultuous history of research and development, *Artificial Intelligence*, the science and engineering of making intelligent machines, still bears the stigma of failed expectations mostly, and perhaps of forbidden knowledge as well. As a result, AI seems to be in a constant need and search of new images and new paradigms, as witnessed from disparaged expert systems in the past, to autonomous multiagents or distributed AI, cognition or cognitive informatics, and computational intelligence in the current studies.

Computational Intelligence was originally defined as the combination of the methods in fuzzy logic, neural networks, and genetic algorithms. Today, CI has emerged from *the fusion of the fields of granular computing, neuro-computing, and evolutionary computing*, and becomes one of the most active research areas and widely used techniques in information science and engineering. As it has been claimed that *AI is the way of the future*, we might say that *CI is the way of the future computing*. Why it is conceived as the way of the future computing can be precisely summarized by paraphrasing Donald Michie's AI definition: *CI is about making computing more fathomable and more under the control of human beings, not less*.

The name of computational intelligence indicates the link to and the difference with artificial intelligence. While some techniques within CI are often counted as AI techniques, there is a clear difference between those techniques and traditional or mainly logic-based AI techniques. In general, typical AI techniques are developed in a top-down fashion by imposing the structure and process of solutions from above, whereas CI techniques are normally implemented in a bottom-up approach through the emergence of order and structure from unstructured beginnings. Another significant contrast between AI and CI can easily be observed from their respective papers and textbooks. As pointed out by many AI researchers, AI papers and textbooks often discuss the big questions, such as *"how to reason with uncertainty," "how to reason efficiently,"* or *"how to improve performance through learning.*" It is more difficult, however, to find descriptions of concrete problems or challenges that are still ambitious and interesting, yet not so open-ended. However, with papers and textbooks in CI, concrete problems and specific algorithms are normally the center and focus of their discussions.

Historically, the areas of interest covered by computational intelligence are also known under the term *soft computing*. Many consider the name of soft computing was chosen to indicate its difference from operations research, also known as hard computing. The two areas are connected by the problem domains they are applied in, but while operations research algorithms usually come with crisp and strict conditions on the scope of applicability and proven guarantees for a solution under certain optimality criteria, soft computing puts no conditions on the problem but also provides no guarantees for success, a deficiency which is compensated by the robustness of the methods. Clearly, the link to soft computing, especially the nature of "soft" characteristics, is still one of the most significant features exhibited by CI techniques.

Yet there is no consensus on what CI exactly is at this point. Attempting to define *Computational Intelligence* is still a challenge that, even after more than a decade of research and development, has not been addressed to everyone's satisfaction. For example, a more broad definition of *Computational Intelligence* is the study of adaptive mechanisms to enable or facilitate intelligent behaviors in complex and changing environments. In this book we have taken a liberal and literal view of *Computational Intelligence:* that is, t*he computational part of the Artificial Intelligence.* As a matter of fact, in his classical introduction to Artificial Intelligence, "*What is Artificial Intelligence?*" John McCarthy, who coined the very name of Artificial Intelligence in 1956, responds to the basic question of "*what is intelligence?*" with "*intelligence is the computational part of the ability to achieve goals in the world.*" In this sense, *AI* and *CI* should be tied more closely and our view of computational intelligence is justified.

This book is edited from the contributions of world-renowned experts, and some of them are pioneers, in the field of computational intelligence. The articles can roughly be grouped into the two categories: theoretical developments and practical applications. The first group deals with issues in granular computing, learning methods, evolutionary optimization, swarm intelligence, linguistic dynamic systems, neurocomputing, and neuro-fuzzy computing; while the second group concerns with applications in motif identification in DNA and protein sequences, adaptive approximation for nonlinear functions, neuromuscular control, reverse engineering of protein and gene networks, biometric problems, and intelligent control.

Granular computing is one of the key components in the current study of computational intelligence. In Chapter 1, Pedrycz investigates the role of granular computing and logic processing in the context of intelligent systems and demonstrates that both of them are organized into a single conceptual and computational framework. This chapter provides a general overview of granular computing that emphasizes a diversity of the currently available concepts and underlines their common features that make the entire pursuit highly coherent. The logic facet of processing is cast in the realm of fuzzy logic and fuzzy sets that construct a consistent processing background necessary for operating on information granules. The synergetic links between granular and logic as well as several main categories of logic processing units or logic neurons have been examined in order to show how they contribute to high functional transparency of granular processing, help capture prior domain knowledge and give rise to a diversity of the resulting models. This work represents a significant step towards the establishment of a framework that uses granular computing as a fundamental environment supporting the development of intelligent systems. Chapter 2 addresses another important issue in granular computing, that is, the abstraction of conventional dynamic systems for the purpose of conducting linguistic analysis based on numerical representation. To this end, the concepts and methods developed in linguistic dynamic systems (LDS) by Wang are utilized. Specifically, conventional dynamic systems are converted to different types of LDS for the purpose of verification and comparison. The evolving laws of a type-I LDS are constructed by applying the fuzzy extension principle to those of its conventional counterpart with linguistic states. In addition to linguistic states, the evolving laws of type-II LDS are modeled by a finite number of linguistic decision rules. Analysis of fixed points is conducted based on point-to-fuzzy-set mappings and linguistic controllers are designed for goals specified in words for type-II LDS. An efficient numerical procedure called α-cuts mapping is developed and applied for simulation studies.

Ever-increasing dataset sizes and ever-larger problems have spurred research into efficient learning methods. Eschrich and Hall present in Chapter 3 a new learning algorithm along this direction, called *slicing*, that embodies the principles of distributed learning, not only to simplify a learning problem overall but also to simplify the individual learning tasks. Slicing, or partitioning and learning, can be seen as a method in which the training set is partitioned into disjoint regions of feature space and treated as a set of learning tasks that can be run in a distributed environment without extensive communication. This chapter examines slicing algorithm with respect to a series of real-world datasets, including a biologically-motivated problem, and shows that it can be used as a general meta-learning technique for distributed learning. Clearly, slicing is more accurate than using a single classifier, can reduce the individual learning task size and provides a mechanism to distribute the data mining task. Chapter 4 discusses margin methods for both supervised and unsupervised learning problems. For supervised learning, a complete framework of posterior probability support vector machines (PPSVMs) is proposed for weighted training samples using modified concepts of risks, linear separability, margin and optimal hyperplane. Within this framework, a new optimization problem for unbalanced classification problems is formulated and a new concept of support vectors is established. Tao and Wang extend the margin idea further to unsupervised learning problems and establish a universal framework for one-class, clustering and PCA (principal component analysis) problems.

Optimization using evolutionary computing is a very important field of studies. In Chapter 5, Yen describes a generic, two-phase framework for solving constrained optimization problems using genetic algorithms. In the first phase of the algorithm, the objective function is completely disregarded and the constrained optimization problem is treated as a constraint satisfaction problem. The genetic search is directed toward minimizing the constraint violation of the solutions and eventually finding a feasible solution. In the second phase the simultaneous optimization of the objective function and the satisfaction of the constraints are treated as a bi-objective optimization problem. The proposed algorithm is analyzed under different problem scenarios using Test Case Generator-2 and its capability to perform well independent of various problem characteristics is demonstrated. Note that Yen's algorithm performs competitively with the state-of-the-art constraint optimization algorithms on eleven test cases which were widely studied benchmark functions in the literature.

Neuro-evolutionary computing, neuro-fuzzy computing, and swarm intelligence are the subject of many current research works in computational intelligence. In Chapter 6, Cai, Venayagamoorthy, and Wunsch investigate a hybrid training algorithm based on particle swarm optimization (PSO) and evolutionary algorithm (EA) for feedforward and recurrent neural networks. Particularly, the applications of the hybrid PSO-EA algorithm for training of a feedforward neural network as a board evaluator for the game Go and for training of a recurrent neural network to predict the missing values from a time series are presented to show its potential. Results indicate that the proposed hybrid algorithm performs better than individual application of PSO or EA algorithm as a training algorithm. In Chapter 7, Lin and Wang present a novel approach to combining wavelet networks and multilayer feedforward neural networks for fuzzy logic control systems. While most of the existing neuro-fuzzy systems focus on implementing the Takagi-Sugano-Kang (TSK) fuzzy inference model, they fail to keep the knowledge structure that is critical in interpreting the learning process and providing insights to the working mechanism of the underlying systems. It is their intention to utilize individual subnets to implement decision-making process of the fuzzy logic control systems based on the Mamdani model. Center average defuzzification has been implemented by a neural network so that a succinct network structure is obtained. More importantly, wavelet networks have been adopted to provide better locality capturing capability and therefore better performance in terms of learning speed and training time. Ant colony algorithms are the pioneers of swarm intelligence. Ant colony optimization (ACO), a new meta-heuristic method based on the observation of real ant colony activities and behaviors, offers a new way to solve many complicated optimization problems. In Chapter 8, Zhang, Xu, and Zhang provide a detailed overview on the principle of ACO and its various applications.

Bioinformatics is an exciting and important area of applications for computational intelligence. Liu and Xiong start applications with the problem of motif discoveries in unaligned DNA and protein sequences in Chapter 9. Current popular algorithms for this problem face two difficulties: high computational cost and the possibility of insertions and deletions. This chapter proposes a self-organizing neural network structure as a new solution. This network contains several subnetworks with each performing classifications at different levels. The top level divides the input space into a small number of regions and the bottom level classifies all input patterns into motifs and non-motif patterns. A low computational complexity is maintained through the use of the layered structure so that each pattern's classification is performed with respect to a small subspace of the whole input space. The introduction of pairwise distance between patterns enables their method to deal with up to two insertions/deletions allowed in a motif, while other existing algorithm can only deal with one insertion or deletion. Note that simulation results show that their algorithm can identify motifs with more mutations than existing algorithms and their algorithm works well for long DNA sequences as well. In Chapter 10, Berman, DasGupta, and Sontag present an interesting investigation on some computational problems that arise in the reverse engineering of protein and gene networks. They discuss the biological motivations, provide precise formulations of the combinatorial questions that follow from these motivations and then describe their computational complexity issue, namely efficient approximation algorithms for these problems.

Similar to bioformatics, biometrics is another focus of current applications of CI techniques. In Chapter 11, Tian, Chen, Zhang, and Yang provide an overview of the advances in automatic fingerprint recognition algorithms and fingerprint applications. More attention has been paid to the enhancement of low-quality fingerprints and the matching of distorted fingerprint images, both issues are considered to be significant and challenging tasks. This chapter also includes an interesting case of application: fingerprint mobile phones that utilize the fingerprint recognition to ensure information security. In addition, two important competitions for fingerprint recognition algorithms, Fingerprint Verification Competition 2004 (FVC2004) and Fingerprint Vendor Technology Evaluation 2003 (FpVTE2003), are introduced and discussed.

Finally, we consider applications in estimation and control. In Chapter 12, He investigates a neuromuscular control system for posture and movement. Valuable lessons on how to design a versatile control system can be learnt from the functional structures of the system and strategies in adaptation to various tasks and environmental conditions. This chapter presents two case studies on how the neuromuscular control system develops adaptive and predictive control strategies to achieve effective stability and performance improvement against perturbations. Results demonstrate that a feedforward predictive control strategy is often developed to achieve efficient control with satisfactory performance, instead of energy demanding stiffness control. Although no current CI techniques have yet been explicitly applied in this study, we believe this is a promising area where CI methods can play a key and natural role. Chapter 13 discusses the use of adaptive approximation methods in estimating unmodeled nonlinear functions to improve tracking performance. Zhao and Farrell focus on methods for the on-line estimation of functions that bound the achievable adaptive approximation accuracy. Whereas the results currently in the literature involved global forgetting, the adaptation laws for the estimated bounds herein involve local forgetting. They show such localized bounds have utility for self-organizing adaptive approximators that adjust the structure of the function approximator during system operations. Last but not least, Yi, Yubazaki, and Hirota propose in Chapter 14 a SIRMs (single input rule modules) dynamically connected fuzzy inference model and apply it to the control of several underactuated systems such as truck-trailer system, ball-beam system, and parallel-type double inverted pendulum system. For each input item, an SIRM is constructed and a dynamic importance degree is introduced and consists of a base value insuring the participation of the input item throughout a control process and a dynamic value changing with control situations The model output is obtained by summarizing the products of the dynamic importance degree and the fuzzy inference result of each SIRM. Simulation results indicate that the proposed model is effective even for very complicated systems.

We have edited this book because we are excited about the emergence of computational intelligence as an integrated science and want to bring the state of the art in both theories and applications in the Series on Intelligent Control and Intelligent Automation. However, a volume in the Series can only provide a rather limited description of the current progress in this new field. Although we have made every effort to include a broad spectrum of approaches and applications in recent CI research and development, the coverage is by no means near balanced or complete.

Realization of this project is due to the generous contributions and assistance of many distinguished researchers. We would like to express our deep appreciation to all contributors for their great support and significant effort. We are also grateful to Senior Editor Steven Patt and other editors at the World Scientific Publishing Company, especially for their patience in numerous extensions of the publication deadline. For assistance with editing, we acknowledge the huge support from many of our associates and assistants, especially Dr. Yanqing Gao, Dr. Sanqing Hu, and Dr. Qinghai Miao.

Fei-Yue Wang
Chinese Academy of Sciences
Beijing, China
The University of Arizona
Tucson, Arizona, USA

Derong Liu
University of Illinois at Chicago
Chicago, Illinois, USA

List of Contributors

Piotr Berman
Associate Professor
Department of Computer Science
and Engineering
Pennsylvania State University
University Park, PA 16802
Phone: (814) 865-1611
Fax: (814) 865-3176
berman@cse.psu.edu

Xinjian Chen
Laboratory of Complex Systems
and Intelligence Science
Institute of Automation
Chinese Academy of Sciences
Beijing 100080, China
Phone: (86) 10-82618465
Fax: (86) 10-62527995
xjchen@fingerpass.net.cn

Steven A. Eschrich
Bioinformatics Staff Scientist
Moffitt Cancer Center and
Research Institute
Tampa, FL 33612-9497
Phone: (813) 745-1303
Fax: (813) 745-6107
eschris@moffitt.usf.edu

Xindi Cai
Department of Electrical and
Computer Engineering
University of Missouri-Rolla
Rolla, MO 65409
Phone: (573) 341-6811
Fax: (573) 341-4532
cai@umr.edu

Bhaskar DasGupta
Associate Professor
Department of Computer Science
University of Illinois at Chicago
Chicago, IL 60607
Phone: (312) 355-1319
Fax: (312) 413-0024
dasgupta@cs.uic.edu

Jay Farrell
Professor
Department of Electrical Engineering
University of California, Riverside
Riverside, CA 92521
Phone: (909) 787-2159
Fax: (909) 787-2425
farrell@ee.ucr.edu

Lawrence O. Hall
Professor
Department of Computer Science
and Engineering
University of South Florida
Tampa, FL 33620
Phone: (813) 971-0129
Fax: (813) 974-4195
hall@csee.usf.edu

Kaoru Hirota
Professor
Interdisciplinary Graduate School
of Science and Engineering
Tokyo Institute of Technology
Yokohama 226-8502, Japan
Phone: (81) 45-9245686
Fax: (81) 45-9245676
hirota@hrt.dis.titech.ac.jp

Derong Liu
Associate Professor
Department Electrical and
Computer Engineering
University of Illinois at Chicago
Chicago, IL 60607-7053
Phone: (312) 355-4475
Fax: (312) 996-6465
dliu@ece.uic.edu

Jiping He
Professor
Harrington Department of
Bioengineering
Arizona State University
Tempe, AZ 85287
Phone: (480) 965-0092
Fax: (480) 965-4292
jiping.he@asu.edu

Yuetong Lin
Assistant Professor
Department of Electronics,
and Computer Technology
Indiana State University
Terre Haute, IN 47809
Phone: (812) 237-3399
Fax: (812) 237-3397
liny@indstate.edu

Witold Pedrycz
Professor
Department of Electrical and
Computer Engineering
University of Alberta
Edmonton, Alberta T6R 2T1
Canada
and
Systems Research Institute
Polish Academy of Sciences
Warswaw
Poland
Phone: (780) 492-3333
Fax: (780) 492-1811
pedrycz@ee.ualberta.ca

Eduardo Sontag
Professor
Department of Mathematics
Rutgers University
New Brunswick, NJ 08854-8019
Phone: (732) 445-3072
Fax: (206) 338-2736
sontag@math.rutgers.edu

Jie Tian
Professor
Laboratory of Complex Systems
and Intelligence Science
Institute of Automation
Chinese Academy of Sciences
Beijing 100080, China
Phone: (86) 10-82618465
Fax: (86) 10-62527995
tian@doctor.com

Fei-Yue Wang
Professor
Laboratory of Complex Systems
and Intelligence Science
Institute of Automation
Chinese Academy of Sciences
Beijing 100080, China
and
Department of Systems and
Industrial Engineering
University of Arizona
Tucson, AZ 85721
Phone: (520) 621-6558
Fax: (520) 621-6555
feiyue@email.arizona.edu

Qing Tao
Professor
Laboratory of Complex Systems
and Intelligence Science
Institute of Automation
Chinese Academy of Sciences
Beijing 100080, China
Phone: (86) 10-82614560
Fax: (86)10-62545229
qing.tao@mail.ia.ac.cn

Ganesh K. Venayagamoorthy
Assistant Professor
Department of Electrical and
Computer Engineering
University of Missouri-Rolla
Rolla, MO 65409
Phone: (573) 341-6641
Fax: (573) 341-4532
gkumar@ieee.org

Jue Wang
Professor
Laboratory of Complex Systems
and Intelligence Science
Institute of Automation
Chinese Academy of Sciences
Beijing 100080, China
Phone: (86) 10-82614560
Fax: (86) 10-62545229
Jue.wang@mail.ia.ac.cn

Donald C. Wunsch II
Mary K. Finley Missouri
Distinguished Professor
Department of Electrical and
Computer Engineering
University of Missouri-Rolla
Rolla, MO 65409
Phone: (573) 341-4521
Fax: (573) 341-4532
dwunsch@umr.edu

Junqin Xu
Associate Professor
Department of Mathematics
Qingdao University
Qingdao 266071, China
Phone: (86) 532-85953712
Fax: (86) 532-85953672
xujunqinqdu@yahoo.com

Gary G. Yen
Professor
School of Electrical and
Computer Engineering
Oklahoma State University
Stillwater, OK 74078-5032
Phone: (405) 744-7743
Fax: (405) 744-9198
gyen@ceat.okstate.edu

Xiaoxu Xiong
Department Electrical and
Computer Engineering
University of Illinois at Chicago
Chicago, IL 60607-7053
Phone: (312) 413-2407
Fax: (312) 996-6465
xxiong@cil.ece.uic.edu

Xin Yang
Associate Professor
Laboratory of Complex Systems
and Intelligence Science
Institute of Automation
Chinese Academy of Sciences
Beijing 100080, China
Phone: (86) 10-82618465
Fax: (86) 10-62527995
yx@fingerpass.net.cn

Jianqiang Yi
Professor
Laboratory of Complex Systems
and Intelligence Science
Institute of Automation
Chinese Academy of Sciences
Beijing 100080, China
Phone: (86) 10-62658815
Fax: (86) 10-62658815
jianqiang.yi@mail.ia.ac.cn

Naoyoshi Yubazaki
CEO
Mycom, Inc.
12 S. Shimobano
Saga Hirosawa
Ukyo
Kyoto 616-8303
Japan
Phone: (81) 75-8823601
Fax: (81) 75-8826531
yubazaki@mycom-japan.co.jp

Siying Zhang
Professor
Institute of Complexity Science
Qingdao University
Qingdao 266071, China
Phone: (86) 532-85953820
Fax: (86) 532-85953672
zhangsiying@qdu.edu.cn

Yuanyuan Zhao
Department of Electrical Engineering
University of California, Riverside
Riverside, CA 92521
Phone: (951) 827-2901
Fax: (909) 787-2425
yzhao@ee.ucr.edu

Jihui Zhang
Professor
Institute of Complexity Science
Qingdao University
Qingdao 266071, China
Phone: (86) 532-85953712
Fax: (86) 532-85953672
zhangjihui@qdu.edu.cn

Yangyang Zhang
Laboratory of Complex Systems
and Intelligence Science
Institute of Automation
Chinese Academy of Sciences
Beijing 100080, China
Phone: (86) 10-82618465
Fax: (86) 10-62527995
zhangyy@fingerpass.net.cn

Contents

Chapter 1

A Quest for Granular Computing and Logic Processing

Witold Pedrycz

Abstract: Granular computing is concerned with processing carried out at a level of coherent conceptual entities – information granules. Such granules are viewed as inherently conceptual entities formed at some level of abstraction whose processing is rooted in the language of logic (especially, many valued or fuzzy logic). The objective of this study is to discuss the role of Granular Computing and logic processing in the context of intelligent systems and demonstrate that both of them are organized into a single conceptual and computational framework. We provide a general overview of Granular Computing emphasizing a diversity of the currently available concepts and underlining their common features that make the entire pursuit highly coherent. The logic facet of processing is embodied in the framework of fuzzy logic and fuzzy sets that constitutes a consistent background necessary for representing and processing information granules. Several main categories of logic processing units (logic neurons) are discussed that support aggregative (and-like and or-like operators) and referential logic mechanisms (dominance, inclusion, and matching). We show how the logic neurons contribute to high functional transparency of granular processing, help capture prior domain knowledge and give rise to a diversity of the resulting models.

1.1 Introduction

The quest for designing and deploying intelligent systems has been with the community of researchers and practitioners for at least several decades. It manifested in different ways and has been realized in the variety of conceptual frameworks

1

with each of them promoting some research agenda and focusing on some philosophical, methodological and algorithmic aspects (such as pattern recognition, symbolic processing, evolutionary mechanisms). One of the recent developments concerns Computational Intelligence (CI) – a unified and comprehensive platform of conceptual and computing endeavors of fuzzy sets, neurocomputing and evolutionary methods. CI promotes and dwells on synergy: it carefully identifies the nature of the individual technologies and exploits their highly complementary character.

While fully acknowledging the benefits and rationale behind CI, in this study, we elaborate on a different point of view that tackles the fundamentals of the development of intelligent systems. Central to our main line of thought is Granular Computing [1, 21] – a cohesive and coherent platform of representing and processing information granules. Such granules are represented in many different ways including sets (intervals), fuzzy sets, and rough sets, to name several dominant alternatives. As being abstract entities, their processing is very much embedded in the framework of logic that is highly consistent with the underlying nature of information granules and supports the transparency of the underlying processing and its semantics.

This chapter is structured into seven sections. First, in Section 2 we discuss the fundamentals of Granular Computing and outline some historical aspects of the main developments along with a brief view at the diversity of the abstraction mechanisms of fuzzy sets, rough sets, and interval analysis. Section 3 is concerned with the unified processing environment of Granular Computing and logic, especially their continuous (multivalued and fuzzy) variants. Subsequently, in Sections 4 and 5 we proceed with the realization of the logic layer of this environment by studying various fuzzy logic neurons and networks. Interpretation aspects of networks are presented in Section 6. Conclusions are covered in Section 7.

1.2   Granular Computing

Information granules permeate human endeavors. No matter what specific task is taken into account, we usually cast it into a certain conceptual setting. This is the framework we formulate generic concepts adhering to some appropriate level of abstraction, carry out further processing, and communicate the results to the environment. Information granules are pivotal to a way in which experimental datasets are perceived and interpreted by humans. Instead of being buried in huge piles of numeric data, the user summarizes such datasets and forms a very limited vocabulary of information granules (for instance, fuzzy sets) that are easily comprehended, come with a well-defined semantics and in this manner help express main relationships and dependencies between individual variables arising at some specific level of information granularity. Even more prominently, granularity implements information abstraction: by moving to a certain level of abstraction (granulation) we hide unnecessary details and concentrate on essentials (main dependencies, general trends) existing in the data. Quite frequently, information granules are formed via data clustering.

Granular Computing is an emerging and highly unifying paradigm of information processing. While we have already noticed a number of important conceptual and computational constructs developed in the domain of system modeling, machine learning, image processing, pattern recognition, and data compression in which various abstractions (and ensuing information granules) came into existence, Granular Computing has become innovative and intellectually proactive in several fundamental ways.

It identifies the essential commonalities between the surprisingly diversified problems and technologies used there, which could be cast into a unified framework we usually refer to as a granular world. This is a fully operational processing entity that interacts with the external world (that could be another granular or numeric world) by collecting necessary granular information and returning the outcomes of the granular computing. With the emergence of the unified framework of granular processing, we get a better grasp as to the role of interaction between various formalisms and visualize mechanisms of interaction between them.

It brings together the existing formalisms of set theory (interval analysis), fuzzy sets, rough sets, etc. under the same roof by clearly visualizing that in spite of their visibly quite distinct underpinnings (and ensuing processing), they also come with fundamental commonalities. In this sense, Granular Computing establishes a stimulating environment of synergy between the individual approaches.

By dwelling on the commonalities of the existing formal approaches, Granular Computing helps build heterogeneous and multifaceted models of processing of information granules by clearly recognizing the orthogonal nature of some of the existing and well established frameworks (say, probability, namely probability density functions and fuzzy sets)

Granular Computing fully acknowledges a notion of variable granularity whose range could cover detailed numeric entities and very abstract and general information granules. It looks at the aspects of compatibility of such information granules and ensuing communication mechanisms of the granular worlds.

With the fundamentals of Granular Computing come a variety of conceptual frameworks and ensuing algorithms; see Figure 1.2.1. These are quite different from the standpoint of the methods, interpretation of results, and interaction between other environments of Granular Computing. The diversity is quite profound; here we briefly comment on some of them and elaborate on the most essential features. By no means we pretend to cover the topic in detail; with the number

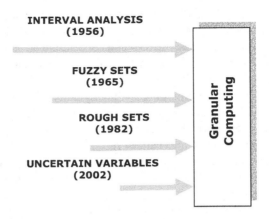

Figure 1.2.1: An overview of Granular Computing with its main contributing conceptual frameworks.

of authoritative publications in the area is not necessary. What becomes perhaps far more needed is a synthetic and comparative analysis of various alternatives of Granular Computing.

Historically, the first framework of Granular Computing was realized in the language of sets and intervals (giving rise to so-called interval analysis). The origin of this interval calculus or interval analysis was rooted in the realm of finite precision numeric computing. These highly conservative models help express important phenomena of propagation and accumulation of computing errors. As the bounded yet finite granules, intervals were quite attractive in describing numerous phenomena ranging from system identification, control under uncertainty, and pattern classification. The concept of fuzzy sets coined in 1965 was critical in re-defining the fundamental idea of dichotomy (yes-no, membership-exclusion, etc.) sets or intervals fully subscribed to. Since then fuzzy sets introduced and supported a notion of partial membership of element to a given concept (information granule), cf. [24, 25]. The applied facets of fuzzy sets is represented across many disciplines including control and decision-making, cf. [10, 20], intelligent systems [11], and pattern recognition [14]. Rough sets [15] allow capture concepts with "rough" boundaries where there we come up with a description of concepts articulated in terms of lower and upper boundaries. Such boundaries arise naturally when we attempt to articulate a complex concept in the language of less descriptive, simpler entities that form our basic vocabulary. Rough sets and fuzzy sets are complementary to a significant extent; we have witnessed emergence of hybrid granules in the form of rough-fuzzy or fuzzy-rough sets. Further generalized concepts of information granules can be found in [1, 2, 3].

Defining and subsequently measuring the size of information granules is of apparent relevance to Granular Computing and particular models developed there. We can come up with a uniform view on this matter in spite of the striking diversity of the underlying formalisms. Intuitively, the size of the granules relates in one way or another to the number of elements they comprise of. The terms such as set cardinality, width of an interval, σ-count of fuzzy set come immediately to mind. They are especially appealing and intuitively meaningful as they relate to the very essence of abstraction: the unified treatment and a cohesiveness of elements drawn together by the information granule under discussion.

1.3 Granular Computing and Logic: Synergistic Links

In essence, Granular Computing promotes the concept of abstraction – a fundamental faculty of human endeavors. The level of abstraction is inherently problem and user oriented. Different levels of abstraction that reflect upon the detail of problem description in terms of basic concepts (entities), primary relationships between them, interaction with the user/analyst. As driven by abstraction, the resulting variables and relationships between them (as being perceived at the particular level of detail) are more abstract entities. This naturally implies that their processing should exploit the language that is in par with the fundamentals of such information granules. Casting the processing in logic comes as a viable and highly justifiable alternative. What type detailed logic is deemed the most appropriate arises as the next specific question. Likewise in Granular Computing, we have at our disposal a variety of logics starting from two-valued one (leading to the well known Boolean calculus, digital systems, etc.), moving to three-valued logic and multivalued logics and including fuzzy logic. The option of fuzzy logic arises here as an interesting alternative owing to the richness of the underlying semantics and useful computing facilities implied primarily by the truth values located in the unit interval.

In a nutshell, the proposed logic kernel (called here fuzzy networks) completes a logic-based processing of input signals and realizes a certain logic-driven mapping between input and output spaces. As the networks interact with a physical world whose manifestation does not usually arise at the level of logic (multivalued) signals, it becomes apparent that there is a need for some interface of the model. Such interfaces are well known in fuzzy modeling [20], cf. also [13, 18, 19]. In the literature they commonly arise under a name of fuzzifiers (granular coders aimed at information granulation) and defuzzifiers (granular decoders with a main function of degranulation of information). The role of the coder is to convert a numeric input coming from the external environment into the internal format of membership grades of the fuzzy sets defined for each input variable. These results of a nonlinear normalization of the input (no matter what original ranges the input variables assume) and a linear increase of the dimensionality of the new logic space in comparison with the original one). On the other hand, the decoder accepts the results of the logic processing and transforms them into some numeric values. The layered architecture of the fuzzy models with clearly distinguished interfaces and the logic-processing core is illustrated in Figure 1.3.1.

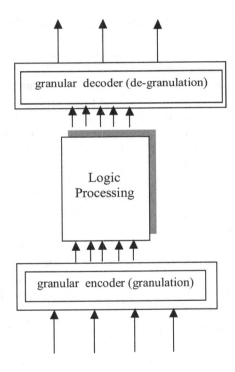

Figure 1.3.1: A general layered structure of fuzzy modeling; the use of granular encoders and decoders are essential in the development of communication mechanisms with the modeling environment.

With the design of the interfaces, we exercise two general approaches (in the literature we encounter far more diversified techniques but they are usually more specialized in the sense of the underlying computing mechanisms):

1. Granulation of individual variables

This mechanism of granulation is quite common in the realm of fuzzy modeling. In essence, we define several fuzzy sets in the universe of discourse of the variable of interest so that any input is transformed via the membership functions defined there and the resulting membership grades are used in further computations by the model. From the design standpoint, we choose a number of fuzzy sets, type of membership functions and a level of overlap between consecutive fuzzy sets. Some general tendencies along this line are thoroughly reported in the literature. By selecting the number of fuzzy sets (usually between 3 and 9 with the magic of 7 ± 2 emphasized throughout the literature), we position modeling activities at some required level of information granularity (a level of modeling details we are interested in). The type of membership functions helps model the semantics of the information granules. Among many possibilities, we commonly encounter triangular fuzzy sets and Gaussian membership functions. These two types come with an extensive list of arguments that help make a suitable selection with respect to the main objectives of the model (e.g., those concerning a tradeoff between interpretation and accuracy of modeling). The overlap level is essential from different points of view, namely (a) semantics of the linguistic terms, (b) nonlinear numeric characteristics of the fuzzy model, and (c) completeness of the model.

2. Nonlinear or linear normalization

Here we transform an original variable defined in some space, say $[a, b]$ (subset of $\mathbb{R}$) is scaled to the unit interval. This could be done with the aid of some mapping $\phi \colon [a, b] \to [0, 1]$ that could be either linear or nonlinear. In any case we consider that f is monotonically increasing with $\phi(a) = 0$ and $\phi(b) = 1$. This transformation does not affect the dimensionality of the problem.

1.4 Main Categories of Fuzzy Logic Processing Units

In this section, we present the main categories of the logic neurons as they were introduced and discussed in [22, 24, 25]. The underlying taxonomy involves aggregative and referential neurons and very much ties up with their logic functionality. The naming of the neurons reflect the underlying processing realized by them. The aggregative neurons concentrate on the logic type of aggregation of the inputs (truth values) while the referential neurons are aimed at logic processing of results of referential transformations of the corresponding truth values.

Aggregative neurons

Formally, these neurons realize a logic mapping from $[0,1]^n$ to $[0,1]$. Two main classes of the processing units exist in this category.

(i) *OR neuron* realizes an *and* logic aggregation of inputs $\mathbf{x} = [x_1, x_2, \cdots, x_n]$ with the corresponding connections (weights) $\mathbf{w} = [w_1, w_2, \cdots, w_n]$ and then summarizes the partial results in an *or*-wise manner (hence the name of the neuron). The concise notation underlines this flow of computing, $y = \mathrm{OR}(\mathbf{x}; \mathbf{w})$ while the realization of the logic operations gives rise to the expression (we commonly refer to it as an s-t convolution)

$$y = \overset{n}{\underset{i=1}{\mathrm{S}}} (x_i \, \mathrm{t} \, w_i).$$

Bearing in mind the interpretation of the logic connectives (t- and s-norms), the OR neuron realizes the following logic expression carried out on the input signals

$$(x_1 \text{ and } w_1) \text{ or } (x_2 \text{ and } w_2) \text{ or } \cdots \text{ or } (x_n \text{ and } w_n).$$

Apparently the inputs are logically "weighted" by the values of the connections before producing the final result. In other words we can treat "y" as a truth value of the above statement where the truth values of the inputs are affected by the corresponding weights. Noticeably, lower values of w_i discount the impact of the corresponding inputs; higher values (especially those being positioned close to 1) do not affect the original truth values of the inputs resulting in the logic formula. In limit, if all connections w_i, $i = 1, 2, \cdots, n$, are set to 1 then the neuron produces a plain *or*-combination of the inputs,

$$y = x_1 \text{ or } x_2 \text{ or } \cdots \text{ or } x_n.$$

The values of the connections set to zero eliminate the corresponding inputs. Computationally, the OR neuron exhibits nonlinear characteristics (that is inherently implied by the use of the t- and s-norms (that are evidently nonlinear mappings). The plots of the characteristics of the OR neuron shown in Figure 1.4.1 visualize this effect (note that the characteristics are affected by the use of some norms). The connections of the neuron contribute to its adaptive character; the changes in their values form the crux of the parametric learning.

(ii) *AND neuron*: The neurons in the category, denoted by $y = \mathrm{AND}(\mathbf{x}; \mathbf{w})$ with $\mathbf{x}$ and $\mathbf{w}$ being defined as in case of the OR neuron, are governed by the expression

$$y = \overset{n}{\underset{i=1}{\mathrm{T}}} (x_i \, \mathrm{s} \, w_i)$$

Here the *or* and *and* connectives are put together in a reversed order: first the inputs are combined with the use of the s-norm and the partial results are aggregated

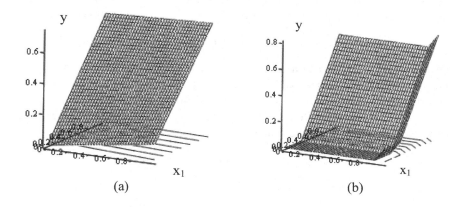

Figure 1.4.1: Characteristics of the OR neuron for selected pairs of t- and s-norms. In all cases the corresponding connections are set to 0.1 and 0.7 with intent to visualize their effect on the input-output characteristics of the neuron: (a) product and probabilistic sum, (b) Lukasiewicz *and* and *or* connectives (b).

and-wise. Higher values of the connections reduce impact of the corresponding inputs. In limit $w_i = 1$ eliminates the relevance of x_i. With all w_i set to 0, the output of the AND neuron is just an *and* aggregation of the inputs

$$y = x_1 \text{ and } x_2 \text{ and } \cdots \text{ and } x_n.$$

Let us conclude that the neurons are highly nonlinear processing units depending upon the specific realizations of the logic connectives. They also come with potential plasticity whose usage becomes critical when learning the networks involving these neurons.

Referential (reference) neurons

The essence of referential computing deals with processing logic predicates. The two-argument (or generally multivariable) predicates such as *similar, included in, dominates* are essential components of any logic description of a system. In general, the truth value of the predicate is a degree of satisfaction of the expression $P(x, a)$ where "a" is a certain reference value (reference point). Depending upon the meaning of the predicate (P), the expression $P(x, a)$ reads as "x is similar to a," "x is included in a," "x dominates a," etc. In case of many variables, the compound predicate comes in the form $P(x_1, x_2, \cdots, x_n,$

$a_1, a_2, \cdots, a_n)$ or more concisely $P(\mathbf{x}; \mathbf{a})$ where $\mathbf{x}$ and $\mathbf{a}$ are two vectors in the n-dimensional unit hypercube. We envision the following realization of $P(\mathbf{x}; \mathbf{a})$

$$P(\mathbf{x}; \mathbf{a}) = P(x_1, a_1) \text{ and } P(x_2, a_2) \text{ and } \cdots \text{ and } P(x_n, a_n)$$

meaning that the satisfaction of the multivariable predicate relies on the satisfaction realized for each variable separately. As the variables could come with different levels of relevance as to the overall satisfaction of the predicates, we represent this effect by some weights (connections) $w_1, w_2, \cdots, w_n$ so that the above expression can be given in the following form

$$P(\mathbf{x}; \mathbf{a}, \mathbf{w}) = [P(x_1, a_1) \text{ or } w_1] \text{ and } [P(x_2, a_2) \text{ or } w_2] \text{ and } \cdots$$
$$\text{and } [P(x_n, a_n) \text{ or } w_n].$$

Taking another look at the above expression and using a notation $z_i = P(x_i, a_i)$, it converts to a certain AND neuron $y = \mathrm{AND}(\mathbf{z}; \mathbf{w})$ with the vector of inputs $\mathbf{z}$ being the result of the computations done for the logic predicate. Then the general notation to be used reads as $\mathrm{REF}(\mathbf{x}; \mathbf{w}, \mathbf{a})$ and using the explicit notation we have

$$y = \mathop{\mathbf{T}}_{i=1}^{n} (\mathrm{REF}(x_i, a_i) \text{ s } w_i)$$

In essence, as visualized in Figure 1.4.2, we may conclude that the reference neuron is a realized in a two-stage construct where first we determine the truth values of the predicate (with a treated as a reference point) and then treat these results as the inputs to the AND neuron.

So far we have exploited the general term of predicate computing not confining ourselves to any specific nature of the predicate. Among a number of possibilities, we discuss the three of them, which tend to occupy an important role. Those are predicates of inclusion, dominance and matching (similarity). As the names stipulate, the predicates return truth values of satisfaction of the relationship of inclusion, dominance and similarity of a certain argument "x" with respect to the given reference "a". The essence of all these calculations is in the determination of the given truth values and this is done in the carefully developed logic framework so that the operations retain their semantics and interpretability. What makes our discussion coherent is the fact that the proposed operations originate from triangular norms. The inclusion operation, denoted by "$\subset$" is modeled by an implication "$\rightarrow$" that is induced by a certain left continuous t-norm [18, 20]

$$a \rightarrow b = \sup\{c \in [0,1] | atc \le b\}, \quad a, b \in [0,1].$$

For instance, for the product the inclusion takes on the form

$$a \rightarrow b = \min(1, b/a).$$

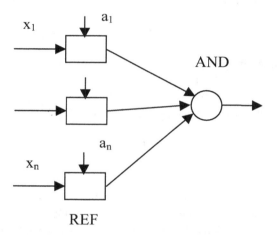

Figure 1.4.2: A schematic view of computing realized by a reference neuron an involving two processing phases (referential computing and aggregation).

The intuitive form of this predicate is self-evident: the statement "x is included in a" and modeled as $\mathrm{INCL}(x, a) = x \to a$ comes with the truth value equal to 1 if x is less or equal to a (which in other words means that x is included in a) and produces lower truth values once x starts exceeding the truth values of "a". Higher values of "x" (those above the reference point "a") start generating lower truth values of the predicate. The dominance predicate acts in a dual manner. It returns 1 once "x" dominates "a" (so that its values exceeds "a") and values below 1 for x lower than the given threshold. The formal model can be realized as $\mathrm{DOM}(x, a) = a \to x$. With regard to the reference neuron, the notation is equivalent to the one being used in the previous case (xx), that is $\mathrm{DOM}(\mathbf{x}; \mathbf{w}, \mathbf{a})$ with the same meaning of $\mathbf{a}$ and $\mathbf{w}$.

The similarity (match) operation is an aggregate of these two,

$$\mathrm{SIM}(x, a) = \mathrm{INCL}(x, a) \; \mathrm{t} \; \mathrm{DOM}(x, a)$$

which is appealing from the intuitive standpoint: we say that x is similar to a if x is included in a *and* x dominates a. Noticeably, if $x = a$ the predicate returns 1; if x moves apart from "a" the truth value of the predicate becomes reduced. The resulting similarity neuron is denoted by $\mathrm{SIM}(\mathbf{x}; \mathbf{w}, \mathbf{a})$ and reads as

$$y = \mathop{\mathbf{T}}_{i=1}^{n} \left(\mathrm{SIM}(x_i, a_i) \; \mathrm{s} \; w_i \right)$$

The reference operations form an interesting generalization of the threshold operations in the sense we admit some partial satisfaction of the constraints (boundaries). The plots of the referential neurons with two input variables are shown in Figures 1.4.3 and 1.4.4; here we have included two realizations of the t-norms to illustrate their effect on the nonlinear characteristics of the processing units.

It is worth noting that by moving the reference point to the origin (**0**) and **1**-vertex of the unit hypercube (with all its coordinates being set up to 1), the referential neuron starts resembling the aggregative neuron. In particular, we have

for $\mathbf{a} = \mathbf{1} = [1\ 1\ 1\ \cdots\ 1]$ the inclusion neuron reduces to the AND neuron

for $\mathbf{a} = \mathbf{0} = [0\ 0\ 0\ \cdots\ 0]$ the dominance neuron reduces to the standard AND neuron

One can draw a loose analogy between some types of the referential neurons and the two categories of processing units we encounter in neurocomputing [11]. The analogy is based upon the *local* versus *global* character of processing realized therein. Perceptrons come with the global character of processing. Radial basis functions realize a local character of processing as focused on receptive fields. In the same vein, the inclusion and dominance neurons are after the global nature of processing while the similarity neuron carries more confined, local processing.

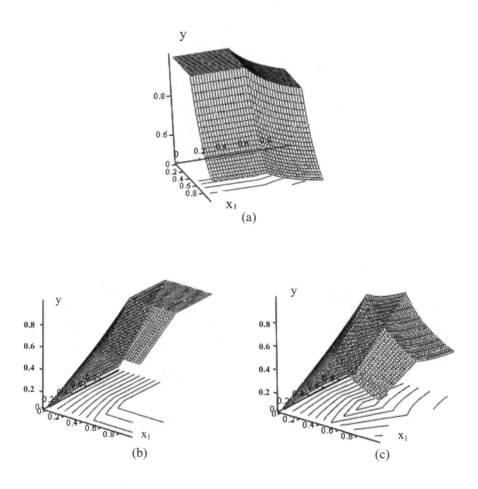

Figure 1.4.3: Characteristics of the reference neurons for the product (t-norm) and probabilistic sum (s-norm). In all cases the connections are set to 0.1 and 0.7 with intent to visualize the effect of the weights on the relationships produced by the neuron. The point of reference is set to (0.5, 0.5): inclusion neuron (a), dominance neuron (b), similarity neuron (c).

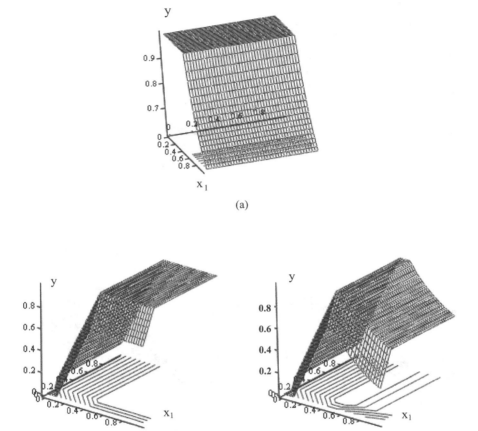

(a)

(b)   (c)

Figure 1.4.4: Characteristics of the reference neurons for the Lukasiewicz t-norm and s-norm (that is at $b = \max(0, a + b - 1)$ and at $b = \min(1, a + b)$). In all cases the connections are set to 0.1 and 0.7 with intent to visualize the effect of the weights. The point of reference is set to (0.5, 0.5): inclusion neuron (a), dominance neuron (b), similarity neuron (c).

1.5   A General Topology of the Network

As we have developed a host of logic processing units, we can use then in the developing a general architecture of the network. In this design we are guided by several requirements. First, we would like to achieve a substantial level of flexibility so that the structure could be easily and effectively adjusted to the experimental data and a variety of problems and their underlying logic settings. Second, we would like to assure a high level of interpretability which is the case here: evidently each neuron comes with a well-defined semantics and our intent is to retain it so at the very end the network can be easily mapped (translated) into a well-structured and transparent logic expression. This quest for interpretability and transparency has been clearly identified and strongly promoted in the most recent literature, cf. [4]; refer also to [5, 6, 13, 20, 23] for additional issues raised with this regard. In the logic description we will dwell upon the well-established components of fuzzy logic: logic operators and linguistic modifiers. Having these requirements in mind, a general architecture of the network is shown in Figure 1.5.1.

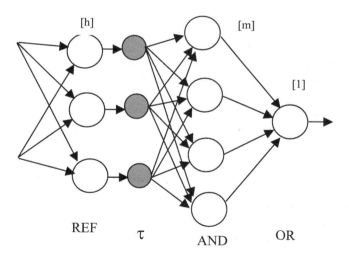

Figure 1.5.1: A general architecture of the network constructed with logic-based neurons; see a detailed description in text. The dimensions of the layers (number of nodes) are marked by numbers placed in brackets (upper part of the figure).

The network comes with several layers where each of them has a clear functional role to play and is highly interpretable. The first layer (carrying out some specialized referential processing) is composed of "h" referential neurons (inclusion, dominance, similarity). The results of this processing are raised to some power (indicated by some small shadowed circle) and then combined *and*-wise in the third layer of the network. The elements there are AND neurons with all connections hardwired to zero. The width of this layer is equal to "m". In the sequel the results are combined by the layer of OR neurons. Let us now move on to the computational details by the same time concentrating on the interpretation of the underlying computations. The truth values generated by the referential neurons reflect the level of satisfaction of the logic predicates

$$z_1 = P_1(\mathbf{x}; \mathbf{a}_1), \ z_2 = P_2(\mathbf{x}; \mathbf{a}_2), \ \cdots, \ z_h = P_h(\mathbf{x}; \mathbf{a}_h)$$

The powers of z_i, denoted here as $\tau_i\,(z_i)$ where τ_i assumes a few discrete values (say 1/2, 0, 1, 2, and 4) are interpreted as linguistic modifiers operating upon z_i and producing some concentration or dilution effect [20, 24]. More specifically, the collection of the modifiers maps on the corresponding powers of the truth values in a usual way we commonly encounter in the literature:

1/2 - *more or less* (dilution effect);

0 - unknown;

1 - true (neutral);

2 - *very* (concentration effect);

4 - *very* (*very*) = *very*^{2} (strong concentration effect).

For instance, the expression

$$\text{INCL}([x_1, x_2], [0.2, 0.7], [0.6, 0.9])^{0.5}$$

that involves the two first layers of the network translates into the following linguistic statement

$$y = \textit{more or less}\{[x_1 \text{ included in } 0.6] \textit{ or } 0.2 \textit{ and } [x_2 \text{ included in } 0.9] \textit{ or } 0.7\}.$$

(Noticeably, the core part of this expression could be extracted by carrying out some additional pruning). In a similar way, by writing down the logic formulas for the AND and OR neurons, we arrive a complete logic expression of the network.

While we have concentrated on the interpretability of the logic networks, it is equally important to discuss its development. As they fall under the umbrella of specialized neural networks, a lot of gradient-based learning techniques designed there are of immediate use to us and they have been used in the past, cf. [8, 11, 16, 17, 18, 19]. In contrast, the structure discussed here is highly heterogeneous. In the optimization we are immediately faced with the problem of structural optimization (e.g., when building the first layer of the referential neurons) that is beyond the reach of the popular gradient based learning; various mechanisms of genetic optimization become of interest [7, 20].

1.6   Interpretation Issues of Logic Networks

The interpretation aspects inherently associated with the logic framework are of significant relevance. The network can be directly translated into a series of logic statements and this translation is immediate owing to the logic expression coming with each neuron. Several detailed logic descriptions are worth describing in more detail.

1. The "standard" rule-based description of data where the rules are of the form
   "if condition$_1$ is A_1 *and* condition $_2$ is B_1 *and* $\cdots$ *and* condition $_n$ is Z_1
   or

   $\vdots$

   or
   then conclusion is F""

maps directly on the first layer of AND neurons that feed a single OR neuron located in the output layer. Evidently as the neurons come with the connections, they play a vital role in the calibration of the conditions and the rules. The calibration alludes to the meaning of the connections of the AND and OR neurons. As already discussed in Section xx, higher values of the connections of the AND neuron lead to the reduction or complete elimination of the corresponding conditions (the conditions become masked and the rule gets then more general). The calibration of the OR neuron modifies the confidence factor (relevance) of each rule. Again, as already discussed, higher values of the connections give rise to the higher relevance of the respective rules. Interestingly, owing to the connections of the AND neurons, each rule can operate in the subspace of the overall condition space; by eliminating some variables of the input space. This feature becomes an important asset as it reduces the dimensionality of the space of interest and contributes to the compactness of the rules. Note that "handcrafted" rules usually are identical as to the condition space and commonly operate in the complete input space.

2. The "standard" rules are homogeneous and involve only *and* connectives of the individual conditions. The elevated richness of the rules comes with reference neurons where we allow the inputs to be "compared" with some reference points as well as involve several types of reference operations that again contribute to the diversity of the rules. In particular, we can envision that the in the same rule there could be several reference operators. For instance, the general form of the rule reads as follows

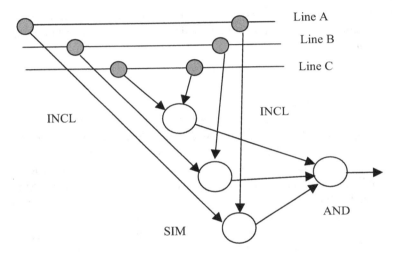

Figure 1.6.1: Distribution of sensors along three lines and their logic relationships between them and the resulting logic network.

"if condition$_1$ is $P(A_1, a_1)$ *and* condition$_2$ is $T(B_1, b_1)$ *and* $\cdots$ *and* condition $_n$ is $G(Z_1, f_1)$

> *or*
>
> $\vdots$
>
> *or*

then conclusion is F"

(where P, T, G, $\cdots$, are logic predicates of interest and a_1, b_1, $\cdots$, form the points of reference).

There could be interesting variations on such rules where we admit the reference points to be dynamic rather than fixed setpoints and involve some other variables. For instance, we can associate signals coming from different sensors where we expressed some prior domain knowledge as to the relationships between the readings of the sensors distributed along some data lines, refer to Figure 1.6.1. The output of the AND neuron indicates a degree of satisfaction of the spatial relationships between the readings of the sensors.

1.7   Conclusions

In this study, we discuss the role of Granular Computing as a fundamental environment supporting the development of intelligent systems. Various conceptual alternatives (such as interval analysis, fuzzy sets, and rough sets) are presented and contrasted. The general topology of granular models is discussed; in particular we emphasize the need of carrying out processing of information granules in the logic-driven framework. Being also motivated by the genuine need of constructing networks that exhibit plasticity while retaining interpretability, we have developed a heterogeneous structure composed of logic neurons. The two main categories of aggregative and reference neurons are deeply rooted in the fundamental operations encountered in fuzzy sets (including logic operations, linguistic modifiers, and logic reference operations). The direct interpretability of the network we addressed in the study helps develop a transparent logic description of data. As the network takes advantage of using various neurons, this imposes an immediate requirement of structural optimization and leads to the utilization of the mechanisms of genetic optimization (genetic algorithms).

Bibliography

[1] A. Bargiela and W. Pedrycz, *Granular Computing: An Introduction*, Dordrecht: Kluwer Academic Publishers, 2002.

[2] Z. Bubnicki, "Uncertain variables and their application to decision making problems," *IEEE Trans. on Systems, Man, and Cybernetics-A*, vol. 31, no. 6, pp. 587–596, 2001.

[3] Z. Bubnicki, *Uncertain Logics, Variables and Systems*, Berlin: Springer Verlag, 2002.

[4] J. Casillas et al. (eds.), *Interpretability Issues in Fuzzy Modeling*, Berlin: Springer Verlag, 2003.

[5] J. A. Dickerson and M. S. Lan, "Fuzzy rule extraction from numerical data for function approximation," *IEEE Trans on System, Man, and Cybernetics-B*, vol. 26, pp. 119-129, 1995.

[6] A. F. Gomez-Skarmeta, M. Delgado, and M. A. Vila, "About the use of fuzzy clustering techniques for fuzzy model identification," *Fuzzy Sets and Systems*, vol. 106, pp. 179–188, 1999.

[7] E. Goldberg, *Genetic Algorithms in Search, Optimization, and Machine Learning*. Reading, MA: Addison-Wesley, 1989.

[8] K. Hirota and W. Pedrycz, "OR/AND neuron in modeling fuzzy set connectives," *IEEE Trans. on Fuzzy Systems*, vol. 2, pp. 151–161, 1994.

[9] K. Hirota and W. Pedrycz, "Fuzzy relational compression," *IEEE Trans. on Systems, Man, and Cybernetics-B*, vol. 29, pp. 407–415, 1999.

[10] J. Kacprzyk, *Wieloetapowe Sterowanie Rozmyte*, Warszawa: PWN, 2001.

[11] B. Kosko, *Neural Networks and Fuzzy Systems*, Englewood Cliffs, NJ: Prentice Hall, 1991.

[12] S. Mitra and S. K. Pal, "Logical operation based fuzzy MLP for classification and rule generation," *Neural Networks*, vol. 7, pp. 353–373, 1994.

[13] S. Mitra and S.K. Pal, "Fuzzy multilayer perceptron, inferencing and rule generation," *IEEE Trans. on Neural Networks*, vol. 6, pp. 51–63, 1995.

[14] S. K. Pal and S. Mitra, *Neuro-Fuzzy Pattern Recognition*, New York: John Wiley, 1999.

[15] Z. Pawlak, *Rough Sets–Theoretical Aspects of Reasoning About Data*, Dordercht: Kluwer Academic Publishers, 1991.

[16] W. Pedrycz, "Neurocomputations in relational systems," *IEEE Trans. on Pattern Analysis and Machine Intelligence*, vol. 13, pp. 289–297, 1991.

[17] W. Pedrycz, "Fuzzy neural networks and neurocomputations," *Fuzzy Sets and Systems*, vol. 56, pp. 1–28, 1993.

[18] W. Pedrycz and A. Rocha, "Knowledge-based neural networks," *IEEE Trans. on Fuzzy Systems*, vol. 1, pp. 254–266, 1993.

[19] W. Pedrycz, P. Lam, and A. F. Rocha, "Distributed fuzzy modelling," *IEEE Trans. on Systems, Man and Cybernetics-B*, vol. 5, pp. 769–780, 1995.

[20] W. Pedrycz and F. Gomide, *An Introduction to Fuzzy Sets: Analysis and Design*, Boston: MIT Press, 1998.

[21] W. Pedrycz (ed.), *Granular Computing. An Emerging Paradigm*, Heidelberg: Physica-Verlag, 2001.

[22] M. Setnes, R. Babuska, and H. Vebruggen, "Rule-based modeling: Precision and transparency," *IEEE Trans on System, Man, and Cybernetics-C*, vol. 28, pp. 165–169, 1998.

[23] T. Sudkamp and R. J. Hammel II, "Rule base completion in fuzzy models," In *Fuzzy Modelling: Paradigms and Practice*, W. Pedrycz (ed.), Dordercht: Kluwer Academic Publishers, 1996, pp. 313–330.

[24] L. A. Zadeh, "Fuzzy logic = computing with words," *IEEE Trans. on Fuzzy Systems*, vol. 4, pp. 103–111, 1996.

[25] L. A. Zadeh, "Towards a theory of fuzzy information granulation and its application in human reasoning and fuzzy logic," *Fuzzy Sets and Systems*, vol. 90, pp. 111–127, 1997.

Chapter 2

Abstraction and Linguistic Analysis of Conventional Numerical Dynamic Systems

Fei-Yue Wang

Abstract: Linguistic dynamic systems (LDS) are dynamic processes involving mainly computing with words instead of numbers for modeling and analysis of complex systems and human–machine interfaces. The goal of studying LDS is to establish a methodology of design, modeling, and analysis of complex decision-making processes bridging the machine world in numbers and the human world in words. Specifically in this chapter, conventional dynamic systems are converted to different types of LDS for the purpose of verification and comparison. The evolving laws of a type-I LDS are constructed by applying the fuzzy extension principle to those of its conventional counterpart with linguistic states. The evolution of type-I LDS represents the dynamics of state uncertainty derived from the corresponding conventional dynamic process. In addition to linguistic states, the evolving laws of type-II LDS are modeled by a finite number of linguistic decision rules. Analysis of fixed points is conducted based on point-to-fuzzy-set mappings and linguistic controllers are designed for goals specified in words for type-II LDS. An efficient numerical procedure called α-cuts mapping is developed and applied to obtain extensive simulation results.

2.1 Introduction

In its traditional sense, modeling and analysis of dynamic systems are based on computing that involves manipulation of numbers and symbols. By contrast, human employs mostly words in computing and reasoning to arrive at conclusions

23

expressed as words from linguistic premises. Therefore, incorporating computing with words into modeling and analysis will be an interesting and important direction for future research. This is true especially when one is dealing with social, political or economical, rather than engineering or physical systems. In this chapter, we outline an approach along this direction based on the theory of linguistic dynamic systems (LDS) developed by Wang [22–26].

Clearly, many efforts have already been taken toward this objective in the past [1, 4, 12, 18]. For example, methods from knowledge based systems, expert systems, linguistic structures, multi-valued logic, fuzzy logic, and many others have been proposed and developed in the past three decades. Although these methodologies have been successfully used to solve many problems in large complex systems, none of them has led to a theoretical framework upon which concepts and methods for system analysis and synthesis parallel to those well known for conventional dynamic systems, such as stability analysis and control design, can be developed. In [22–26], Wang have used Kosko's interpretation [10] of fuzzy sets to consider LDS as mappings on fuzzy hypercubes; and by introducing cellular structures on hypercubes using equi-distribution lattices developed in number theory [8], these mappings can be approximated as cell-to-cell mappings in a cellular space [6, 7], in which each cell represents a linguistic term (a word) defined by a family of membership functions of fuzzy sets; in this way, LDS can be studied in the cellular space, and thus, methods and concepts of analysis and synthesis developed for conventional nonlinear systems, such as stability analysis and design synthesis, can be modified and applied for LDS; while cell-to-cell mappings provide us with a very general numeric tool for studying LDS, it is not the most effective method to handle the special cases of type-I and type-II LDS to be studied here.

In this chapter, a LDS is called a type-I LDS if its states are linguistic and its evolving laws are constructed based on conventional functions by using the fuzzy extension principle [10, 28], while a LDS is called a type-II LDS if its states are linguistic and its evolving laws are modeled by a finite number of linguistic rules. The difference between type-I and type-II LDS is two-folds: First, the evolving laws of type-I LDS are of crisp forms and the fuzzy extension principle is used to assign linguistic meanings of new states based on previous states. On the other hand, the evolving laws of type-II LDS are directly approximated by a finite set of linguistic rules. Second, the computational mechanisms behind type-I and type-II LDS are different. In type-I LDS, the evolving of states is calculated accurately using the fuzzy extension principle, while in type-II LDS, the evolving of states is computed approximately using fuzzy compositions.

One special kind of LDS is fuzzy dynamic systems [19]. As addressed by Layne and Passino [11], fuzzy dynamic systems can be generated from conventional dynamic systems by using two methods. In their first method, the fuzzy extension principle is used to transform a conventional equation into a fuzzy model. From the LDS point of view, the state space of such fuzzy dynamic systems is transformed into a word space while the evolving laws are described by conventional equations. Therefore, this kind of fuzzy dynamic systems is a special case of type-I LDS. There are a few analytic results on the fixed points of type-I LDS [3, 13, 20, 21]. In [11], the Lyapunov stability of a first-order fuzzy dynamic system with Gaussian membership functions was studied. However, in general it is difficult to study the behaviors of type-I LDS analytically. In this chapter, we present a numerical method called α-cut mapping to study type-I LDS. Their second method of generating fuzzy dynamic systems is to use fuzzy compositions. In this case, the corresponding fuzzy dynamic systems are a special case of type-II LDS of which the evolving laws are represented by fuzzy relational matrices. An energetic function was introduced in [9] to study the stability of this kind of type-II LDS and later its controllability [5]. A more comprehensive form of type-II LDS will be presented here based on fuzzy mappings, along with an efficient numerical method for studying their behaviors.

Most recently, Zadeh has proposed "fuzzy LOGIC = computing with words" and the generalized constraint language (GCL) to deal with propositions of perception-based information [30, 31]. In addition, Pedrycz's group has developed the concepts and methods of information granularity [14], fuzzy granular models [15], granular world [16], and algorithms and models of generalization and specialization of information granules [17], which are quite useful in hierarchical information representation and processing. For interested readers, a good reference of the current work can be found in Wang's edited book [27]. To a certain extend, those recent works are closely related to the work on LDS described in the present chapter, however, our focus here is more on dynamic and computational aspects of computing with words.

The purposes of this chapter are three-folds: First, a systematic framework for LDS representing conventional dynamic systems will be established. Second, the conclusions derived for behaviors of LDS are verified by comparing the corresponding behaviors of conventional dynamical systems. Our purpose here is not to use LDS to model simple conventional systems, but to demonstrate the similarities and differences between conventional dynamic systems and LDS. Since the behaviors of conventional dynamic systems are well known, a benchmark can be easily set up for the purpose of comparison. Third, we try to use numerical simulation results to discover and show the unique properties of LDS.

The organization of this chapter is as follows. In Section 2.2, the method of converting conventional dynamic systems to type-I LDS based on the fuzzy extension principle and the numerical method for solving type-I LDS called α-cut mapping are constructed. In Section 2.3, the structure, numerical procedure and existence of fixed-points of type-II LDS are discussed. In Section 2.4, the LDS controller design principles for controlling type-II LDS are addressed. Finally, in Section 2.5, the chapter is concluded with remarks for future works.

2.2 Type-I Linguistic Dynamic Systems

The procedure of converting a conventional dynamic system into a type-I LDS is called *abstracting process*, namely, extracting linguistic dynamic models in words from conventional dynamic models in numbers. To transfer a conventional dynamic system into a type-I LDS, the fuzzy extension principle is used to specify the computational procedure for the current linguistic states to the next linguistic states. Note in this case, the system model has the same form as that of the corresponding conventional dynamic system.

2.2.1 General Type-I Linguistic Dynamic Systems

Let us consider a conventional nonlinear autonomous discrete-time system

$$x(k+1) = f(x(k)), \quad k \in N, \tag{2.2.1}$$

where $x(k) \in R^n$ is state vector, and $f: R^n \to R^n$ the evolving law. The union of all interested states is denoted as Ω_x. The question is "if the initial state of this system is given by a word or a linguistic term, what is the procedure to calculate the evolving process of 'words' through the conventional dynamical system?" In terms of type-I LDS modeling, this process for system (2.2.1) is represented as

$$X(k+1) = F(X(k)), \quad k \in N, \tag{2.2.2}$$

where $X(k)$ is a word defined on $W(\Omega_x)$ and $F: W(\Omega_x) \to W(\Omega_x)$ is a mapping in the word domain, where N and $W(\Omega_x)$ are the sets of all integers and words defined on Ω_x. Mapping F is computed using the fuzzy extension principle as follows:

$$\mu_{X(k+1)}(w) = \bigvee_{w=f(x)} \mu_{X(k)}(x), \quad \forall w \in \Omega_x. \tag{2.2.3}$$

In general, it is difficult to find the closed-form solution for (2.2.3). Therefore, a numerical procedure should be developed to reveal its properties.

Numerical procedure for solving type-I linguistic dynamic systems

For numerical calculation of (2.2.3), a simple sampling of the universe of discourse with the uniform resolution is not appropriate since in the vicinity of asymptotically stable equilibrium points, the sampling resolution should be much higher than those used in other regions. However, it is difficult to know where the asymptotically stable equilibrium points are without numerical analysis. To overcome this dilemma, one can trace the evolution of α-cuts of linguistic states. Based on this consideration, a numerical procedure, called α-cuts *mapping*, is introduced to analyze (2.2.3).

Step 1: Initialization

Set $k = 0$, $i = 1$ and initial linguistic state $X(0) = X_0$. Choose a set of α values $\{\alpha_1, \cdots, \alpha_S\}$, where $\alpha_{l+1} > \alpha_l$, $l = 1, \cdots, S - 1$, for the α-cuts.

Step 2: The evolving law

Calculate the α_i-cut of the evolving law using the following equation:

$$X(k + 1)_{\alpha_i}(w) = \bigcup_{w=f(x)} X(k)_{\alpha_i}. \tag{2.2.4}$$

This equation shows that a point $w \in X(k + 1)_{\alpha_i}$ if $w = f(x)$ and $x \in X(k)_{\alpha i}$, where $X(k)_{\alpha i}$ denotes the α_i-cuts of $X(k)$.

Step 3: α_i-loop

If $i < S$, let $i \leftarrow i + 1$, go to Step 2. Else, find $X(k + 1)$ by combining α-cuts into a fuzzy membership function using a discrete form of resolution of identity [29] as

$$X(k + 1)_{\alpha_i}(w) = \bigcup_{\alpha_i} \alpha_i X(k + 1)_{\alpha_i}$$

where $\alpha_i X(k + 1)_{\alpha i}$ is the fuzzy set whose membership function is

$$\mu_{\alpha_i X(k+1)_{\alpha_i}}(x) = \begin{cases} \alpha_i, & \text{if } x \in X(k + 1)_{\alpha_i}, \\ 0, & \text{otherwise.} \end{cases}$$

Step 4: k-loop

If k is less than a prescribed value then let $k \leftarrow k + 1$, $i = 1$ and go to Step 2. Else, stop. □

By using the above procedure, we can find the corresponding type-I LDS for nonlinear discrete-time control systems:

$$\begin{aligned} x(k + 1) &= f(x(k), u(k)), \\ y(k) &= h(x(k)), \\ u(k) &= g(y(k), r(k)). \end{aligned} \tag{2.2.5}$$

The behaviors of the corresponding type-I LDS in many cases are qualitatively similar to those of its conventional counterpart. However, in many other cases, its behaviors have no conventional counterparts. In order to reveal the unique behaviors of type-I LDS, examples of linear type-I LDS is investigated next.

2.2.2 Linear Type-I Linguistic Dynamic Systems

Assume that all linguistic states are represented by fuzzy numbers, for a linear control system,

$$x(k + 1) = Ax(k) + Bu(k),$$
$$y(k) = Cx(k), \qquad (2.2.6)$$
$$u(k) = Ky(k) + r(k).$$

The corresponding type-I LDS is

$$X(k + 1) = AX(k) + BU(k),$$
$$Y(k) = CX(k), \qquad (2.2.7)$$
$$U(k) = KY(k) + R(k).$$

Note that in (2.2.7), A, B, C, and K represent fuzzy operations based on the extension principle. Here the boldface is used to distinguish them from conventional matrices in (2.2.6). When system (2.2.6) is stable, the supports of linguistic states $X(k)$ of the type-I LDS (2.2.7) can be either kept in finite regions or become unbounded during the evolving processes as shown in the following examples:

$$A = \begin{pmatrix} 0.01 & 0.03 & 1.3 \\ 0 & -0.01 & 2 \\ 1 & 0 & -0.01 \end{pmatrix}, \quad B = \begin{pmatrix} -1.5 \\ -1 \\ 2 \end{pmatrix},$$

$$C = (0, 1, 0.5), \quad K = -0.3.$$

The eigenvalues of A are 1.1624, -1.1162, and -0.0563, and the original plant is unstable. The eigenvalues of $(A + BCK)$ are 0.9680, $-0.4890 + i0.6185$, and $-0.4890 - i0.6185$. Thus, the controlled plant is stable. Figure 2.2.1(a) shows the waveform of $x_1(k)$ with initial conditions $x_1(0) = x_2(0) = x_3(0) = 2$ and $r = 1$. To demonstrate that the conventional dynamic system is a special case of the corresponding type-I LDS, in Figure 2.2.1(b) the support of $X(0)$ is chosen to be very close to the crisp initial state in Figure 2.2.1(a). Figure 2.2.1(b) shows $X_1(k)$ with initial conditions $X_1(0) = X_2(0) = X_3(0) = $ "very very close to 2", represented by triangle fuzzy numbers with supports of size 2×10^{-10}. $R = $ "very very close to 1", represented by a triangle fuzzy number with support of size 2×10^{-10}. Note that in this case the behavior of the type-I LDS is almost the same as that of its conventional counterpart. However, although the center of area of $X_1(k)$ is still equal to $x_1(k)$, the support of $X_1(k)$ grows exponentially with respect to k. As one can see from Figure 2.2.1(b), the support of $X_1(k)$ grows from the order of 10^{-10} to the order of 10^{-1} in 29 iterations. To make this more

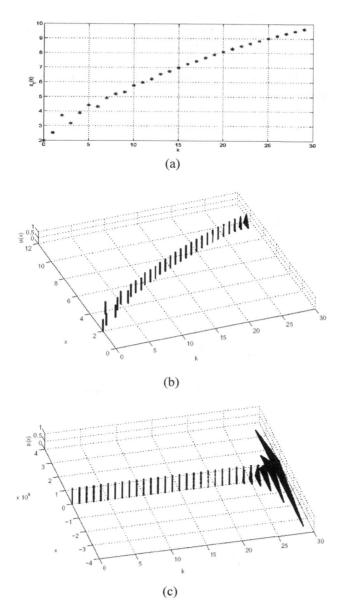

Figure 2.2.1: Dynamics of uncertainty: (a) $x_1(k)$, (b) $X_1(k)$ with very small amount of initial uncertainty, (c) $X_1(k)$ with big amount of initial uncertainty.

clear, in Figure 2.2.1(c) the sizes of supports of $X_1(0), X_2(0), X_3(0)$ and R are all chosen to 2. Observe that the support of $X_1(k)$ expands to the size of the order of 10^9 in 29 iterations in this case, however, the center of area of $X_1(k)$ is still equal to $x_1(k)$ no matter how big and how fast the supports grow.

This example indicates that the uncertainty of the initial condition grows exponentially for a stable linear system. Is this a general case? To address this question, let us analyze the evolving processes of supports of linguistic states. For a fuzzy number X, let us use $\text{supp}(X)$ to denote the size of its support. It follows from (2.2.7) that

$$\text{supp}(X_i(k+1)) = \sum_{j=1}^{n} |a_{ij}| \, \text{supp}(X_j(k)) + \sum_{j=1}^{m} |b_{ij}| \, \text{supp}(U_{ij}(k)),$$

$$\text{supp}(Y_l(k)) = \sum_{q=1}^{n} |c_{lq}| \, \text{supp}(X_q(k)),$$

$$\text{supp}(U_{ij}(k)) = \sum_{l=1}^{p} |k_{jl}| \text{supp}(Y_l(k)) + \text{supp}(R_j(k)).$$

From which we have

$$\begin{aligned}
\text{supp}(X_i(k+1)) &= \sum_{j=1}^{n} |a_{ij}| \, \text{supp}(X_j(k)) \\
&\quad + \sum_{j=1}^{m} \sum_{l=1}^{p} \sum_{q=1}^{n} |b_{ij}| \, |k_{jl}| \, |c_{lq}| \, \text{supp}(X_q(k)) \\
&\quad + \sum_{j=1}^{m} |b_{ij}| \, \text{supp}(R_j(k)) \\
&= \sum_{q=1}^{n} \left[|a_{iq}| + \sum_{j=1}^{m} \sum_{l=1}^{p} |b_{ij}| \, |k_{jl}| \, |c_{lq}| \right] \text{supp}(X_q(k)) \\
&\quad + \sum_{j=1}^{m} |b_{ij}| \, \text{supp}(R_j(k)) \\
&\leq \Delta_k \sum_{q=1}^{n} \left[|a_{iq}| + \sum_{j=1}^{m} \sum_{l=1}^{p} |b_{ij}| \, |k_{jl}| \, |c_{lq}| \right] + \Gamma_k \sum_{j=1}^{m} |b_{ij}|
\end{aligned}$$

where

$$\Delta_k \overset{\triangle}{=} \max_{1 \leq q \leq n} \text{supp}(X_q(k)), \qquad \Gamma_k \overset{\triangle}{=} \max_{1 \leq j \leq m} \text{supp}(R_j(k)).$$

When

$$\sum_{q=1}^{n} \left[|a_{iq}| + \sum_{j=1}^{m} \sum_{l=1}^{p} |b_{ij}| \, |k_{jl}| \, |c_{lq}| \right] + \sum_{j=1}^{m} |b_{ij}| < 1, \qquad (2.2.8)$$

we have

$$\lim_{k \to \infty} \operatorname{supp}(R(k)) = 0.$$

Then

$$\lim_{k \to \infty} \operatorname{supp}(X_i(k+1)) = 0.$$

For the example presented in Figure 2.2.1, clearly condition (2.2.8) is not satisfied. However, condition (2.2.8) is only sufficient for preventing the initial uncertainties from growing exponentially. In order to have a better understanding of the propagation of initial uncertainties in type-I LDS, let us assume that reference signal $r(k) = 0$, then system (2.2.6) can be rewritten as

$$x(k+1) = (A + BKC)x(k)$$

of which the corresponding type-I linear LDS is given by

$$X(k+1) = (A + BKC)X(k) \stackrel{\triangle}{=} MX(k)$$

where $M = \{m_{ij}\} \stackrel{\triangle}{=} A + BKC$. The dynamics of the center of area of $X(k)$ is determined by the eigenvalues of M matrix while the dynamics of $\operatorname{supp}(X(k))$ is characterized by the eigenvalues of matrix $\tilde{M} = \{|m_{ij}|\}$. This is due to the fact that $\operatorname{supp}(X(k))$ satisfies the following relationship:

$$\operatorname{supp}(X(k+1)) = \tilde{M}\operatorname{supp}(X(k)).$$

We present two examples to illustrate this conclusion. Assume

$$M_1 = \begin{pmatrix} 0.8 & 0 & 0 \\ 0 & 0.9 & 0 \\ 0 & 0 & 0.7 \end{pmatrix}$$

and let $M = TM_1T^{-1}$. Then M has the same eigenvalues as M_1. If we choose T as

$$T = \begin{pmatrix} 1 & 2 & 3 \\ 20 & 300 & 100 \\ 10 & 10000 & 100 \end{pmatrix}$$

then

$$M = TM_1T^{-1} = \begin{pmatrix} 0.9556 & -0.0079 & 0.0002 \\ 5.2493 & 0.5325 & 0.0100 \\ 7.7953 & -0.4457 & 0.9118 \end{pmatrix}.$$

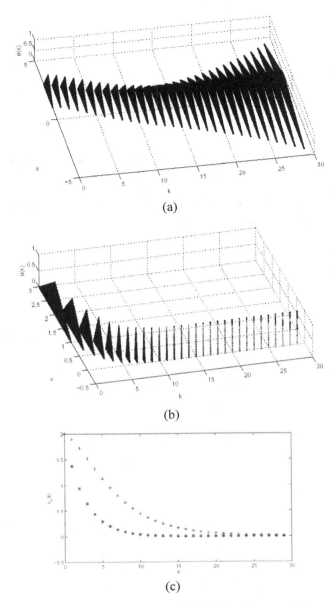

(a)

(b)

(c)

Figure 2.2.2: Dynamics of uncertainty: (a) $X_1(k)$ with expanding supports, (b) $X_1(k)$ with converging supports, (c) the center of area of $X_1(k)$ for (a) and (b) (marked by "+" and "*", respectively.

In this case, the eigenvalues of $\tilde{M}$ are given by 0.4459, 1.0649 and 0.8892. Since one of the eigenvalues of $\tilde{M}$ is outside of the unit circle, $\text{supp}(X(k))$ will increase exponentially. The simulation result is shown in Figure 2.2.2(a), the initial conditions for all linguistic states are chosen as a triangle fuzzy number with parameter [left, center, right] = [1, 2, 3].

If we choose T as

$$T = \begin{pmatrix} 1 & 2 & 30 \\ 20 & 300 & 100 \\ 10 & 10000 & 100 \end{pmatrix}$$

then

$$M = TM_1T^{-1} = \begin{pmatrix} 0.6803 & 0.0061 & -0.0001 \\ -0.4050 & 0.8190 & 0.0025 \\ -0.6015 & -0.0203 & 0.9007 \end{pmatrix}.$$

The eigenvalues of $\tilde{M}$ are given as 0.6643, 0.8332 and 0.9025. Since all eigenvalues of are within the unit circle, $\text{supp}(X(k))$ converges to zero. The simulation result is shown in Figure 2.2.2(b). The centers of area of $X_1(k)$, $c(k)$, for the above two cases are shown in Figure 2.2.2(c) as two curves marked by "+" and "*", respectively. Observe that in both cases the centers of area approaches to 0. The evolving process of $c_1(k)$ is the same as the conventional states $x_1(k)$ with $x_1(0) = c_1(0)$.

2.3 Type-II Linguistic Dynamic Systems

In a type-II LDS, the numeric state space is converted to a linguistic one, and the evolving laws are transformed from numerical functions to linguistic rules. Type-II LDS can be generated by using various kinds of mechanisms. Still, the most convenient one is based on the fuzzy extension principle.

Let us use the first equation (i.e., the plant) in control system (2.2.5) as an example to show the method of transforming a conventional dynamic system into a type-II LDS. Assume that the state and control spaces Ω_x and Ω_u are covered by word sets $\{X_1, \cdots, X_{N_x}\}$, $\{U_1, \cdots, U_{N_u}\}$, respectively. Then the corresponding type-II LDS model for (2.2.5) can be defined by $N_x \times N_u$ rules, expressed as a rule base,

$$R_{\text{plant}} \triangleq \left\{ \begin{array}{cccc} R_p(X_1, U_1), & R_p(X_1, U_2), & \cdots & R_p(X_1, U_{N_u}), \\ R_p(X_2, U_1), & R_p(X_2, U_2), & \cdots & R_p(X_2, U_{N_u}) \\ \vdots & \vdots & \ddots & \vdots \\ R_p(X_{N_x}, U_1), & R_p(X_{N_x}, U_2) & \cdots & R_p(X_{N_x}, U_{N_u}) \end{array} \right\} \quad (2.3.1)$$

where $R_p(X_i, U_j)$, $1 \le i \le N_x$, $i \le j \le N_u$ is a linguistic rule of the form

$$\text{IF } X(k) \text{ is } X_i, \ U(k) \text{ is } U_j, \ \text{THEN } X(k+1) \text{ is } F(X_i, U_j). \qquad (2.3.2)$$

Using the fuzzy extension principle, word $F(X_i, U_j)$ is calculated by

$$\mu_{F(X_i, U_j)}(w) = \bigvee_{w=f(x,u)} \mu_{X_i}(x) \wedge \mu_{U_j}(u), \forall w \in \Omega_x. \qquad (2.3.3)$$

Therefore, $R_p(X_i, U_j)$ is a fuzzy implication with a membership function specified by

$$\mu_{R_p(X_i, U_j)}(v, u, x_{k+1}) = \mu_{X_i}(v) \wedge \mu_{U_j}(u) \wedge \mu_{F(X_i, U_j)}(x_{k+1}).$$

Intuitively, as the number of words covering Ω_x and Ω_u increases, the accuracy of a type-II LDS approximating its conventional counterpart improves. However, this statement is not quite true because the approximation accuracy depends on not only the number of words used but also the nature of fixed points of the type-II LDS. In many cases, the fixed points of a type-II LDS are not in the vicinity of those of its conventional counterpart even though $N_x \times N_u$ are quite large.

2.3.1 Examples

As an example, we examine the well-known logistic system controlled by a simple feedback rule:

$$x(k+1) = -\gamma \left[x(k) - \frac{1}{2}(1 + \frac{1}{\gamma}) \right]^2 + \gamma \left[\frac{1}{4}(1 + \frac{1}{\gamma})^2 \right] + u(k),$$
$$y(k) = x(k), \qquad\qquad\qquad\qquad\qquad\qquad\qquad\qquad (2.3.4)$$
$$u(k) = -1.5y(k).$$

When $\gamma = 1$, this controlled system asymptotically approaches 0. Since logistic system can become chaotic through a period-doubling bifurcation route, parameter perturbations can easily induce periodic and even chaotic behaviors. When we transfer system (2.3.4) into a type-II LDS, the parameter perturbation may be such that the logistic system becomes chaotic. In this case, the corresponding type-II LDS may oscillate irregularly.

In Figure 2.3.1(a), the rule base for the corresponding type-II LDS with 5×5 words is presented. The center of area $c(k)$ of $X(k)$ is shown in Figure 2.3.1(b). Note that in this case $c(k)$ locates exactly at the equilibrium point of the conventional control system. This is just a special case where the type-II LDS happens to have a fixed point of which the center of area is at 0. Generally speaking, an increase in the number of words does reduce the approximation errors. Let us

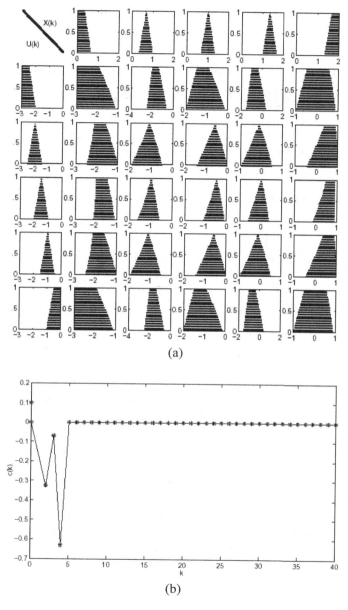

(a)

(b)

Figure 2.3.1: (a) Rule base for type-II LDS with 5×5 words covering the state and control spaces; (b) Centers of area of $X(k)$.

Table 2.3.1: Relation between number N_x and center of area of linguistic state, a chaotic system

N_x	1	2	3	4	5	6
$\bar{c}$	0	−0.91	0	−0.58	0	−0.34
Remark	S	S	S	P2	S	P2
N_x	7	8	9	10	11	
$\bar{c}$	−0.17	−0.27	−0.14	−0.07	−0.04	
Remark	S	P2	PM	P2	PM	

Table 2.3.2: Relation between N_x and center of area of linguistic state, a stable system

N_x	1	2	3	4	5	6	7
$\bar{c}$	−0.48	−0.44	−0.35	−0.30	−0.26	−0.21	−0.18
N_x	8	9	10	11	15	20	
$\bar{c}$	−0.17	−0.15	−0.135	−0.13	−0.09	−0.08	

choose the average of $c(k)$ (denoted by $\bar{c}$) between $k = 30$ and 40 as the average approximation of type-II LDS to the conventional system under various situations, then list $\bar{c}$ as a function of $N_x = N_u$. The results are presented in Table 2.3.1, where symbols S, P2 and PM denote steady, period-2 and irregular oscillating solutions, respectively. From this table we first observe that $\bar{c}$ shows a decreasing trend when N_x becomes large. Second, $c(k)$ becomes more and more likely to oscillate irregularly around its average value as N_x increases. This is because when N_x increases, the number of fixed points of the LDS increases so that the solution is more likely to jump from one fixed point to another.

Since logistic system is chaotic under different parameter perturbations, it is not very clear how much its chaotic behaviors contribute to the approximation errors of the type-II LDS to the conventional counterpart. To clarify the effects of number of words to the approximation error, examine a very simple control system

$$x(k + 1) = 1.1x(k) + u(k),$$
$$y(k) = x(k), \qquad (2.3.5)$$
$$u(k) = -1.5y(k).$$

This system converges to 0 with arbitrary initial conditions. Using the similar process as in the previous example, the relation between $\bar{c}$ and N_x is presented in Table 2.3.2, from which one can observe that $\bar{c}$ approaches to zero monotonically as N_x increases. This is consistent to the intuition that the more words a LDS uses to model a conventional system, the more accurate the approximation will be.

2.3.2 Fixed Points of Type-II Linguistic Dynamic Systems

To simplify our analysis, here we only study the fixed points of autonomous type-II LDS. Assume a set of words $\{X_1, \cdots, X_{Nx}\}$ is used to cover Ω_x, and that each word is represented by a fuzzy set:

$$X_i \triangleq \mu_{X_i}(x), x \in \Omega_x, i = 1, \cdots, N_x.$$

The evolving law consists of N_x rules in a rule base,

$$R_p \triangleq \{R_p(X_1), R_p(X_2), \cdots, R_p(X_x)\} \tag{2.3.6}$$

where $R_p(X_i)$, $1 \leq i \leq N_x$, are linguistic rules of the form

$$\text{IF } X(k) \text{ is } X_i, \text{ THEN } X(k+1) \text{ is } W(X_i),$$

where X_i and $W(X_i)$ are two fuzzy sets. Using the extension principle, for a conventional autonomous dynamic system,

$$x(k+1) = f(x(k)),$$

the membership function of W can be calculated as

$$\mu_{W(X_i)}(w) = \bigvee_{w=f(x)} \mu_{X_i}(x), \ x \in \Omega_x.$$

Therefore, $R_p(X_i)$ is an implication whose membership function is given by

$$\mu_{R_p(X_i)}(v, x_{k+1}) = \mu_{X_i}(v) \wedge \mu_{W(X_i)}(x_{k+1}).$$

Now, given a linguistic state $X(k)$, it follows from the linguistic model (2.3.6) that the state at $(k+1)$st iteration, $X(k+1)$, is found by the following sup-* composition:

$$\mu^{[i]}_{X(k+1)}(x_{k+1}) = \bigvee_v \left[\mu_{X(k)}(v)\right] \wedge \left[\mu_{X_i}(v) \wedge \mu_{W(X_i)}(x_{k+1})\right], \forall x_{k+1} \in \Omega_x$$

and

$$\mu_{X(k+1)}(x_{k+1}) = \bigvee_{1 \leq i \leq N_x} \mu^{[i]}_{X(k+1)}(x_{k+1}). \tag{2.3.7}$$

Assume all fuzzy sets used for words are convex, without loss of generality, for a given state $X(k)$, consider first only the contribution of the ith rule $R_p(X_i)$ to the next state $X(k+1)$:

$$\mu_{X(k+1)}(x_{k+1}) = (\underset{v}{\vee} \mu_{X(k)}(v) \wedge \mu_{X_i}(v)) \wedge \mu_{W(X_i)}(x_{k+1}). \qquad (2.3.8)$$

This corresponds to the case when the ith rule is active and all the other rules are not activated. Define $x_{kv} \in \Omega_x$ as a point satisfying

$$\mu_{X(k)}(x_{kv}) \wedge \mu_{X_i}(x_{kv}) = \underset{v}{\vee} \mu_{X(k)}(v) \wedge \mu_{X_i}(v).$$

Then we can construct a point-to-fuzzy-set mapping $\psi \colon \Omega_x \to W(\Omega_x)$ as

$$\mu_{\psi(x_{kv})}(x) \overset{\triangle}{=} [\mu_{X_k}(x_{kv}) \wedge \mu_{X_i}(x_{kv})] \wedge \mu_{W(X_i)}(x), \forall x \in \Omega_x.$$

It follows from the definition of the fixed points of point-to-fuzzy-set mappings that a fixed point of ψ, x^* must satisfies the condition [2]

$$\mu_{\psi(x^*)}(x^*) \geq \mu_{\psi(x^*)}(x), \forall x \in \Omega_x. \qquad (2.3.9)$$

Therefore, (2.3.8) can be represented in the form of a fuzzy mapping

$$X(k+1) = \psi(x_{kv})$$

from which and (2.3.9), if x_{kv} is a fixed point of ψ then $\psi(x_{kv})$ is a fixed point of the ith rule. Therefore, from Theorem 2.4 of [2], if $C \subseteq \Omega_x$ is a nonempty convex compact subset of a real locally convex topological vector space, and ψ is a convex and closed fuzzy mapping over C, then ψ has a fixed point in C. The fuzzy mapping ψ is *closed* if and only if it is upper semicontinuous, and it is *convex* if and only if for any $x \in \Omega_x$, fuzzy set $\psi(x)$ is a convex fuzzy set. Therefore, a fuzzy set X^* is a fixed point of the ith rule (2.3.8) if the following conditions are satisfied:

C1: X_i and $W(X_i)$ are convex.

C2: The supports of X_i and $W(X_i)$ are overlapped to each other.

C3: $\mu_{X^*}(w) = (\vee_x \mu_{X^*}(x) \wedge \mu_{X_i}(x)) \wedge \mu_{W(X_i)}(w)$.

Once the fixed points of each rule are found, the fixed points of the entire rule base R_p are simply the union of fixed points of all rules. When different initial conditions are applied to a type-II LDS, the LDS can settle at different equilibrium states which are the unions of fixed points of certain rules.

Remark 2.3.1. Conditions C1–C3 for the existence of fixed points of the ith rule have important applications in the design of linguistic controllers for type-II LDS.

For example, assume that we have a plant whose dynamics is represented by a type-II LDS and there are no fixed points or the fixed points are in undesirable regions. In this situation, we can use linguistic control laws to manipulate the dynamics of the plant such that conditions C1–C3 are satisfied in the desired regions. It follows from condition C3 that a fixed point X^* must be in $W(X_i)$. □

To verify that conditions C1–C3 are sufficient to the existence of fixed points of type-II LDS, consider the fixed points of the following rule:

IF $X(k)$ is "close to 1", THEN $X(k+1)$ is "close to 2".

Note that triangle fuzzy numbers are used to represent both "close to 1" and "close to 2", with corresponding supports as (0,2) and (1,3), respectively. Thus, the overlapped region of two supports is interval (1,2), hence conditions C1 and C2 are satisfied. Analytically, the membership functions of X_i and $W(X_i)$ can described as

$$\mu_{X_i}(x) = \begin{cases} x, & x \in (0,1], \\ 2-x, & x \in (1,2), \\ 0, & \text{otherwise}, \end{cases} \qquad \mu_{W(X_i)}(x) = \begin{cases} x-1, & x \in (1,2], \\ 3-x, & x \in (2,3), \\ 0, & \text{otherwise}. \end{cases}$$

Since both membership functions are symmetric with respect to $x = 1.5$, it follows that $x_{kv} = 1.5$ and $\mu_{X_i}(x_{kv}) = 0.5$. Therefore, the "biggest" fixed point, X_1^*, is given by

$$\mu_{X_1^*}(x) = \begin{cases} x-1, & x \in (1,1.5], \\ 0.5, & x \in (1.5,2.5], \\ 3-x, & x \in (2.5,3), \\ 0, & \text{otherwise} \end{cases}$$

which is a trapezoid fuzzy set. Other fixed points have the general form

$$\mu_{X^*}(x) = \begin{cases} x-1, & x \in (1,1+a]. \\ a, & x \in (1+a,3-a]. \\ 3-x, & x \in (3-a,3). \\ 0, & \text{otherwise}. \end{cases}$$

However, to which fixed point the type-II LDS converges depends on initial states. If the support of the initial state is not overlapped with that of X_i, then the type-II LDS does not approach to fixed points.

The numerical results are shown in Figure 2.3.2. The α-cuts mapping with $S = 40$ is used in the computation. In Figure 2.3.2(a) shows the result when initial state $X(0) = $ "close to 1.5". Note that when $k = 1$, the state is not at the fixed point. At $k = 2$, the state arrives at the "biggest" fixed point X_1^*. In Figure

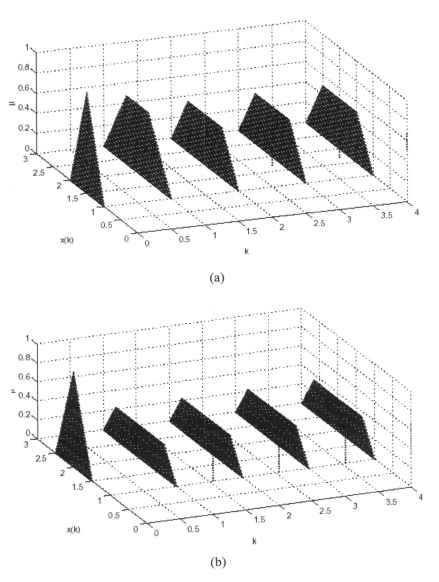

(a)

(b)

Figure 2.3.2: Fixed points of a simple type-II LDS: (a) the "biggest" fixed point X_1^*; (b) the second fixed point which is a subset of X_1^*.

2.3.2(b) presents the result when initial state $X(0) =$ "close to 2". Note that when $k = 2$, the state arrives at a fixed point whose height is smaller than that of X^* given in Figure 2.3.2(a).

2.4 Linguistic Control Design for Goal States Specified in Words

In many cases, one desires to drive a LDS to a prescribed state using *linguistic control laws*, so called problem of designing linguistic controllers for LDS. In this section, we present a simple procedure for constructing linguistic controllers for type-II LDS along with several numerical examples.

2.4.1 Design Procedure

The following procedure is proposed to construct a linguistic controller for a type-II LDS:

1. Assume LDS model (2.3.1) has been founded using the procedure described in the previous section or other methods.

2. Specify the control goal by word X^*. It is impossible to make the controlled LDS to arrive at X^* unless for some i, j and U, X^* satisfies

$$\mu_{X^*}(x) = \bigvee_{v,u} \left[\mu_{X^*}(v) \wedge \mu_U(u)\right] \wedge \left[\mu_{X_i}(v) \wedge \mu_{U_j}(u) \wedge \mu_{F(X_i,U_j)}(x)\right],$$
$$\forall x \in \Omega_x$$

3. In order to make the controlled LDS settle at X^*, find all i and j such that $\text{supp}(X^*)$ and $\text{supp}(F(X_i,U_j))$ are overlapped, namely,

$$\Lambda \overset{\triangle}{=} \{(i,j)|\text{supp}(X^*) \cap \text{supp}(F(X_i, U_j)) \neq \phi\}.$$

If $\Lambda = \phi$, then the LDS cannot be controlled to state X^* and one has to go back to Step 2 and choose a new control goal.

4. Although it is impossible to control an LDS without introducing additional uncertainties along the control process, our goal is to find an optimal control policy that would minimize the uncertainty in the final state. However, it is extremely difficult to conduct such optimization over the entire control process at this point. Instead, a stepwise minimization of uncertainty for the next state is performed on the control policy here. Define the loss of certainty as the amount of decrease in the maximum truth value of the membership of the next state, and then the following procedure is used to determine the control word at iteration k:

4.1) Let (p, q) point to the first element of the set Λ and set $\Gamma = 0$.

4.2) Calculate

$$x_{\max} = \bigvee_x \mu_{X(k)}(x) \wedge \mu_{X_p}(x).$$

4.3) Choose control $U_{p,q}(k)$ satisfying the relation

$$x_{\max} \leq \bigvee_u \mu_{U_{p,q}(k)}(u) \wedge \mu_{U_q}(u).$$

Obviously, the simplest way to choose $U_{p,q}(k)$ is to make $U_{p,q}(k) = \mu_{U_q}(u)$. Another possible way to find $U_{p,q}(k)$ is to use equation

$$x_{\max} = \bigvee_u \mu_{U_{p,q}(k)}(u) \wedge \mu_{U_q}(u).$$

Note that there are infinite many solution for this equation and so it provides us an additional degree of freedom for choosing control words. In our simulation, we simply set $U_{p,q}(k)$ to

$$\mu_{U_{p,q}(k)}(u) = x_{\max} \wedge \mu_{U_q}(u).$$

4.4) Calculate

$$F_{\max} = \bigvee_x \mu_{X^*}(x) \wedge \mu_{F(X_p, U_q)}(x).$$

4.5) If $\Gamma < \min(x_{\max}, F_{\max})$ go to Step 4.6. Otherwise, let

$$\Gamma = \min(x_{\max}, F_{\max}) \text{ and } U(k) = U_{p,q}(k).$$

4.6) If (p,q) is not the last element in Λ, then let (p,q) point to the next element in Λ and go to Step 4.2.

4.7) Choose $U(k)$ as the control policy at iteration k. $U(k)$ guarantees that the loss of certainty in the state is minimized from k to $k + 1$.

5. Calculate $X(k + 1)$ using the fuzzy composition and rule base in (2.3.1).

6. If stopping conditions are not met, go to Step 4. Else, stop.   □

Remark 2.4.1. Notice that this linguistic control algorithm is to emulate a dynamic policy maker by minimizing the loss of certainty at each iteration. It is possible to construct different linguistic controllers if goals or criteria for optimization are different. Since LDS are most likely applied to deal with "soft" decision-making processes and used by human policy makers, the standard "defuzzification" procedure in fuzzy control systems may not be necessary here.   □

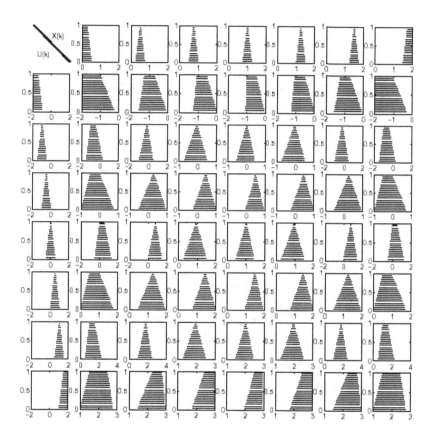

Figure 2.4.1: Evolving relations of a type-II LDS from logistic system with 7×7 words.

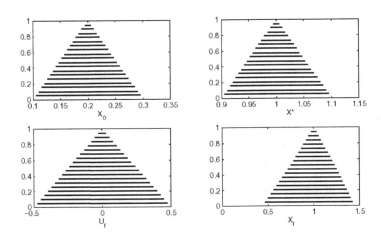

Figure 2.4.2: Simulation 1 – initial state, control goal, final control and state.

2.4.2 Numerical Examples

Figure 2.4.1 presents the evolving relations $R_p(X_i, U_j)$ of a type-II LDS that has been produced from the controlled logistic system using the extension principle with parameter $\gamma = 1$ when 7 words are used to cover Ω_x and Ω_u, respectively. The first row of Figure 2.4.1 lists all state words X_i, $i = 1, \cdots, 7$. while the first column lists all input or control words U_j, $j = 1, \cdots, 7$. Each membership function is represented by its 20-level α-cuts. The membership function in the $(j+1)$st row and $(i+1)$st column represents $F(X_i, U_j)$. Our goal is to control this type-II LDS to a given word X^* when its initial linguistic state is word X_0. The design procedure presented in Section 2.4.1 is used to find control words or policies at each iteration. The following numerical examples are used to illustrate the proposed procedure for designing LDS controllers.

In the first simulation, the initial condition is given by $X_0 = $ "close to 0.2" with a membership function given at the upper-left corner of Figure 2.4.2. The control goal is word $X^* = $ "close to 1" with a membership function given at the upper-right corner of Figure 2.4.2. The evolving process of the state is shown in Figure 2.4.3(a). Note that after one iteration, the controlled LDS settles at a word very close to the control goal. Therefore the control policy works well to drive the LDS to "close to 1". This can clearly be seen from the center of area $c(k)$ of $X(k)$ as shown in Figure 2.4.3(b), where $c(k)$ is only a little bit less than 1 for $k \geq 2$. The evolving process of the control policy is illustrated in Figure 2.4.3(c). For the purpose of comparison, the membership function of the final state $X_f = X(6)$ is

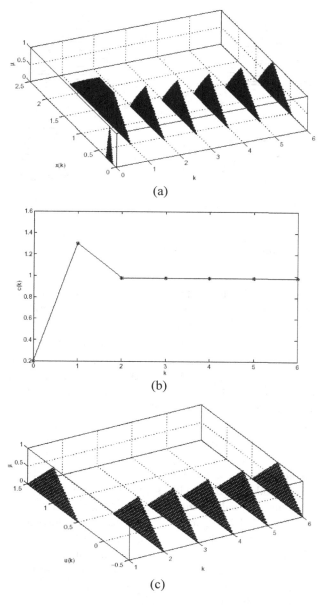

Figure 2.4.3: Simulation 1: (a) linguistic state $X(k)$, (b) COA of $X(k)$, (c) control $U(k)$.

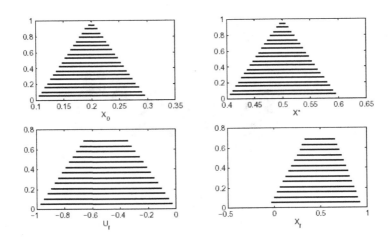

Figure 2.4.4: Simulation 2 – initial state, control goal, final control and state.

presented in Figure 2.4.2. Clearly, the support of X_f is much bigger than that of X^*, the controlled result is only a fuzzy approximation to the control goal.

In the second simulation, the initial condition is given by $X_0 =$ "close to 0.2" and the control goal is $X^* =$ "close to 0.5", as shown in Figure 2.4.4.

The evolving process of the state is illustrated in Figure 2.4.5(a). Note that after two iterations, the controlled LDS settles to a period-2 solution near the control goal X^*. This can be clearly seen from the center of area of $X(k)$ in Figure 2.4.5(b). The evolving process of the control policy is presented in Figure 2.4.5(c). Observe that the control policy also has a periodic-2 oscillation due to the low accuracy of the LDS plant model. To eliminate it, one needs to increase the accuracy of the LDS plant model by using more words to cover Ω_x and Ω_u.

In the third simulation, instead of 7 words, 10 words are used to cover Ω_u and a new LDS plant model is produced as shown in Figure 2.4.6. Assume that the initial state and control goal are the same as in Figure 2.4.4, the control result is presented in Figure 2.4.7. Figure 2.4.7(a) indicates that after the first iteration the system arrives at an equilibrium word that takes X^* as its subset. However, as shown in Figure 2.4.8, the support of the equilibrium word is almost 7 times as bigger as that of X^*. Therefore, the uncertainty in the equilibrium word is much higher than that in X^*. But the center of area of $X(k)$, as shown in Figure 2.4.7(b), indicates that the equilibrium word is reasonably "close to 0.5" and thus the control goal is accomplished. The evolving process of the control policy is described in Figure 2.4.7(c).

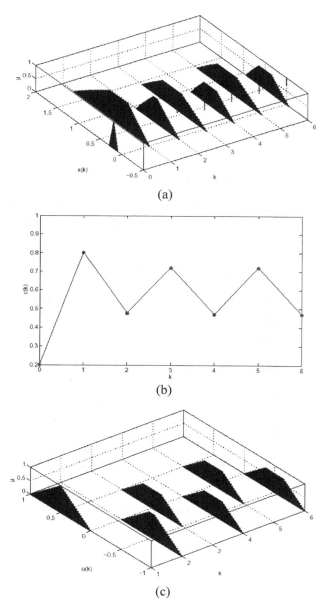

Figure 2.4.5: Simulation 2: (a) linguistic state $X(k)$, (b) COA of $X(k)$, (c) control $U(k)$.

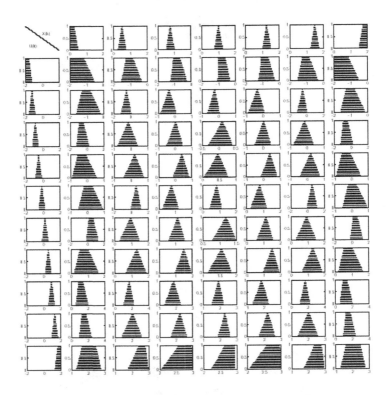

Figure 2.4.6: Evolving relations of a type-II LDS from logistic system with 7×10 words.

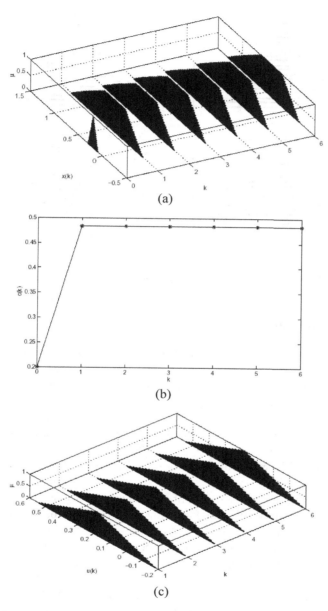

Figure 2.4.7: Simulation 3: (a) linguistic state $X(k)$, (b) COA of $X(k)$, (c) control $U(k)$.

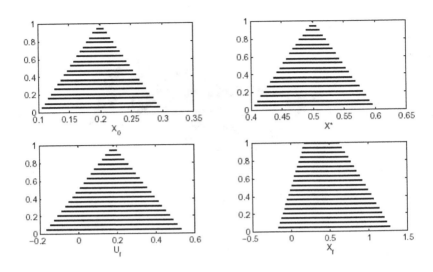

Figure 2.4.8: Simulation 3 – initial state, control goal, final control and state.

2.5   Conclusions

In this chapter we present the concepts and structures of type-I and type-II LDS for abstracting linguistic dynamic models from conventional dynamic systems described by differential or difference equations. A numerical procedure called α-cuts mapping for analyzing type-I and type-II LDS is developed and behaviors of the fixed points of type-II LDS are investigated. In addition, to drive a type-II LDS to given goal states specified by words or linguistic, a simple method of constructing linguistic controllers for type-II LDS is proposed and simulation results indicate that linguistic controllers perform well if LDS plant models are accurate enough.

In actual applications, the design of an LDS usually begins with raw data and one needs to use different technologies to "extract" LDS models from the data. Our future work will focus on such LDS whose control strategies would be management policies for complex systems or processes, such as social and economical systems and their decision-making processes. While LDS controllers may not be "good" or "accurate" enough to control conventional systems whose structures and parameters are normally known with great accuracy, they may still have great potential in decision making and decision analysis for complex systems whose structures and parameters are virtually unknown and are extremely difficult to find or even identify. Of course, the current work is only a small step towards such objectives.

Acknowledgements

This work is supported in part by the Outstanding Young Scientist Research Fund (60125310) and a grant on LDS research from the National Natural Science Foundation of China. The author would also like to express his appreciation to his former research assistant, Mr. Yang Tao, for his assistance and help in conducting all numerical simulation examples and in formulating part of analysis in this chapter.

Bibliography

[1] M. Braae and D. A. Rutherford, "Theoretical and linguistic aspects of the fuzzy logic controller," *Automatica*, vol. 15, no. 5, pp. 553–577, 1979.

[2] D. Butnariu, "Fixed points for fuzzy mappings," *Fuzzy Sets Syst.*, vol. 7, no. 2, pp. 191–207, 1982.

[3] C. A. Cabrelli, B. Forte, U. M. Molter, and E. R. Vrscay, "Iterated fuzzy set systems: a new approach to the inverse problem for fractals and other sets," *J. Math. Anal. Appl.*, vol. 171, no. 1, pp. 79–100, 1992.

[4] B. Cleveland and A. Meystel, "Dynamic predictive planning + fuzzy logic CONTROL=intelligent control," *Proc. of IEEE Int'l Symp. Intelligent Control*, Philadelphia, PA, 1990, pp. 101–106.

[5] M. M. Gupta, G. M. Trojan, and J. B. Kiszka, "Controllability of fuzzy control systems," *IEEE Trans. Systems, Man and Cybernetics*, vol. SMC-16, no. 4, pp. 576–582, 1986.

[6] C. S. Hsu, "A generalized theory of cell-to-cell mapping for nonlinear dynamical systems," *ASME J. Appl. Mech.*, vol. 48, pp. 634–842, 1981.

[7] C. S. Hsu, *Cell-to-Cell Mapping*, New York: Springer-Verlag, 1987.

[8] L. K. Hua and Y. Wang, *Applications of Number Theory to Numerical Analysis*, New York: Springer-Verlag, 1981.

[9] J. B. Kiszka, M.M . Gupta, and P. N. Nikiforuk, "Energetic stability of fuzzy dynamic systems," *IEEE Trans. Systems, Man and Cybernetics*, vol. SMC-15, no. 8, pp. 783–792, 1985.

[10] B. Kosko, *Neural Networks and Fuzzy Systems: a Dynamical Systems Approach to Machine Intelligence*, Englewood Cliffs, NJ: Prentice Hall, 1992.

[11] J. R. Layne and K. M. Passino, "Lyapunov Stability Analysis of Fuzzy Dynamic Systems," in *Fuzzy Control: Synthesis and Analysis*, New York: Wiley, 2000, pp. 95–111.

[12] E. H. Mamdani and S. Assilian, "Applications of fuzzy logic to approximate reasoning using linguistic synthesis," *IEEE Trans. Comput.*, vol. 26, no. 12, pp. 1182–1191, 1977.

[13] H. T. Nguyen, "A note on the extension principle for fuzzy sets," *J. Math. Anal. Appl.*, vol. 64, no. 2, pp. 369–380, 1978.

[14] W. Pedrycz and G. Bortolan, "Reconstruction problem and information granularity, fuzzy sets and systems," *IEEE Trans. Fuzzy Syst.*, vol. 5, no. 2, pp. 234–248, 1997.

[15] W. Pedrycz and A. V. Vasilakos, Linguistic models and linguistic modeling," *IEEE Trans. Systems, Man and Cybernetics*, vol. 29, no. 6, pp. 745–757, 1999.

[16] W. Pedrycz and G. Vukovich, Granular words: representation and communication problems," *Int. J. Intell. Syst.*, vol. 15, pp. 1015–1026, 2000.

[17] W. Pedrycz and G. Vukovich, Abstraction and specialization of information granules," *IEEE Trans. Systems, Man and Cybernetics*, vol. 31, no. 1, pp. 106–111, 2001.

[18] T. J. Procyk and E. H. Mamdani, A linguistic self-organizing process controller," *Automatica*, vol. 14, pp. 15–30, 1978.

[19] J. B. Pu, *A Cell-to-Cell Mapping Based Analysis and Design of Fuzzy Dynamic Systems and its Applications*, Ph.D. Thesis, Systems and Industrial Engineering Department, University of Arizona, 1995.

[20] H. Roman-Flores, L. C. Barros, and R.C. Bassanezi, "A note on Zadeh's extensions," *Fuzzy Sets Syst.*, vol. 117, no. 3, pp. 327–331, 2001.

[21] H. Roman-Flores and A. Flores-Franulic, "Discrete fuzzy dynamical systems," *Proc. 18th International Conference of the North American Fuzzy Information Processing Society-NAFIPS*, 1999, pp. 75–76.

[22] F.-Y. Wang, "Modeling, analysis and synthesis of linguistic dynamic systems: a computational theory," *Proc. of IEEE International Workshop on Architecture for Semiotic Modeling and Situation Control in Large Complex Systems*, Monterey, CA, Aug. 1995, pp. 173–178.

[23] F.-Y. Wang, "Outline of a computational theory for linguistic dynamic systems: towards computing with words," *Int. J. Intell. Control Syst.*, vol. 2, no. 2, pp. 211–224, 1998.

[24] F.-Y. Wang and Y. Lin, "Linguistic dynamic systems and computing with words for complex systems," *Proc. IEEE International Conference on Systems, Man, and Cybernetics*, Nashville, TN, 2000, vol. 4, pp. 2399–2404.

[25] F.-Y. Wang, Y. Lin, and J. B. Pu, "Linguistic Dynamic Systems and Computing with Words for Modeling and Simulation of Complex Systems," in *Discrete Event Modeling and Simulation Technologies*, H. S. Sarjoughian and F. E. Cellier, Editors, New York: Springer-Verlag, 2001, pp. 75–92.

[26] F.-Y. Wang, "A framework for computing with words and linguistic dynamic systems," *Journal of Pattern Recognition and Artificial Intelligence*, vol. 14, no. 5, pp. 337–384, 2001.

[27] P. Wang, *Computing with Words*, New York: Wiley, 2001.

[28] L. A. Zadeh, "Fuzzy sets," *Inform. Control*, vol. 8, pp. 338–353, 1965.

[29] L. A. Zadeh, "The concept of linguistic variable and its application to approximate reasoning," *Inform. Sci.*, vol. 8 & 9, pp. 199–249, 301–357, 43–80, 1975.

[30] L. A. Zadeh, "Fuzzy LOGIC = computing with words," *IEEE Trans. Fuzzy Syst.*, no. 4, pp. 103–111, 1996.

[31] L. A. Zadeh, "From computing with numbers to computing with words–From manipulation of measurements to manipulation of perceptions," *IEEE Trans. Circuits Syst.*, vol. 45, no. 1, pp. 105–119, 1999.

Chapter 3

Slicing: A Distributed Learning Approach

Steven A. Eschrich and Lawrence O. Hall

Abstract: Data mining is an active area of research involving many large-scale machine learning issues. Ever-increasing dataset sizes and ever-larger problems have spurred research into efficient learning methods [4, 5, 7, 32, 37, 41, 45], including the traditional computer science method of divide and conquer. Distributed learning involves multiple learners working in a distributed fashion to solve the same problem. Distributed learning can lead to better scalability and is generally well-suited to the parallel and massively distributed architectures provided by the modern Internet. Slicing, or partitioning and learning, is a new learning algorithm that embodies the principles of distributed learning, not only to simplify a learning problem overall but also to simplify the individual learning tasks. It can be seen as a method in which the training set is partitioned into disjoint regions of feature space. These disjoint regions can be treated as a set of learning tasks that can be run in a distributed environment, without extensive communication. Classification takes place by assigning a test example to the classifier built from the corresponding region of feature space. Both the base learner and the partitioning strategy can be varied in this approach. Thus, slicing can be used as a general meta-learning technique for distributed learning. The slicing algorithm is examined with respect to a series of real-world datasets, including a biologically-motivated problem. Potential difficulties can occur with the slicing algorithm, however, solutions exist within the literature for mitigating the instability that arises when learning from less data. Slicing is demonstrated to be more accurate than using a single classifier, can reduce the individual learning task size and provides a mechanism to distribute the data mining task.

3.1   Introduction

Distributed learning, in part, attempts to reduce the individual computational requirements of individual learners. Common approaches to lowering the size of a learning task involve subsampling or random disjoint partitioning [32]. These approaches lower the number of examples used for constructing a classifier. Typically the smaller subsample or disjoint partition increases the likelihood of a poor classifier, therefore they are used within an ensemble. Slicing is a meta-learning technique that attempts to address this difficulty by not only partitioning the data into subsets, but doing so in an intelligent fashion so as to reduce the overall learning task size while maintaining overall accuracy.

We consider an example of a typical distributed algorithm that also reduces the individual learning task size. The technique is known as an ensemble of small subsamples [11]. An ensemble of classifiers is constructed by repeatedly choosing small sub-samples of data from the training set (typically without replacement). It has been shown [11] that the number of ensemble members and the subsample size must both be taken into account when using this technique. The individual learning task sizes are reduced through the use of subsamples, thereby making it an attractive distributed learning algorithm.

Several learning techniques are relevant to slicing. Perhaps the most significant similarity is to radial basis functions (RBF networks) in neural networks [36, 40]. RBF networks create a collection of basis functions throughout feature space, possibly through the use of k-means clustering. Output responses (classification) are learned as weighted combinations of the distances to each kernel. Slicing uses a concept similar to radial basis functions, however the use of arbitrary classifiers within a cluster (or kernel) lends further flexibility to the approach. Additionally, slicing is explicitly defined for a distributed framework in such a way as to reduce the individual learning task size.

The neural network community has long utilized a mixture of experts (ME) architecture [23, 27] for creating localized expertise (see Figure 3.1.1 for a graphical description as adapted from [19]). In the ME method, a series of neurons are specialized for particular subsets of training examples. Each neuron is presented with each training pattern and a typical neural network learning method (e.g., backpropagation [44]) is used to update the weights. Due to random weight initializations, the individual neurons will diverge in terms of predictions. Thus each neuron learns a classification model independently. A gating network (or arbiter) is used to learn an appropriate mixture of each neuron's output. The gating network is itself a neuron, using only the training example as input. For each pattern, it is expected to learn the appropriate mixture of each expert neuron. Thus certain expert neurons focus on particular subsets of examples and are given large weights

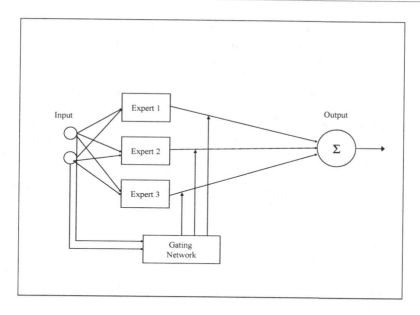

Figure 3.1.1: Mixture of experts.

in the overall mixture, for those examples. However, each neuron is expected to learn using all of the available data.

The mixture of experts approach is similar to broader ensemble combination methods. These include approaches such as stacking [48] and meta-learning [6–8]. Both methods learn to select the appropriate classifier or combination of classifiers for a given test example as a second-level machine learning problem. The mixture of experts is unique in that the arbiter does not train on the output of individual experts. Slicing addresses scalability by defining regions in which individual classifiers are expected to be expert, in advance of training.

Another similar approach is the Dynamic Classifier Selection by Local Accuracy (DCS-LA) algorithm [49]. In this method, a group of classifiers is created from the entire training set. The individual classifiers can either be generated by the same learner (e.g., a neural network using different random initializations) or by a group of different learners (e.g., neural networks, decision trees, etc.). Given a particular test example, the most accurate classifier for the surrounding local region of feature space is determined. The simplest accuracy estimation is the training set accuracy within the local region, as determined by the K-nearest neighbors in the training set. The approach improved classification accuracy in all datasets used in their experiments. The DCS-LA method is similar in spirit to the mixture of experts approach in which classifiers are considered to be "expert" in only certain regions of feature space. Thus the final meta-classifier is essentially a piece-wise set of classifiers.

Both the mixture of experts and the DCS-LA methods for learning require all classifiers to be constructed from the entire training set. In addition, they are expected to produce reasonably accurate predictions across the entire feature space even if the prediction will not be used. Slicing is directed toward limiting the coverage of an individual classifier, thereby reducing the task size associated with learning. Partitioning occurs in advance and thus pre-determines the regions of local expertise. The partitioning occurs according to similarity in feature space, thus allowing us to realize computational savings in the learning process.

3.2 Slicing

Slicing is a meta-learning technique that divides a dataset into contiguous partitions of feature space. Both the partitioning process and the learning method are variables for this technique.

The term slicing is used to denote the method of scaling the learning process through partitioning, specifically clustering. The process of slicing consists of several key steps that are shown in Figure 3.2.1. This illustration uses two features (dimensions) of the Iris dataset, a small dataset from the UCI Machine Learning Repository [3]. Each ellipse represents a cluster of data and the three classes in the dataset are represented by different symbols. Partitioning of the dataset can happen in a variety of ways; clustering is one possible approach. Once the partition is created, a learner can then be used to construct models on each subset of data. In Figure 3.2.1, this notion is illustrated by creating decision trees or neural networks from each cluster of data. In the simplest form, slicing requires no communication among subsets of data. Classification is done by choosing the most appropriate classifier from the possible slices. In Figure 3.2.1, this corresponds to identifying the cluster in which the test example falls then using the corresponding classifier.

Slicing reduces individual learning tasks by reducing the required coverage of their classification models. Slicing can be accomplished by creating non-overlapping clusters of data. This approach not only creates disjoint subsets of data, but limits a subset to a particular region of feature space. A variety of clustering algorithms exist within the literature; the simple k-means algorithm [18, 33] was chosen for this work. The scalability of the clustering algorithm can be a key weakness in its adoption for slicing. The viability of the approach through use of simple clustering is demonstrated and suggestions are provided later for other partitioning alternatives.

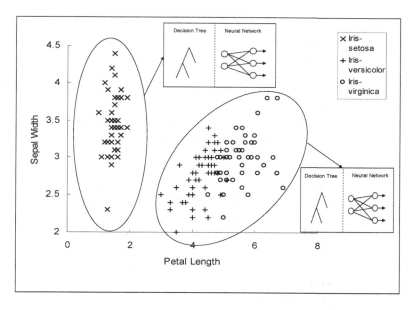

Figure 3.2.1: Slicing example using Iris dataset. Clusters are defined by the ellipse shapes and classes are denoted by different symbols. A classifier (e.g., a decision tree or neural network) is created from each cluster of data.

The k-means clustering algorithm proceeds by initializing a set of cluster centroids to random elements of the dataset. The training set is then assigned to a particular cluster, as represented by its mean (centroid) according to the smallest distance in feature space. The Euclidean distance measure is generally assumed, as seen in Equation 3.2.1. The distance is denoted $D(\mathbf{x}, \mathbf{y})$ for two examples, $\mathbf{x}$ and $\mathbf{y}$. Each example consists of s features.

$$D(\mathbf{x}, \mathbf{y}) = \sqrt{\sum_{i=1}^{s} (x_i - y_i)^2}. \tag{3.2.1}$$

Once all examples are assigned to clusters, the cluster centroids can be recomputed. This process continues using an iterative optimization approach until no examples change cluster membership, or a limit on the number of iterations is reached. The simplistic clustering method tends to create spherical clusters with the Euclidean distance metric.

Once partitions of the training set are created, either through k-means clustering or other partitioning methods, classifiers are constructed on the individual clusters of data. This is done in a distributed fashion with no communication, since each cluster of data is independent. Several different classifier types are considered, including decision trees, naive Bayes and neural networks.

Once all classifiers are constructed, test examples are classified. First, the test example is assigned to the appropriate cluster. When using k-means, this is done by simply storing the cluster centroids from the training set and choosing the closest centroid with respect to the distance to the test example. The corresponding classifier, that was built from the given cluster of data, is considered the "expert" classifier for the particular region of feature space. Thus, this classifier is used to identify the classification of the test example.

3.3   Variance Reduction in Slicing

One key finding from the mixture of experts literature is that learning from smaller subsets of data generally leads to higher variance in classifier error [27]. This is clear from the standpoint of learning theory; more hypotheses are consistent when there are fewer data points to fit. More precisely, researchers have decomposed classifier error into bias and variance terms. Several techniques exist for reducing the variance of an unstable learning method, as discussed earlier. Bagging and the randomized C4.5 algorithm [10] are considered as techniques for variance reduction in slicing. In addition, we discuss a method that is integrated with the slicing approach: overlapping, or soft, partitions. By relaxing the disjoint partition requirement for our slicing algorithm, classifier accuracy is improved. Below several methods of creating overlapping clusters of data, as first presented in [12], are described.

Once overlapping partitions are defined, classifiers no longer use disjoint subsets of data. This allows a classifier to gain a broader view of the region of feature space it is responsible for learning. More importantly, it allows several classifiers to overlap their regions of expertise. A given test example can then be classified by an ensemble of classifiers, rather than by a single classifier. All classifiers deemed "capable" of classifying a test example can be used. Soft partitions create more accurate ensembles than disjoint partitioning which can also outperform learning from the entire dataset.

Clustering is a method of creating a partition of a dataset within the slicing algorithm. Simple partitional k-means clustering is used to generate a set of cluster centroids representing a partition of a dataset. Given a set of cluster centroids and a particular clustering algorithm, the partition is immediately defined. However, several different partitions can be generated from the same set of cluster centroids. The simplest approach is using the k-means method of cluster assignment – an example is assigned to the closest cluster. Soft partitions can be created by assigning (replicating) an example to all clusters that are "sufficiently close", according to some membership criteria. Three different criteria have been considered for replication strategies [12]: the Fuzzy C-Means membership function, the Possibilistic C-Means membership function and a simple distance ratio method. Each of these strategies creates overlapping partitions of the dataset.

An important distinction should be made in the process of creating soft partitions. These partitions are generated by considering the cluster assignment problem as a soft decision problem. However, the resulting clusters are not fuzzy clusters; examples are either included or not within a cluster. Each cluster of data becomes a training set for a classifier. If weights (or membership values) are included with each example, a classifier capable of manipulating this information would be required. For simplicity, this issue is avoided by replicating the example with full weight into each "appropriate" cluster.

It should also be noted that the Fuzzy C-Means and Possibilistic C-Means membership functions are not used as methods of creating cluster centroids (i.e., the typical use of these functions). Rather, they are used as decision functions for replication purposes. As detailed below, a simpler method of making replication decisions can also be used and will often do very well. Here, clustering should be seen merely as a partitioning method that is completely unsupervised.

3.3.1 Soft Replication

Soft replication of examples into clusters is done by a simple distance ratio measure. Given the minimum distance from an example to the closest cluster centroid ($d_{\min}$), an example is simply replicated to all clusters whose distance (to the cluster centroid) is within $n \times d_{\min}$, for some value n. For our experiments we simply chose $n = 2$. In other words, replication is done for an example into all clusters that are within twice the distance of the closest cluster. More formally the distance ratio R is defined in Equation 3.3.1. R describes the ratio R_i of cluster distance to minimum cluster distance ($d_{\min}$).

$$R_i = \frac{d_i}{d_{\min}} \tag{3.3.1}$$

The decision to replicate an example to cluster C_i can then be made when R_i is less than R_*. In our experiments, $R_* = 2$. Although this method is simple our results indicate it is a very useful approach.

Classification of test examples is done in an analogous manner. A test example is considered to be in all clusters that are within $n \times d_{\min}$ and the corresponding classifiers are queried for predictions. The predictions are then combined using a weighted vote. More formally, consider a classification decision C according to Equation 3.3.2. In the case of classification, each classifier outputs a classification

$$V_c = \sum_{i=1}^{k} \frac{1}{R_i} \times V_{ic} \tag{3.3.2}$$

(vote) V_{ic} weighted by a value of $1/R_i$ (where R_i is the ratio of cluster distance to minimum cluster distance from above). Classifiers corresponding to clusters near the threshold distance are weighted by approximately $1/2$. The classifier corresponding to the closest cluster is weighted by 1. Again, the approach is simplistic however experimental results show it works well.

3.3.2 Fuzzy Replication

A more sophisticated method of soft partitioning is a fuzzy membership-based replication strategy. An example is replicated into clusters for which its fuzzy membership is sufficiently high. Of course, we could as easily include each example and its associated membership in every cluster. Section 3.6 discusses the advantages of the simple replication approach. Classification of test examples is done by fuzzy weighting of the relevant classifiers (corresponding to relevant clusters).

The standard Fuzzy C-Means membership function is given by Equation 3.3.3 (see [2] for a full description of FCM). For simplicity, we choose A as the identity matrix, $m = 2.0$, Euclidean distance as the norm, and $c = K$, where K was chosen initially in the k-means clustering.

$$u_{ik} = \left[\sum_{j=1}^{c} \left(\frac{\|\mathbf{x_k} - \mathbf{v_i}\|_A}{\|\mathbf{x_k} - \mathbf{v_j}\|_A} \right)^{\frac{2}{m-1}} \right]^{-1}, 1 \leq i \leq c; 1 \leq k \leq n; \qquad (3.3.3)$$

Membership values are constrained to the interval $[0, 1]$ and all memberships for an example must sum to one. The cluster membership of an example is related to the distances to all cluster centroids (V_i). This is likely to be an advantage, since replication should only occur in clusters in which an example has significant membership. For example, if we consider only the Euclidean distance between a point and a cluster centroid, we cannot measure the degree of similarity to other clusters directly. With the fuzzy membership, we have this measure.

Three additional constraints were added to help implement the idea of including only clusters in which the example has significant membership. First, all membership comparisons are differential in nature. In other words, we consider membership values only in relation to the highest membership value ($u_{\max}$) computed for an example. Second, thresholding is done at a cutoff membership value. Replication occurs only for clusters with membership values above this threshold, which correlates to clusters in which the example has significant membership. The membership threshold value is $Th = u_{\max} - \epsilon$. For our experiments, $\epsilon = 0.2$ provided a reasonable tradeoff between too many and too few replicates. Conditions did exist in which an example was very fuzzily assigned to all clusters, in which case the example is then replicated across all clusters (reasonable behavior in such a situation).

More formally, for each example we calculate the membership values u_i for every cluster, using Equation 3.3.3. We then find the maximum membership value $u_{\max}$ (i.e., the membership in the closest cluster). The replication decision is a binary function $F_{\text{replicate}}$ as given in the following equation

$$F_{\text{replicate}} = \left\{ \begin{array}{ll} 1 & u_i \geq u_{\max} - \epsilon \\ 0 & \text{otherwise} \end{array} \right\} \qquad (3.3.4)$$

Classification using fuzzy replication occurs as follows. The membership values for a test example T are calculated according to Equation 3.3.3 from the existing cluster centroids. We threshold the memberships at the value of $Th = u_{\max} - \epsilon$. For each membership value u_i (representing cluster i which was above the threshold), the corresponding classifier is queried for a prediction. The predictions are combined via weighted voting; the weights are the computed membership values. The membership threshold that was applied reduces the overall computational effort since fewer classifiers need to be queried. In addition, only classifiers that can reasonably predict a classification for T are queried. This forms a group of experts around each test example (in feature space), with varying levels of expertise (as given by the membership function).

One difficulty that arises in the use of the fuzzy membership function from FCM is the constraint that memberships must sum to one. With a large number of clusters (large K) the membership function must distribute the membership values accordingly. This leads to n-way ties in which the largest membership value is rather small. Consider the simple case of five very close clusters. An example may have membership values of approximately 0.2 in each cluster. While this is an appropriate decision for FCM, the example will likely be replicated in all clusters in the partition including remote (e.g., low membership) clusters. Thus constraining the membership values causes problems in this context. These arguments against the membership function in FCM are well-detailed in [29].

3.3.3 Possibilistic Replication

Possibilistic C-Means (PCM) clustering [29] is a soft approach in which membership values are not constrained to sum to one. Thus the second soft approach to partitioning was using the possibilistic clustering membership function. The general form of the membership is given in Equation 3.3.5. For simplicity, we chose η_i to be equal (with a value of 1.0). The value of η is described as a scaling parameter that describes the "size" of the cluster [30] and methods for estimating η exist. By also choosing the value of $m = 2.0$, we are able to simplify the membership formula significantly, to the form in Equation 3.3.6.

$$u_{ij} = \frac{1}{1 + \left(\frac{d_{ij}^2}{\eta_i}\right)^{\frac{1}{m-1}}} \qquad (3.3.5)$$

$$u_{ij} = \frac{1}{1 + d_{ij}^2} \qquad (3.3.6)$$

Given the possibilistic interpretation of training examples in different clusters, a threshold membership value must be defined for replication to occur. Since the membership values u_i are not constrained to sum to 1.0 for all clusters, the following approach is used. Cluster memberships greater than $\epsilon \times u_{max}$ indicated replication should occur to that cluster. In experiments, $\epsilon = 0.5$ was empirically determined to be a reasonable value.

Classification using possibilistic replication occurs in much the same way as with fuzzy replication. The possibilistic membership of the test example T in each of the existing clusters is calculated. A threshold of $\epsilon \times u_{max}$ is used to zero out small membership values. The remaining non-zero memberships correspond to "close" clusters, therefore the classifiers built from examples within these clusters are queried for predictions. A final classification decision is made via weighted voting, this time the weights are the possibilistic membership values.

The possibilistic approach to overlapping partitions provides a more robust strategy for replicating (and weighting) examples than the other two approaches. If we consider the clustering approach to simply provide data partitions, then the number of clusters can be varied according to several concerns including problem size and feature space density. Any number of cluster memberships may be large, corresponding to "closeness" to any number of clusters. This contrasts to FCM in which ties will dilute the maximum membership value seen.

3.3.4 Summary

We believe these overlapping partition strategies create a "localized" bagging effect. Experiments have shown that when over-clustering of a dataset occurs, slicing remains effective when overlapping partitions are allowed. This is due to many similar clusters of data created in regions of feature space. Due to the nature of overlap, many of the clusters are very nearly identical. Each cluster then produces similar classifiers and are weighted similarly in a voted classification. This achieves, in effect, a bagging effect for localized regions. A key accomplishment of this approach is that the "bag" of data is generally much smaller than the full dataset.

3.4 Experiments

The effectiveness of the slicing algorithm can be seen through a series of experiments on several benchmark datasets from the UCI Machine Learning repository [3] and other locations. Four datasets were chosen for their large size and continuous features. The initial clustering using k-means requires continuous features; see Section 3.6 for further discussion of this point. Table 3.4.1 describes the dataset characteristics. For all experiments, a ten-fold cross-validation is done to estimate classifier accuracy. Accuracies are reported as the mean across the ten folds. Statistical significance (when reported) is done by a paired t-test between folds, at the $\alpha = 0.05$ level.

A protein secondary-structure prediction dataset was also used in our experiments. The dataset consists of 1,219 protein chains for training and 63 protein chains in the test set [26]. These protein chains were used in [26] to develop and validate the feed-forward back-propagation neural network that won the CASP-3 secondary structure prediction contest. The protein chains are submitted to PSI-BLAST [1] and a scoring matrix representing the log-likelihood of each of the amino acids being substituted is returned. This matrix for the protein is split into windows of 15 amino acids and each window is used as an example, together with

Table 3.4.1: Dataset details

Dataset	Source	Number of examples	Number of features	Number of classes
phoneme	Elena	5,404	5	2
satimage	UCI	6,435	35	6
pendigits	UCI	10,992	16	10
letter	UCI	20,000	16	26
Jones	other	209,529	315	3

the simplified protein secondary structure (helix, strand or coil). In addition, an N/C terminus bit is added for each amino acid to signify whether or not the chain exceeds a terminus. This Jones training set one consists of 209,529 amino acid examples, of dimension 315. The test set consists of 17,731 examples representing 63 protein chains.

For the experiments, the learners used include the C4.5R8 [42] decision tree learner, a neural network and a naive Bayes algorithm, implemented within the Weka learning environment [47]. The neural network is trained using a validation set. Training stops after no improvement in validation accuracy occurs within 15 epochs. The best set of weights found (relative to validation set accuracy) are used for test set predictions.

Modifications were made to the C4.5R8 code to produce randomized C4.5 as described in [10] for the variance reduction step. The default parameters require building 200 trees by choosing among the top 20 tests. When using bagging for variance reduction, 50 bags were generated. Clustering was done with a simple k-means clustering algorithm on normalized features, scaled in the range $[0, 1]$. The values of K were chosen in order to examine the range of clusters from 2 to 26 (the maximum number of classes in the letter dataset).

Many researchers advocate the use of unpruned trees to maintain diversity when creating ensembles. Pruning can often remove the variations in decision trees that may limit the overall diversity. Dietterich [10] chose pruned or unpruned trees based on classification accuracy when comparing randomization, bagging and boosting. In 17 of 33 datasets used in the analysis of randomization, the unpruned trees were more accurate classifiers. Therefore, for randomization we generate only unpruned trees. We use pruned trees for bagging, following the convention of Quinlan [43] when using bagging within C4.5. It should be noted, however, that the diversity argument also applies to bagging. Again considering experiments performed by Dietterich [10], pruned trees were more accurate in bagging for only 10 of the 33 datasets used in that work. Therefore, future work could investigate more thoroughly the use of pruning within slicing.

3.4.1 Slicing

First, the effectiveness of the slicing approach without variance-reduction is examined. Tables 3.4.2–3.4.4 show a summary of the results from using each classifier on each dataset. The accuracy reported for slicing is the best across the range of clusters tested. Also included is the number of clusters at which the maximum accuracy was found. Slicing using the naive Bayes and neural network classifiers outperformed a single classifier using the entire dataset. In the case of slicing using C4.5, classification accuracy improved in only two of the four datasets.

It is interesting to note that with a single exception, the maximum accuracy using C4.5 occurs with large numbers of clusters. In the case of the naive Bayes algorithm, the best performance always occurs at the maximum number of clusters tested (26). The neural network did not exhibit this particular phenomenon. A decision tree learner can adjust the size of the decision tree based on the amount of data provided [39] and the parametric form for the naive Bayes classifier tends to be stable in estimating normal distributions. However, a neural network has many free variables in the form of network weights. The total number of weights does not change per dataset in our experiments, regardless of the size of training data. Hence, a neural network is likely to vary in the accuracy vs. the size of the training subset used.

Figures 3.4.1–3.4.12 show the detailed results with each classifier type on each dataset, when used with slicing. Several interesting results are evident. In two of the four datasets, slicing using C4.5 tends to be more accurate than using C4.5 on the entire dataset. However, slicing with C4.5 performs poorly on the other two datasets (letter and phoneme). This is likely due to the increased variance as discussed earlier. In the case of neural networks and naive Bayes, the performance is better when using slicing. Generally, the larger the number of clusters the more accurate slicing becomes. This is likely due to the reduction in the size of the individual learning task. There are clearly limits in the reduction of size in the training sets, however the results clearly indicate that creating a subspace of feature space creates a simpler task for the learner.

3.4.2 Large Feature Space: Satimage

One notable issue exists in the overall slicing results. Slicing using neural networks generally performs more accurately than a single neural network for a small number of clusters. However, in the case of the satimage dataset the accuracy of slicing drops sharply when using more than three clusters. The satimage

Table 3.4.2: Slicing results using C4.5. Shown are results from learning over the entire training set, slicing into K clusters. The accuracy reported is the highest accuracy from the range of K.

Dataset	Baseline C4.5	Slicing (C4.5)	K
phoneme	86.92	86.75	3
satimage	86.51	87.41	22
pendigits	96.33	97.11	26
letter	88.09	87.12	26

Table 3.4.3: Slicing results using a feedforward, backpropagation neural network. Shown are results from learning over the entire training set, slicing into K clusters. The highest accuracy is reported across all values of K tested.

Dataset	Neural network	Slicing (neural network)	K
phoneme	82.31	83.88	12
satimage	90.22	90.40	3
pendigits	94.75	98.57	7
letter	82.48	91.31	26

Table 3.4.4: Slicing results using a naive Bayes classifier. Shown are results from learning over the entire training set, slicing into K clusters. The highest accuracy is reported across all values of K tested.

Dataset	Bayes	Slicing (Bayes)	K
phoneme	76.13	82.57	26
satimage	79.53	86.32	26
pendigits	85.77	97.07	26
letter	64.20	87.71	26

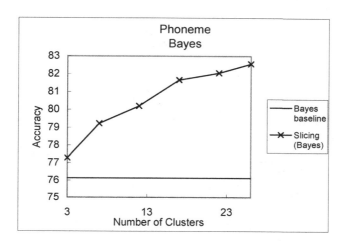

Figure 3.4.1: Naive Bayes slicing (phoneme).

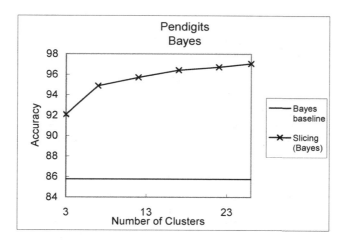

Figure 3.4.2: Naive Bayes slicing (pendigits).

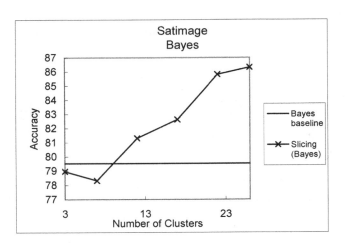

Figure 3.4.3: Naive Bayes slicing (satimage).

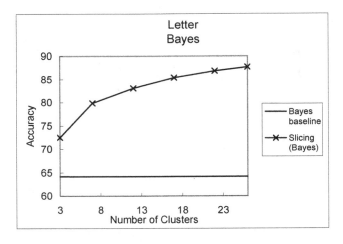

Figure 3.4.4: Naive Bayes slicing (letter).

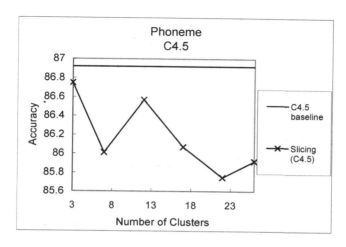

Figure 3.4.5: C4.5 slicing (phoneme).

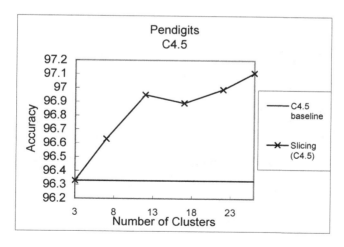

Figure 3.4.6: C4.5 slicing (pendigits).

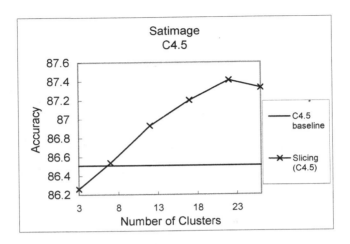

Figure 3.4.7: C4.5 slicing (satimage).

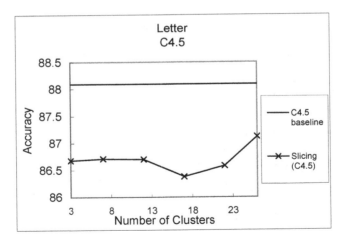

Figure 3.4.8: C4.5 slicing (letter).

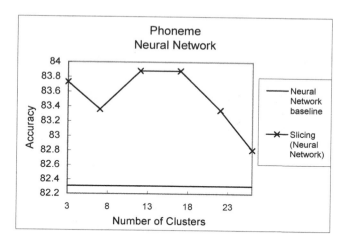

Figure 3.4.9: Neural network slicing (phoneme).

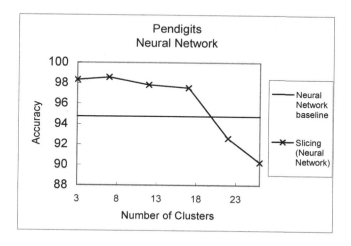

Figure 3.4.10: Neural network slicing (pendigits).

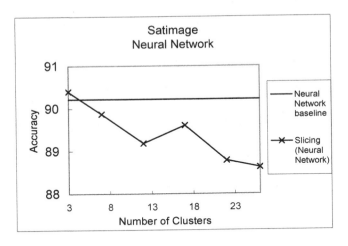

Figure 3.4.11: Neural network slicing (satimage).

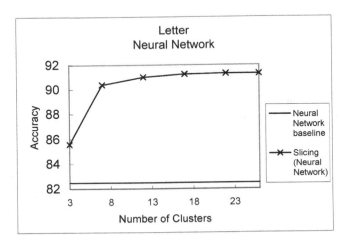

Figure 3.4.12: Neural network slicing (letter).

dataset consists of 35 features – many more features than in any of the other small datasets. The neural network architecture is chosen based on the number of inputs and outputs (35 and 7, respectively). As is done in the Weka [47] software package and elsewhere, the default number of hidden units is one-half the number of inputs plus outputs (21 in this example). Therefore we believe that slicing the input space creates partitions which are too sparse for effective neural network learning. In other words, there are too few examples for the large number of free parameters that must be trained in the neural network. An additional possibility is that the large number of features includes many irrelevant ones. The existence of irrelevant features can skew the clustering algorithm and thus lead to partitions from which classifiers are difficult to construct.

If the dimensionality of the satimage dataset leads to sparse partitions, we can consider methods for reducing the dimensionality. Many such techniques exist, including wrapper-based approaches [28] that utilize classifiers to determine a "good" subset of features to use. An alternate approach is to construct composite features that represent linear combinations of the existing features. The principal components approach [25] is a popular method for reducing the number of features required using this approach. This method involves calculating the covariance matrix from the dataset and determining the associated eigenvectors of the matrix. The eigenvectors are orthogonal and are used as the axes for a transformed feature space. Feature reduction occurs by choosing the eigenvectors with the highest eigenvalues. A sufficient number of eigenvectors can be chosen so as to represent a given percentage of the variance in the dataset. Thus the feature space is transformed so that the axes are linear combinations of existing features and represent the direction of largest variance.

We used the implementation of principal components as provided within the Weka [47] software to reduce the feature space. Linear combinations of the existing 35 features were created that explain 95% of the variance in the dataset. The software created 6 features (from the original 35). Using this constructed feature set, the dimensionality is dramatically reduced and it is hoped that slicing will perform similarly to other datasets.

Figure 3.4.13 shows the results of slicing using the six-feature version of the satimage dataset. The figure clearly demonstrates that the reduced feature set does indeed boost the classification accuracy from slicing relative to a single neural network. However, the classification accuracy of slicing peaks when using 7 clusters and decreases for a larger numbers of clusters. Slicing using 26 clusters is as accurate as a single neural network, however our experiments on other datasets indicate that accuracy often rises with increasing numbers of clusters.

We believe the principal components analysis (PCA) method reduces the sparseness of feature space relative to the number of dimensions. Although the classifi-

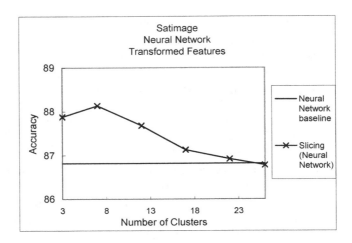

Figure 3.4.13: Neural network slicing results (satimage) with 6 transformed features.

cation accuracy of slicing with six features improves over a single neural network, we note that the classification accuracies are lower than using the full set of 35 features. More sophisticated feature selection or feature creation can boost the accuracy of learning from a reduced set of features. It may provide additional boosts in classification accuracy when using large numbers of clusters within slicing. Here we explain the earlier results more completely rather than attempting to optimize the feature selection method.

3.4.3 Variance Reduction: Bagging and Randomization

As discussed earlier, it is expected that unstable learners will have difficulties when using smaller subsets of data. Several different approaches are considered for lowering the variance in classifier error and hence lowering the classifier error rate. Experiments performed on the neural network indicate that classification accuracy improved on the phoneme and satimage datasets only for $k = 3$. Increasing the number of clusters led to further decreases in classification accuracy, below the level of a single neural network over the entire dataset. Neural network bias assumes a smoothly continuous decision function therefore decomposing feature space through slicing clearly avoids much of inherent instability of neural networks. Stable learning methods do not tend to improve when used with bagging. The naive Bayes approach is generally considered to be stable therefore no variance reduction was performed.

Table 3.4.5: Variance reduction in slicing with C4.5. Shown is the best classification accuracy between bagging and randomization methods. The number in parenthesis indicates the number of clusters.

Dataset	Slicing (C4.5)	Slicing (C4.5) with variance reduction	Technique
phoneme	86.75	89.08	B (12)
satimage	87.41	91.92	R (3)
pendigits	97.11	98.98	R (3)
letter	87.12	96.29	R (3)

The improved classification accuracy in slicing is demonstrated using C4.5 with two different types of variance reduction: bagging and C4.5 randomization. The individual results are displayed in Figures 3.4.14–3.4.17. In both cases improvements are seen, although the C4.5 randomization method tends to outperform bagging within C4.5. This is consistent with [10] in which randomized C4.5 outperformed bagging over the entire dataset. Clearly both techniques increase the overall classification accuracy. Table 3.4.5 summarizes the results from variance reduction. We conclude that the use of variance reduction techniques are effective within slicing. In addition, the scalability of slicing is maintained since the individual learning task when using variance reduction remains small.

3.4.4 Methods for Overlapping Clusters

The use of overlapping clusters of data is introduced above and is expected to provide variance reduction. We performed several experiments examining the impact of overlapping clusters of data on the slicing algorithm. First we consider the three different techniques for creating overlapping clusters using C4.5 as the base classifier. Figures 3.4.18–3.4.20 show the detailed accuracy of each overlap method. In each instance the fairly smooth curves tend to increase with a larger number of clusters. With the exception of the very extreme (e.g., 2–5 clusters) learning from overlapping partitions performed better than C4.5 using the entire dataset.

Table 3.4.6 provides a summary of results for the datasets. The table describes the number of times a particular replication strategy (fuzzy, possibilistic and soft) ranked the highest in classification accuracy. Ties are included as ranking the highest. The table shows that the soft and possibilistic replication methods are often similar in top ranking. This result is useful in that the simplest replication method

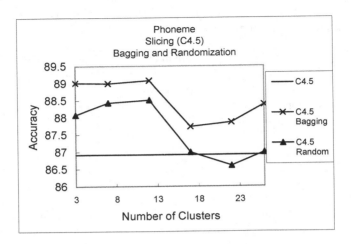

Figure 3.4.14: C4.5 slicing with variance reduction results (phoneme).

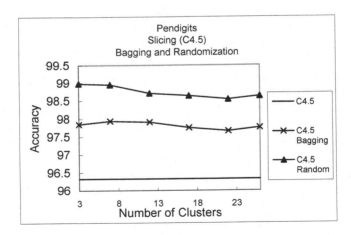

Figure 3.4.15: C4.5 slicing with variance reduction results (pendigits).

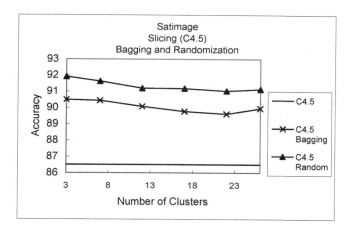

Figure 3.4.16: C4.5 slicing with variance reduction results (satimage).

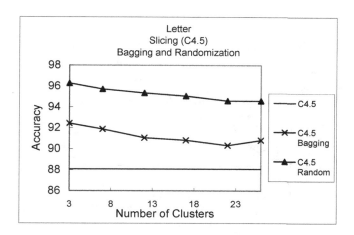

Figure 3.4.17: C4.5 slicing with variance reduction results (letter).

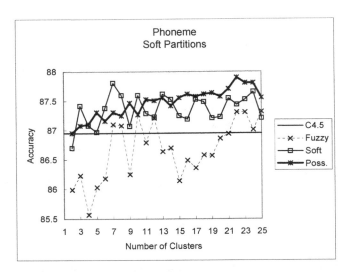

Figure 3.4.18: C4.5 overlapping clusters (phoneme).

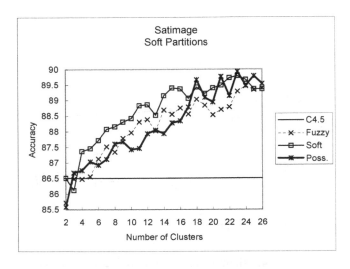

Figure 3.4.19: C4.5 overlapping clusters (satimage).

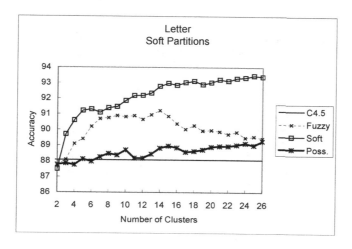

Figure 3.4.20: C4.5 overlapping clusters (letter).

Table 3.4.6: Number of times replication strategy ranked highest in accuracy (including ties), as K varies from 2 to 26

Dataset	Fuzzy	Soft	Possibilistic
phoneme	0	7	17
satimage	0	19	5
pendigits	1	23	0
letter	0	23	1

(soft) provides as accurate an ensemble as the more sophisticated approaches.

Table 3.4.7 shows the overall best accuracy of any replication method for any number of clusters tested (2–26). The most obvious detail in this table is that fact that both the soft and possibilistic replication strategies won on two datasets. The other significant finding is that in all cases learning from soft partitions outperformed C4.5 using the entire dataset, sometimes by rather large margins.

3.4.5 Variance Reduction: Overlapping Clusters

Next the impact of using the soft replication approach as a variance reduction technique is examined. As can be seen in Figures 3.4.21–3.4.24, classification accuracy improves when using C4.5 as the base classifier. What is surprising is in fact the accuracy also improves when using overlapping clusters in conjunction

Table 3.4.7: Overall best classification accuracy for replication method (any cluster size). winners are denoted by replication method (S = soft, P = possibilistic).

Dataset	Baseline C4.5	Overall best accuracy
phoneme	86.95	87.92 (P)
satimage	86.51	89.93 (P)
pendigits	96.33	98.27 (S)
letter	88.08	93.47 (S)

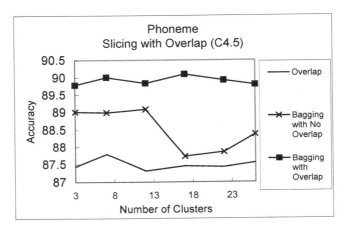

Figure 3.4.21: C4.5 slicing using overlap (phoneme).

with the randomized C4.5 technique. This indicates that the overlapping cluster method is providing additional variance reduction via localized context.

Due to the nature of variance reduction, the overlapping partitions may be expected to hurt (or at least not effect) the more stable learning methods. In the case of the naive Bayes learner, performance did in fact increase above the baseline method of using one classifier on the entire dataset. However the accuracy was in fact lower than learning from disjoint partitions, hence the details of this experiment are not included. This result again reinforces the notion that the overlapping partitions are decreasing variance in classifier error, however this generally impacts only unstable learners.

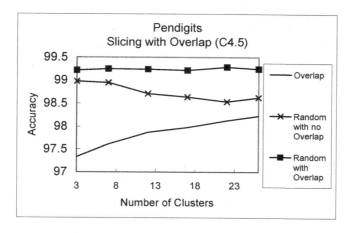

Figure 3.4.22: C4.5 slicing using overlap (pendigits).

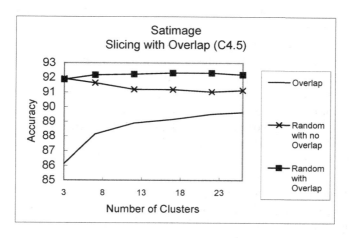

Figure 3.4.23: C4.5 slicing using overlap (satimage).

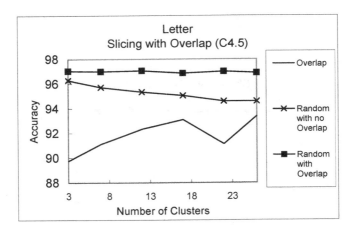

Figure 3.4.24: C4.5 slicing using overlap (letter).

3.4.6 Protein Secondary Structure

We next apply slicing to the Jones dataset (train and test set one) from the protein secondary structure prediction domain. This dataset consists of 315 real-valued features that are normalized in the range $[0, 1]$. Recall that neural networks perform poorly on the satimage dataset due to the large number of features. We believe the same difficulties arise in the Jones dataset, therefore feature selection was performed as a pre-processing step. All results within this section are relative to a reduced-feature dataset, described below.

The principle components analysis method, described earlier for use on the satimage dataset, creates linear combinations of features and results in the reduction of the total number of features used for learning. The PCA method was attempted on the Jones dataset. However, the reduction requires retaining 240 eigenvectors (principal components) of the total 315 features in order to explain 95% of the variance in the dataset. This reduction is not sufficient for our investigation into the effects of dimensionality on slicing and is therefore not considered further.

An alternative feature selection technique is the Correlation-based Feature Selection (CFS) method [17]. It involves the selection of a subset of features that are highly correlated with the class label. In addition, the algorithm attempts to find the subset of correlated features that are minimally correlated with each other. When used in conjunction with a feature selection mechanism such as the best-first search strategy, this approach can select a small subset of features efficiently. The best-first search strategy for feature selection starts by selecting the best pair-wise

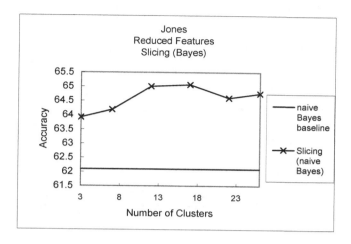

Figure 3.4.25: Slicing using naive Bayes (Jones reduced features).

combination of features. In an iterative fashion, a subset of n features is found by first starting with the previous best subset of $n - 1$ features and evaluating all possible ways of adding a single feature.

We utilize the CFS Subset approach using the best-first search strategy on the Jones dataset. The algorithm is implemented within the Weka [47] learning environment. A subset of 50 features was identified by the selection algorithm. The features are near the center of the protein chain window, which agrees with the physical interpretation of secondary structure in proteins. The accuracy of C4.5 on the Jones test set using the 50 feature subset is slightly better (1.5% higher in accuracy) than using all 315 features. The neural network performs slightly worse with the 50 features (3.3% lower in accuracy). Figures 3.4.25 and 3.4.26 show the improvement in accuracy when slicing is used compared to a single classifier using the entire dataset. Slicing with either the naive Bayes or C4.5 classifier results in improvements of about 2% in accuracy.

The combination of bagging within the slicing framework provides a increase of 13% in classification accuracy over slicing without bagging, when used with C4.5. This result is very similar to that achieved by an ensemble of small subsamples method [11]. In effect, both methods are providing averaging over simpler decision trees. We compare the two approaches more precisely in Subsection 3.5.1 below.

The results of using neural networks on the Jones dataset are included in Figure 3.4.27. A neural network was initially used in this problem [26] and results in a very accurate classifier. It is clear from the figure that neural networks do not

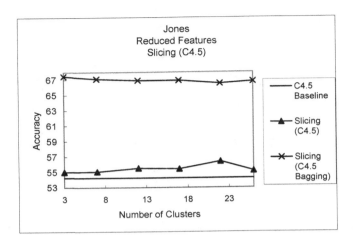

Figure 3.4.26: Slicing using C4.5 (Jones reduced features).

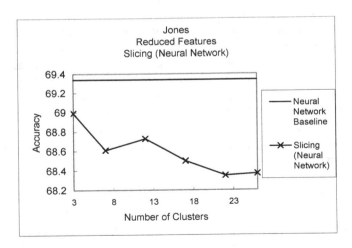

Figure 3.4.27: Slicing using neural networks (Jones reduced features).

perform well within the slicing framework on this problem. However, we believe this result suggests that not only is the learning task size reduced for an individual learner, but in some sense the complexity of the learning task is reduced. Decision tree learners are biased toward finding shorter trees and therefore simpler hypotheses [34]. The naive Bayes classifier also provides this bias through the use of simple statistical models. The neural network, in contrast, is given a particular architecture for the given learning problem. In the case of the reduced-feature Jones dataset, 26 hidden units are used for learning. However, we believe the decreasing performance of slicing using neural networks arises from too many free variables (hidden units), given the ideal decision surface required to accurately classify examples. We believe the use of clustering not only reduces the number of examples but also reduces the complexity of the decision surface needed, however our algorithm does not account for this phenomenon at the present time.

In summary, the slicing method can be used effectively on the Jones dataset with both the C4.5 decision tree learner and the naive Bayes learner. The amount of data per learner is reduced and learning can occur in a distributed fashion, without a loss in classification accuracy. However, neural networks do not perform well when slicing in this domain. Neural networks do not have a bias for simpler hypotheses, which is likely to be important in this domain. More work must be done to fully understand this phenomenon within the protein secondary structure domain.

3.5 Analysis

The experiments presented above indicate that slicing is an effective distributed learning technique. We further investigate the use of slicing as compared to a scalable ensemble of subsamples. The slicing approach can be more accurate than the subsampling approach while meeting all of the criteria for distributed learning. Also, we compare the accuracy of slicing with several techniques reported in the literature.

3.5.1 Small Subsamples

The ensemble of small subsamples approach is scalable in that no single classifier uses the entire dataset and has been demonstrated to be more accurate than a single classifier using the entire dataset [11]. Experiments in [11] indicate that the use of 25% subsamples with 20 classifiers generally results in very accurate ensembles. We use this particular combination of subsample size and number of ensemble members in the comparison below. In the case of slicing, we generally see that overall accuracy increases up to at least 26 clusters. If we assume 26 clusters with an equal number of members, then clustering represents $1/26$, or

Table 3.5.1: Comparison of classifier accuracy in slicing and ensemble methods. The † signifies statistical significance.

Dataset	Subsamples	Slicing
phoneme	85.20	89.08†
satimage	89.08	91.92†
pendigits	97.52	98.98†
letter	90.44	96.29†
Jones	67.67	67.46

approximately 4%, of the dataset. It should be noted, however, that the use of overlap for variance reduction increases the individual subset size above 4%. In the comparisons below, the number of slices is chosen to be 26 (except for Jones, in which 3 clusters were chosen).

Table 3.5.1 summarizes the comparison between the ensemble of subsamples approach with slicing. The slicing results shown are from Table 3.4.5, in which C4.5 was used with variance reduction (randomization or bagging). As can be clearly seen, a statistically significant increase in accuracy occurs when slicing is used for all datasets except Jones. In each case, less than 100% of the dataset is used per classifier however it is clear that slicing creates more accurate partitions. Using the Jones dataset, slicing produces a slightly less accurate classifier despite using only 50 of the 315 total features.

Slicing produces more accurate classification accuracies than the ensemble of subsamples. We believe this is due to the "intelligent" nature of the partitioning method. The subsamples are required to create models with coverage including the entire feature space range. In contrast, slicing reduces the extent of the model coverage for any individual learner. We believe slicing provides a mechanism to enable localized learning. The results presented above clearly indicate that slicing provides "better" subsets of data from which to learn.

3.5.2 Additional Comparisons

The Dynamic Classifier Selection by Local Accuracy (DCS-LA) [49] approach was discussed earlier as an alternative method of selecting experts in regions of feature space. In the paper that describes the algorithm, several experiments were performed to analyze the accuracy of the method. The authors used the satimage and phoneme datasets considered in this dissertation. The experiments performed in [49] are not completely comparable to our experimental setup. A 50% train and test split of the data was used and accuracy results were reported as an average

Table 3.5.2: Comparison: Slicing vs. DCS-LA. DCS-LA accuracy reported as average of a 50% train/test split. Slicing accuracy reported as average of 10-fold cross validation.

Algorithm	satimage	phoneme
DCS-LA local class accuracy	88.49	89.38
DCS-LA overall accuracy	87.64	88.57
Slicing (C4.5 with randomization)	89.08	91.92

from using each half as the training set and testing on the other half of data. In addition, the data was normalized such that the values for each feature had a mean of zero and unit variance. Our normalization method simply scaled the features in the range $[0, 1]$. The comparison is summarized in Table 3.5.2. It is clear from this comparison that our approach is comparable to or slightly more accurate than the DCS-LA method. More significant is the fact that slicing achieves this accuracy without requiring any single classifier to use the entire dataset for training, as is required by the DCS-LA approach.

Next we compare to the mixture of experts (ME) method from the neural network literature. There are fewer empirical studies using the mixture of experts, however an extensive study was done by Moerland in [35]. Several variations of the mixture of experts were used in the experiments. Recall that the gating network within the mixture of experts determines the appropriate mixture of "expert" output per training example. This gating network was initially specified as a single neuron (perceptron) [23]. However, a more sophisticated multi-layer perceptron (MLP) was also considered in the experiments. Table 3.5.3 presents a summary of the results from [35] and two slicing approaches. Slicing using neural networks is slightly worse in accuracy than the mixture of experts approach on the pendigits dataset. However, slicing outperforms both ME methods on the letter dataset. Performance on both pendigits and letter significantly improved using the variance reduction techniques. Thus the slicing algorithm performs at least as well as the mixture of experts (and often better) while at the same time reducing the individual learning task size. Recall that each expert is expected to use the entire training set in the mixture of experts approach.

Finally we compare slicing with the radial basis function approach in the neural network literature. Experiments were performed by Likas, et al. [31] using standard radial basis functions (RBFs) when comparing with a new probabilistic radial basis function. Table 3.5.4 shows the results from their RBF experiments vs. slicing. The authors in [31] used a five-fold cross-validation technique, therefore

Table 3.5.3: Comparison: Slicing vs. mixture of experts. The numbers in parentheses indicate the number of clusters or number of experts used in the respective algorithm.

Algorithm	pendigits	letter
Mixture of experts: perceptron gate	98.90 (10)	90.10 (10)
Mixture of experts: MLP gate	98.80 (10)	87.50 (10)
Slicing (neural network)	98.57 (7)	91.31 (26)
Slicing with overlap (C4.5 with randomization)	99.25 (26)	97.06 (12)

Table 3.5.4: Comparison: Slicing vs. radial basis functions. The numbers in parentheses indicate the number of clusters or kernels used in the respective algorithm.

Algorithm	phoneme	satimage
RBF	76.00 (12)	85.30 (24)
Slicing (C4.5 with randomization)	89.08 (12)	91.92 (3)

the comparison cannot be made precisely. However, it is clear from the results that significant increases in classification accuracy occur when using slicing.

3.6   Discussion

A number of issues remain open in the use of slicing; only the feasibility of this approach has been demonstrated. One significant issue is the scalability of the clustering step of the algorithm. We describe below several alternatives to the simple k-means clustering method. Other open problems in slicing include determining the correct number of clusters to use and the extension of slicing to datasets with discrete features.

3.6.1 Cluster Scalability

The slicing approach can be a scalable and accurate learning method. This method can be applied to learning on very large datasets since each individual classifier is not required to use the entire dataset. k-means clustering was used due to its simplicity. However, a simple and efficient method of creating cluster centroids is essential to the feasibility of this approach. A review of many approaches to scalable clustering is included below; any of these algorithms can be considered as viable (and efficient) replacements for k-means.

A variety of researchers have developed scalable clustering algorithms that could be considered for use in slicing. VFKM [22] is an approach based on the Hoeffding statistical bounds [21]. The approach involves choosing a bound on the error allowed in partitioning, then determining the number of samples required to achieve this bound. A clustering iteration occurs on only the sample of data, rather than entire dataset. Other methods include a single scan of the dataset [4, 13]. These methods work by aggregating points and storing only sufficient statistics from the examples. A classic example of this approach is in BIRCH [50]. A distributed version of k-means was proposed in [14] in the form of communicating sufficient statistics. Fractionation was used [46] to randomly split the dataset and merge the results using a model-based clustering clustering technique. Other notable scalable clustering methods include CLARANS [38] and DENCLUE [20].

Regardless of the clustering method, the final partitions can still be softened through the use of the soft decision strategies outlined earlier. Therefore loss of precision in clustering can be offset through overlapping partitions. We propose that slicing using overlapping partitions is a technique that provides efficiency in clustering (by making hard decisions while clustering) and flexibility in partitioning (by using soft decisions).

3.6.2 Number of Clusters

The most significant question remaining to be addressed is the number of clusters to choose in this process. The graphs and tables in the previous section were based on tests including values of K from 2 to 26. We used the Davies-Bouldin cluster validity index [9] as a measure of the validity of the clustering partition. Figure 3.6.1 shows the plot of validity vs. the number of clusters for both the satimage and letter datasets. One would choose K where the first minimum exists (2 and 3, respectively). Comparing the validity results with the previous classification accuracies indicates no correlation between the two measures. We believe the number of clusters must be a parameter to the algorithm, which must be set either empirically or by the desired amount of distributed processing. This is not unlike the parameter tweaking required to use neural networks. One possible approach is to simply equate the number of clusters with the desired training set size, assuming the clustering algorithm creates relatively balanced clusters.

The increasing accuracy in soft partition learning as the number of clusters increases is an ensemble phenomenon. In ensemble experiments, an increase in the number of members of an ensemble generally produces an overall increase in accuracy. Using large numbers of clusters in soft partitioning has a "bagging-

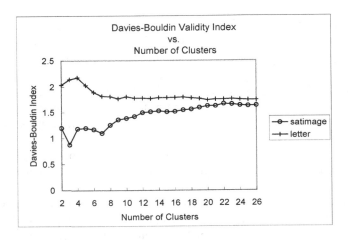

Figure 3.6.1: Cluster validity vs. number of clusters.

like" effect. Over-clustering with hard k-means will inevitably lead to cluster centroids close in proximity. The use of soft boundaries will replicate many of the examples from neighboring clusters. Slicing with variance reduction generates similar (although not identical) effects. Therefore the exact number of clusters is not as crucial, though as is the case in bagging a point of diminishing returns can be expected.

3.6.3 Continuous Features

The use of the simple k-means clustering algorithm limits the availability of the approach to continuous-valued features. A variety of clustering algorithms and distance metrics [15, 16, 24] have been developed to handle discrete, or nominal, features. This issue can be explored further in later work. We believe slicing is relatively insensitive to the partitioning process, therefore in the case of mixed features a partial solution could involve simply clustering on the continuous features. Within each cluster the full set of features would be considered for learning [15, 16].

3.7   Conclusions

Slicing is a distributed learning algorithm that provides a scalable solution while maintaining classification accuracy. We describe the slicing of a dataset in terms of clustering the dataset using the simple k-means clustering algorithm. Several similar, existing techniques provide some insight into the potential difficulty of increased variance when learning from less data. This problem occurs in unstable classifiers, in which the removal of a few examples can significantly alter the constructed classifier. Several techniques exist for reducing the variance in classifier response through ensembles, including bagging and C4.5 randomization. An additional method for variance reduction, using overlapping clusters of data, was also described. Several different alternatives for creating overlapping clusters were also listed. Empirical results suggest that a simple distance-based measure creates overlapping clusters which allow significant boosts in classification accuracy when classifiers are trained on the data from the cluster.

Experimental results were shown using three representative learners from the field of machine learning: a naive Bayes learner, a neural network and a decision tree. The naive Bayes learner uses statistical estimation of a normal distribution and therefore is a stable learning algorithm. Slicing using the naive Bayes learner provided significant increases in classification accuracy as the number of clusters increased. The neural network learner produced similar results in three of the four benchmark datasets. This result demonstrates a surprising level of stability from the neural network learner, although classification accuracy was seen to drop with a larger number of clusters (and correspondingly smaller dataset sizes). The neural network did not perform as expected on the satimage dataset; the poor results were explained and a remedy in the form of dimensionality reduction through principal components analysis was described. The corresponding results from the Jones dataset indicate that slicing with neural networks has other limitations within the secondary structure domain. Finally, slicing using the decision tree learner demonstrated the need for variance reduction methods. Without such methods, the performance of slicing using decision trees was significantly worse than using a single decision tree over the entire dataset.

Variance reduction techniques lower the variance portion of classifier error. Experiments have demonstrated that slicing with decision trees improves significantly when bagging or randomized C4.5 techniques are employed. The classification accuracy on all datasets improved beyond that of a single decision tree built from the entire training set. The use of overlapping clusters also provides boosts to classification accuracy. Surprisingly, the use of overlapping clusters in combination with bagging or randomized C4.5 increased the classification accuracy further. This indicates that overlapping clusters are creating a "localized" bagging effect. This type of effect can be important in scalable, distributed learning algorithms in which the use of bagging over the entire dataset is impossible.

Finally we compared the slicing algorithm to an ensemble of subsamples approach [11]. In each case the slicing method outperformed the ensemble approach, with statistical significance. Next we compared slicing with three different algorithms from the literature: the mixture of experts, radial basis functions and dynamic classifier selection. Slicing performed as well or better than each of these approaches. More significantly, slicing generates accurate classifiers in a scalable manner.

Bibliography

[1] S. Altschul, T. Madden, A. Schaffer, J. Zhang, Z. Zhang, W. Miller, and D. Lipman, "Gapped BLAST and PSI-BLAST: A new generation of protein database search programs," *Nucl. Acids Res.*, vol. 25, no. 17, pp. 3389–3402, Sept. 1997.

[2] J. C. Bezdek and S. K. Pal, Eds., *Fuzzy Models for Pattern Recognition*, Piscataway, NJ: IEEE Press, 1991.

[3] C. Blake and C. Merz, UCI repository of machine learning databases, http://www.ics.uci.edu/~mlearn/~MLRepository.html, Department of Information and Computer Sciences, University of California, Irvine.

[4] P. Bradley, U. Fayyad, and C. Reina, "Scaling clustering algorithms to large databases," *Proc. 4th International Conference on Knowledge Discovery and Data Mining*, AAAI Press, 1998, pp. 9–15.

[5] L. Breiman, "Pasting small votes for classification in large databases and on-line," *Machine Learning*, vol. 36, pp. 85–103, 1999.

[6] P. Chan and S. Stolfo, "Meta-learning for multistrategy and parallel learning," *Proc. 2nd International Workshop on Multistrategy Learning*, 1993, pp. 150–165.

[7] P. Chan and S. Stolfo, "Learning arbiter and combiner trees from partitioned data for scaling machine learning," *Proc. Intl. Conf. on Knowledge Discovery and Data Mining*, 1995, pp. 39–44.

[8] P. Chan and S. Stolfo, "Scaling leaning by meta-learning over disjoint and partially replicated data," *Proc. 9th Florida Artificial Intelligence Research Symposium*, 1996, pp. 151–155.

[9] D. L. Davies and D. W. Bouldin, "A cluster separation measure," *IEEE Transactions on Pattern Analysis and Machine Intelligence*, vol. 1, no. 4, pp. 224–227, 1979.

[10] T. G. Dietterich, "An experimental comparison of three methods for constructing ensembles of decision trees: Bagging, boosting, and randomization," *Machine Learning*, vol. 40, no. 2, pp. 139–158, 2000.

[11] S. Eschrich, N. V. Chawla, and L. O. Hall, "Generalization methods in bioinformatics," *Proc. Workshop on Data Mining in Bioinformatics, Knowledge Discovery and Data Mining*, 2002.

[12] S. Eschrich and L. O. Hall, "Soft partitions lead to better learned ensembles," *Proc. 2002 Annual Meeting of the North American Fuzzy Information Processing Society (NAFIPS2002)*, J. Keller and O. Nasraoui, Eds., IEEE, 2002, pp. 406–411.

[13] F. Farnstrom, J. Lewis, and C. Elkan, "Scalability for clustering algorithms revisited," *SIGKDD Explorations*, vol. 2, no. 1, pp. 51–57, 2000.

[14] G. Forman and B. Zhang, "Distributed data clustering can be efficient and exact," *SIGKDD Explorations*, vol. 2, no. 2, pp. 34–38, 2000.

[15] D. Gibson, J. Kleinberg, and P. Raghavan, "Clustering categorical data: An approach based on dynamical systems," *Proc. 24th International Conference on Very Large Databases*, 1998, pp. 311–322.

[16] S. Guha, R. Rastogi, and K. Shim, "Rock: A robust clustering algorithm for categorical attributes," *Information Systems*, vol. 25, no. 5, pp. 345–366, 2000.

[17] M. A. Hall, *Correlation-based Feature Selection for Machine Learning*, PhD Thesis, Waikato University, Department of Computer Science, Hamilton, NZ, 1998.

[18] J. A. Hartigan, *Clustering Algorithms*, New York: Wiley, 1975.

[19] S. Haykin, *Neural Networks: A Comprehensive Foundation*, 2nd Ed., Englewood Cliffs, NJ: Prentice Hall, 1999.

[20] A. Hinneburg and D. A. Keim, "An efficient approach to clustering in large multimedia databases with noise," *Proc. 4th International Conference on Knowledge Discovery and Data Mining*, New York: AAAI Press, 1998, pp. 58–65.

[21] W. Hoeffding, "Probability inequalities for sums of bounded random variables," *Journal of the American Statistical Association*, vol. 58, pp. 13–30, 1963.

[22] G. Hulten and P. Domingos, "A general method for scaling up machine learning algorithms and its application to clustering," *Proc. 18th International Conference on Machine Learning*, Williamstown, MA: Morgan Kaufmann, 2001, pp. 106–113.

[23] R. A. Jacobs, M. I. Jordan, S. J. Nowlan, and G. E. Hinton, "Adaptive mixtures of local experts," *Neural Computation*, vol. 3, pp. 79–87, 1991.

[24] A. K. Jain, M. N. Murty, and P. J. Flynn, "Data clustering: A review," *ACM Computing Surveys*, vol. 31, no. 3, pp. 264–323, 1999.

[25] I. T. Jolliffe, *Principle Component Analysis*, New York: Springer Verlag, 1986.

[26] D. T. Jones, "Protein secondary structure prediction based on decision-specific scoring matrices," *Journal of Molecular Biology*, vol. 292, pp. 195–202, 1999.

[27] M. I. Jordan and R. A. Jacobs, "Hierarchical mixtures of experts and the EM algorithm," *Neural Computation*, vol. 6, pp. 181–214, 1994.

[28] R. Kohavi, *Wrappers for Performance Enhancement and Oblivious Decision Graphs*, PhD Thesis, Stanford University, Computer Science Department, 1995.

[29] R. Krishnapuram and J. M. Keller, "A possibilistic approach to clustering," *IEEE Transactions on Fuzzy Systems*, vol. 1, no. 2, pp. 98–110, 1993.

[30] R. Krishnapuram and J. M. Keller, "The possibilistic c-means algorithm: Insights and recommendations," *IEEE Transactions on Fuzzy Systems*, vol. 4, no. 3, pp. 385–393, 1996.

[31] A. Likas and M. Titsias, "A probabilistic RBF network for classification," *Proc. INNS-IEEE International Joint Conference on Neural Networks*, 2000, vol. 4, pp. 238–243.

[32] H. Liu and H. Motoda, Eds., *Instance Selection and Construction for Data Mining*, Boston: Kluwer Academic Publishers, 2001.

[33] J. MacQueen, "Some methods for classification and analysis of multivariate observations," *Proc. 5th Berkeley Symposium on Mathematical Statistics and Probability*, Berkeley, CA, 1967, pp. 281–297.

[34] T. Mitchell, *Machine Learning*, New York: McGraw-Hill, 1997.

[35] P. Moerland, "Classification using localized mixtures of experts," *Proc. International Conference on Artificial Neural Networks (ICANN'99)*, London: IEE, 1999, vol. 2, pp. 838–843.

[36] J. E. Moody and C. J. Darken, "Fast learning in networks of locally-tuned processing units," *Neural Computation*, vol. 1, no. 2, pp. 281–294, 1989.

[37] A. W. Moore and M. S. Lee, "Cached sufficient statistics for efficient machine learning with large datasets," *Journal of Artificial Intelligence Research*, vol. 8, pp. 67–91, 1998.

[38] R. Ng and J. Han, "Efficient and effective clustering methods for spatial data mining," *Proc. 20th International Conference on Very Large Databases*, Santiago, Chile: Morgan Kaufmann, 1994, pp. 144–155.

[39] T. Oates and D. Jensen, "The effects of training set size on decision tree complexity," *Proc. 14th International Conference on Machine Learning*, D. Fisher, Ed., Morgan Kaufmann, 1997, pp. 254–262.

[40] M. J. D. Powell, "Radial basis functions for multivariable interpolation: A review," in *Algorithms for Approximation*, J. Mason and M. Cox, Eds., Oxford: Clarendon Press, 1987, pp. 143–167.

[41] F. Provost, D. Jensen, and T. Oates, "Efficient progressive sampling," *Proc. 5th International Conference on Knowledge Discovery and Data Mining (KDD-99)*, 1999, pp. 23–32.

[42] J. R. Quinlan, *C4.5: Programs for Machine Learning*, San Mateo, CA: Morgan Kaufmann, 1992.

[43] J. R. Quinlan, "Bagging, boosting, and C4.5," *Proc. 13th National Conference on Artificial Intelligence*, 1996, pp. 725–730.

[44] D. E. Rumelhart and J. L. McClelland, *Parallel Distributed Processing: Exploration in the Microstructure of Cognition*, vols. 1 and 2, Cambridge, MA: MIT Press, 1986.

[45] J. Shafer, R. Agrawal, and M. Mehta, "Sprint: A scalable parallel classifier for data mining," *Proc. 22nd VLDB Conference*, Mumbai (Bombay), India, 1996, pp. 1–12.

[46] J. Tantrum, A. Murua, and W. Stuetzle, "Hierarchical model-based clustering of large datasets through fractionation and refractionation," *Proc. 8th ACM SIGKDD International Conference on Knowledge Discovery and Data Mining*, Edmonton, Canada, 2002, pp. 183–190.

[47] I. H. Witten and E. Frank, *Data Mining: Practical Machine Learning Tools and Techniques with Java Implementations*, San Francisco, CA: Morgan Kaufmann, 1999.

[48] D. H. Wolpert, "Stacked generalization," *Neural Networks*, vol. 5, pp. 241–259, 1992.

[49] K. Woods, W. P. Kegelmeyer, and K. Bowyer, "Combination of multiple classifiers using local accuracy estimates," *IEEE Transactions of Pattern Analysis and Machine Intelligence*, vol. 19, pp. 405–410, 1997.

[50] T. Zhang, R. Ramakrishan, and M. Livny, "Birch: An efficient data clustering method for very large databases," *ACM SIGMOD Record*, vol. 25, no. 2, pp. 103–114, 1996.

Chapter 4

Marginal Learning Algorithms in Statistical Machine Learning

Qing Tao and Jue Wang

Abstract: It is now commonly known that the margin-driven algorithms play a crucial role in statistical machine learning. In this chapter, we summarize our investigation on margin methods for both supervised and unsupervised learning problems. As for supervised learning, we first describe the nature of SVMs (support vector machines) in terms of margin-based generalization bound. Based on this nature, a complete framework of posterior probability support vector machines (PPSVMs) is proposed for weighted training samples using modified concepts of risk, linear separability, margin and optimal hyperplane. Within this framework, a new optimization problem for unbalanced classification problems is formulated and a new concept of support vectors is established. Furthermore, a soft PPSVM with an interpretable parameter ν is obtained which is similar to the ν-SVM developed by Schölkopf *et al.*, and an empirical method for determining the posterior probability is proposed as a new approach to determine ν. The main advantage of a PPSVM classifier lies in that fact that it is closer to the Bayes optimal without knowing the distributions. To validate the proposed method, two synthetic classification examples are used to illustrate the logical correctness of PPSVMs and their relationship to regular SVMs and Bayesian methods. Compared with fuzzy support vector machines (FSVMs), the proposed PPSVM is a natural and analytical extension of regular SVMs based on the statistical learning theory. As for supervised learning, to extend the margin idea to unsupervised learning problems and establish a universal framework for one-class, clustering and PCA (principal component analysis) problems, an unsupervised learning problem with predefined threshold η is formally described and the intuitive margin is introduced. Then, one-class, clustering and PCA are formulated as three specific η-unsupervised learning problems. By defining a specific hypothesis space in η-one-class problems, the significant minimal sphere algorithm for regular one-class problems is proved to be a maximum margin algorithm. Furthermore, some new one-class, clustering and PCA marginal algorithms can be achieved in different hypothesis spaces. Since the nature in SVM is employed successfully, the proposed algorithms have robustness, flexibility and high performance. To verify our formulation, some experiments are conducted. They demonstrate that the proposed framework is not only of theoretical interest, but they also have potentials in the family of practical unsupervised learning techniques.

4.1 Introduction

In the last few years there have been very significant developments in the understanding of statistical learning theory and SVM (support vector machine) [5, 47, 48]. Statistical learning theory, which focuses on the induction and statistical inference from data to distribution, provides theoretical guarantees for good generalization ability of learning algorithms. As the first theoretically motivated statistical learning algorithm, SVM can minimize the expected risk in PAC (probably approximately correct [45]) frame. The wonderful statistical learning nature in SVM inspires us to reconsider many learning problems in pattern recognition [7]. In this chapter, we describe the nature of SVMs in terms of margin and report our investigation on application of margin methods to solve both supervised and unsupervised learning problems, specifically, unbalanced classification, one-class, clustering and PCA (principal component analysis) problems.

Within the last decade, both theory and practice have pointed to the concept of the margin of a classifier as being central to the success of a new generation of learning algorithms. This is explicitly true for SVMs, which in their simplest form implement maximal margin hyperplanes, but has also been shown to be the case for boosting algorithms such as Adaboost [29]. The generalization bound in terms of the margin has been clearly described and increasing the margin has been shown to improve the generalization performance [5, 27] in PAC framework.

In machine learning, the term "unbalanced" is mainly referred to as an unequal number of examples of the data set in each class. Under this circumstance, the role of samples is different. Since SVMs need no distribution information, all training points in each class are equally treated. However, in many real applications, certain samples may be outliers and some may be corrupted by noise, thus the influences of different samples may be unbalanced, and SVMs may not be robust enough and sometimes their performance could be affected severely by few samples with small probabilities. How can we adapt SVMs to such situations? Inspired by the Bayesian decision theory, we try to solve the problem by weighing the label of each sample using posterior probability $P(\text{class}|\text{sample})$. Unlike those in regular SVMs, these labels may not be 1 or -1 and thus data are in fact unbalanced now. Such a classification problem is called unbalanced in this chapter. To effectively deal with the unbalanced classification problem, the concept of margin in SVMs has to be modified for the use by posterior probability. Clearly, weighing the regular margin by posterior probabilities is a straightforward and the most natural way for this purpose. This idea motivates us to establish a complete framework for weighting the margin, which is called posterior probability support vector machines (PPSVMs) in this chapter.

The formal description of an unsupervised learning problem has been presented in [16, 34]. The objective functions of unsupervised learning problems are essentially in terms of distributions. This implies that the generalization ability is the main concern in the performance of unsupervised learning algorithms. But it seems that there are no available clustering and PCA algorithms like SVMs, in which the good generalization ability is obtained by the data-dependent algorithms. Can we generalize the idea in SVM to get some clustering and PCA algorithms with robustness, flexibility and high generalization performance? On the other hand, one-class problem can be viewed as a particular unsupervised problem. Its generalization follows directly from unsupervised problems. Can the available one-class learning algorithms minimize the expected risk in PAC frame? Thirdly, in many classifications, PCA is usually used to make computation easier and more effective. However, the patterns to be recognized are unknown before we use PCA. This kind of classification tasks obviously needs the generalization ability of PCA. To use the nature of statistical learning theory to investigate unsupervised learning problems, we define an η-unsupervised learning problems with a intuitive margin and use the margin to design algorithms for general unsupervised learning, especially one-class, clustering and PCA problems, and further analyze the relations between our marginal algorithms and the available unsupervised algorithms. This may be the most important contribution of this chapter.

The remainder of this chapter is arranged as follows. The classification problems are formally described in Section 4.2. Some marginal algorithms in SVMs and its theoretical analysis have been stated in Section 4.3. In Section 4.4, the unbalanced classification problems and the framework of weighted maximal margin algorithms are presented. The unsupervised learning problems and our η-unsupervised learning problems are formally described in Section 4.5. Some marginal algorithms for one-class problems are presented and analyzed theoretically in Section 4.6. In Section 4.7, some available clustering algorithms are described and some new clustering algorithms are presented. In Section 4.8, a complete SVM-like framework of linear PCA for deciding the projection direction is constructed. And some conclusions are summarized in the last section.

4.2   Classification Problems and Margin

There are mainly two kinds of supervised learning problems, i.e. classification and regression. In this chapter, we only discuss the classification problems. Consider the following two-category classification problem:

$$(x_1, z_1), (x_2, z_2), \cdots, (x_l, z_l) \in R^N \times Z, \ Z = \{-1, 1\}. \tag{4.2.1}$$

where $(x_i, z_i) \in R^N \times \{1, -1\}$ is independently drawn and identically distributed according to a density function $p(x, z)$, and z_i is the label of x_i. Let $H = \{f(x, \alpha) | f(x, \alpha) \colon R^N \times \Lambda \to \{1, -1\}, \alpha \in \Lambda\}$ be a hypothesis space of problem (4.2.1).

Definition 4.2.1 (Loss function). Let $f \in H$. Let L be defined as follows:

$$L(z, f(x, \alpha)) = \begin{cases} 0, & \text{if } f(x, \alpha) = z; \\ 1, & \text{otherwise.} \end{cases} \tag{4.2.2}$$

L is called the *loss function* of problem (4.2.1). Based on this loss function, the expected and empirical risks can be defined in terms of distributions and data respectively.   □

Definition 4.2.2 (Empirical risk). Let $f(x) \colon R^N \to R$ and $f \in H$. Then

$$err_{emp}(f) = \sum_{i=1}^{l} L(z_i, f(x_i))$$

$$err_D(f) = \int L(z, f(x)) p(x, z) dx dz$$

is called the empirical and expected risk of f about problem (4.2.1) respectively.
□

From Definition 4.2.2, it is easy to know that the expected risk is the total probability that a point is misclassified. In fact, the expected risk reflects the averaged error of all the samples.

Definition 4.2.3 (The Bayesian optimal classifier). Let $f_0 \in H$. If

$$err(f_0) = \min\{\int L(z, f(x)) p(x, z) dx dz, f \in H\}.$$

Then f_0 is called the optimal classifier.   □

In the following, we will restrict the hypothesis space H to $H_0 = \{w^T x + b \colon w \in R^N, b \in R\}$.

Definition 4.2.4 (Separable). If there exists a pair $(w, b) \in R^N \times R$, such that

$$\begin{cases} w^T x_i + b > 0, & \text{if } z_i > 0; \\ w^T x_i + b < 0, & \text{if } z_i < 0; \end{cases}$$

Then the classification problem (4.2.1) is called linearly separable and $f(x) = w^T x + b$ is called a linear classifier. □

Definition 4.2.5 (Margin). Let $w^T x + b$ be a linear classifier of linearly separable problem (4.2.1) and $\|w\| = 1$.

$$\rho(w, b, x_i) = z_i(w^T x_i + b).$$

is called the margin of x_i.

$$\rho(w, b) = \min\{z_i(w^T x_i + b), 1 \le i \le l\}$$

is called the margin of linear classifier $w^T x + b$. If there exists (w_0, b_0), such that,

$$\rho(w_0, b_0) = \max\{\rho(w, b) \colon (w, b) \in R^N \times R, \|w\| = 1\}$$

The hyperplane $w_0^T x + b_0 = 0$ is called the maximal margin hyperplane for classification problem (4.2.1), and $\rho(w_0, b_0)$ is called the maximum margin. □

4.3 Maximal Margin Algorithm in SVM

Finding a maximal margin can now be formulated as solving the following optimization problem:

$$\begin{cases} \max\{\rho(w, b)\}, \\ z_i(w^T x_i + b) \ge \rho(w, b), \ 1 \le i \le l, \ \|w\| = 1. \end{cases}$$

As discussed in [48], the above optimization problem is equivalent to

$$\begin{cases} \min \dfrac{1}{2}\|w\|^2, \\ z_i(w^T x_i + b) \ge 1, \quad 1 \le i \le l. \end{cases} \tag{4.3.1}$$

In the following, the generalization ability of the above maximal margin algorithm will be analyzed. To this end, the well-known VC dimension bound theorem in [5, 47, 48] is introduced.

Theorem 4.3.1 (cf. [5]). *Let h denote the VC dimension of the function class H. For any probability distribution on $R^N \times Z$, with probability of $1 - \delta$ over l random examples, any $f \in H$ that is consistent with the training set has error no more than*

$$R[f] \le \epsilon = \frac{2}{l}\left[h\log\frac{2el}{h} + \log\frac{2}{\delta}\right]$$

provided $h \le l$ and $l > \epsilon/2$, where $R[f]$ is the expected risk. □

Let the hypothesis space $H_0 = \{w^T x + b, w \in R^N, b \in R\}$ be the set of all linear classifiers, and h_0 be the corresponding VC dimension. The following inequality is given in [47, 48],

$$h_0 \leq R_D^2 \Theta^2 + 1, \text{ if } ||w|| \leq \Theta$$

where R_D is the radius of the smallest ball around the samples.

Roughly speaking, SVMs attempt to minimize the expected risk using PAC framework by minimizing the VC dimension of H and the empirical risk simultaneously [3, 24]. Note that in optimization problem (4.3.1), the constraints $(z_i(w^T x_i + b) \geq 1, 1 \leq i \leq l)$ imply the empirical risk is zero. On the other hand, if we bound the margin of a function class, from the above inequality, we can control its VC dimension. Hence, the maximal margin algorithm is in fact to minimize a bound of the expected risk on condition that the corresponding empirical risk is zero. Therefore, the maximal margin algorithm here captures the insight of statistical learning theory. However, there are ramifications of such analysis that go beyond the scope of this chapter. In fact, applying Theorem 3.1 to SVMs requires *a-priori* structuring of the hypothesis space. Bounds that rely on an effective complexity measure rather than the a-priori VC dimension have been proposed by Shawe-Taylor *et al.* in [5, 30]. These bounds rely only on the geometric margin. A nice and direct proof of this theorem is systematically described in [5]. For the purpose of a condensed explanation, we only need to quote a bound on the generalization error in terms of the margin and covering numbers. To this end, we introduce the definition of covering numbers and margin bound theorem in [5].

Definition 4.3.1 (γ-cover of a function class). Let F be a class of real-valued functions on a domain X. A γ-cover of F with respect to a sequence of inputs $S = (x_1, x_2, \cdots, x_l)$ is a finite set of functions A such that for all $f \in F$, there exists $g \in A$, such that $\max_{1 \leq i \leq l}(|f(x_i) - g(x_i)|) < \gamma$. The size of the smallest such cover is denoted by $N(F, S, \gamma)$, while the covering numbers of F are the values

$$N(F, l, \gamma) = \max_{S \in X^l} N(F, S, \gamma). \qquad \square$$

Theorem 4.3.2. *Consider thresholding a real-valued function space F and fix $\gamma > 0$. For any probability distribution D on X, with probability $1 - \delta$ over l random examples S, any hypothesis $f \in F$ that has margin $m(f, S, \eta) \geq \gamma$ on S has error no more than*

$$err_D(f, \eta) \leq \epsilon(l, F, \delta, \gamma) = \frac{2}{l}(log N(F, 2l, \frac{\gamma}{2}) + log\frac{2}{\delta})$$

provided $l > 2/\epsilon$. $\qquad \square$

Based on Theorem 4.3.2, generally, a large margin implies good generalization performance.

Define Lagrangian function as, $L = \frac{1}{2}\|w\|^2 - \sum_{i=1}^{l} \alpha_i(z_i(w^T x_i + b) - 1)$. Then, the dual representation of optimization problem (4.3.1) is

$$
\begin{cases}
\min \ \dfrac{1}{2} \displaystyle\sum_{i,j=1}^{l} \alpha_i \alpha_j z_i z_j \langle x_i, x_j \rangle - \sum_{i=1}^{l} \alpha_i \\[2mm]
\alpha_i \geq 0, \ 1 \leq i \leq l, \ \displaystyle\sum_{i=1}^{l} \alpha_i z_i = 0
\end{cases}
\tag{4.3.2}
$$

Let $\{\alpha_i^0, 1 \leq i \leq l\}$ be a solution of problem (4.3.2). Using the same inference as that in SVMs, optimal w_0 can be written as

$$
w_0 = \sum_{i=1}^{l} \alpha_i^0 z_i x_i
$$

Definition 4.3.2. If $\alpha_i^0 \neq 0$, the corresponding x_i is called a support vector. □

Geometrically speaking, the margin of a classifier is the minimal distance of training points from the decision boundary. The maximal margin classifier is the one with the maximum distance from the *nearest patterns* to the boundary, called *support vectors*. Informally speaking, support vectors are the patterns most informative and difficult for the classification task.

For problems which allow that patterns may be misclassified or in the margin stripe, Cortes and Vapnik proposed soft algorithms by introducing slack variables in constraints ([4, 47, 48])

$$
\begin{cases}
\min \ \dfrac{1}{2}\|w\|^2 + C \displaystyle\sum_{i=1}^{l} \xi_i \\[2mm]
\xi_i \geq 0, \ z_i(w^T x_i + b) \geq 1 - \xi_i, \ 1 \leq i \leq l.
\end{cases}
\tag{4.3.3}
$$

where C is a predefined positive real number and ξ_i's are slack variables.

Similarly, the dual representation of optimization problem (4.3.3) is

$$
\begin{cases}
\min \ \dfrac{1}{2} \displaystyle\sum_{i,j=1}^{l} \alpha_i \alpha_j z_i z_j \langle x_i, x_j \rangle - \sum_{i=1}^{l} \alpha_i \\[2mm]
C \geq \alpha_i \geq 0, \ 1 \leq i \leq l, \ \displaystyle\sum_{i=1}^{l} \alpha_i z_i = 0.
\end{cases}
\tag{4.3.4}
$$

Obviously, the choice of linear classifiers seem to be very limiting. But there is a way to have both linear models and a very rich set of nonlinear classifiers, by using the kernel techniques. The interesting fact about kernel functions is that the scalar product can be implicitly computed in the sample space. Note that the inner product is only concerned in optimization problem (4.3.2) and (4.3.4). So, kernel allows to achieve a nonlinear classifier by computing scalar products in feature space, where one could otherwise hardly perform any computation. As such, SVM is now usually called kernel method in machine learning community. A direct consequence from kernel methods is that: every (linear) algorithm that only uses scalar products can implicitly be executed by using kernels, i.e., one can very elegantly construct a nonlinear version of a linear algorithm. In fact, kernel representations offer an alternative solution by embedding the data to a high dimensional feature space to increase the computational power of linear algorithms. In 2002, a new technique was introduced in [30] allowing us to shift the C-SVM to a specific maximum margin algorithm in an auxiliary function space.

In order to be intuitive geometrically and analyzed easily, the designation of almost all the learning algorithms is divided into two phases: 1) representation, i.e. to look for a fundamental representation space, where the solution can be described in a simpler and more intuitive form; 2) embedding, i.e. to embed the data into a properly selected representation space and search for the simple and intuitive solutions therein based on the embedded data points. It now becomes clear that the representation and embedding of SVMs are solved successfully and thoroughly, and both theories and practices have pointed out that the maximal margin algorithm for linearly separable cases is central to the success of complete framework of SVMs ([5, 30]). Compared with neural networks, the mechanism of SVMs is clear and easy to understand.

In order to effectively control the number of support vectors through a single parameter, Schölkopf et $al.$ constructed a new SVM algorithm called ν-SVM [33]. In addition, ν-SVMs have the advantage of enabling us to eliminate the regularization parameter C in C-SVMs.

Specifically, ν-SVMs are formulated as

$$\begin{cases} \min \dfrac{1}{2}\|w\|^2 - \nu\rho + \dfrac{1}{l}\sum_{i=1}^{l}\xi_i, \\ \xi_i \geq 0,\ z_i(w^T x_i + b) \geq \rho - \xi_i,\ 1 \leq i \leq l,\ \rho > 0, \end{cases} \qquad (4.3.5)$$

where $\nu \geq 0$ is a constant and z_i is 1 or -1. To understand the role of ρ, note that for $\xi = 0$, the above constraints imply that the two classes are separated by the margin $2\rho/\|w\|$.

Theorem 4.3.3. $Assume$ $that$ the $solution$ of $(4.3.5)$ $satisfies$ $\rho > 0$, $then$ the $following$ $statements$ $hold:$

(i) ν is an $upper$ $bound$ on the $fraction$ of $margin$ $errors.$

(ii) ν is a lower bound on the fraction of support vectors. $\qquad\qquad\square$

4.4 Unbalanced Classification Problems and Weighted Maximal Margin Algorithms

The Bayesian decision theory is a fundamental statistical approach to classification problems [7], and its power, coherence, and analytical nature applied in pattern recognition make it among the elegant formulations in science. A Bayesian approach achieves the exact minimum probability of error based entirely on evaluating the posterior probability. However, in order to perform required calculations, a number of assumptions including the availability of *a prior* probability and the class-conditional probability must be made. Clearly, the knowledge of density functions would allow us to solve whatever problems that can be solved on the basis of available data. But in line with Vapnik's principle of never solving a problem that is more general than you actually need to solve [47, 48], one should try to avoid estimating any density when solving a particular classification problem. Hence, in machine learning, the minimization of the expected risk should be achieved only in terms of samples available. Fortunately, the statistical learning theory provides distribution-free conditions and guarantees for good performance of generalization for learning algorithms [47, 48]. It attempts to minimize the expected risk using the PAC (probably approximately correct [45]) framework by simultaneously minimizing the empirical risk and the model complexity [5, 47, 48]. The most significant difference between Bayesian decision and statistical learning theory might be that the former is based on distributions called deductive inference, while the latter is based on data called inductive inference.

As far as the relationship between SVMs and the posterior probability is concerned, some researchers suggest that the output of SVM should be a calibrated posterior probability to enable post-procession [48]. Under this circumstance, the output of SVM will not be binary. In [48], Vapnik proposes fitting this probability with a sum of cosine functions, where the coefficients of the cosine expansion will minimize a regularized function. To overcome some limitations in [48], Platt applies a sigmoid regression to the output of SVMs to approximate the posterior probability [26] and demonstrates that the SVM+sigmoid combinations produce probabilities that are of comparable quality to the regularized maximum likelihood kernel methods. Since the output of SVM is used in the posterior probability estimation, the desired sparsity is achieved for Bayesian classifiers in [26, 48]. On the other hand, Sollich has described a framework interpreting SVMs as maximizing a posterior solution to inference problems [35]. It should be pointed out that, in all the previous papers, the training of SVM is in fact investigated in the balanced cases. In order to solve unbalanced classification problems, Y. Lin *et al.* have presented a modified SVM [20], and C. Lin *et al.* proposed a fuzzy SVM [19]. They all have extended the soft margin algorithm in SVMs by weighing the punishment terms of error. Unfortunately, the linearly separable problem, which forms the main building block for more complex SVMs, is difficult to be discussed in their formulations. Obviously, how to reformulate the entire SVM framework in terms of posterior probabilities for solving unbalanced classification problems is an interesting issue.

In this section, a complete framework of posterior probability support vector machines (PPSVMs) is presented by weighing unbalanced training samples and introducing new concepts for linear separability, margins and optimal hyperplanes. The style of PPSVMs is almost the same as that of SVMs, but the binary output of PPSVMs is based on the posterior probability now. This might constitute an interesting attempt to train a SVM to behave like a Bayesian optimal classifier. In reformulating SVMs, our main contributions come from defining the weighted margin properly and formulating a series of weighted optimization problems to determine hard and soft margins and ν-SVM (as proposed and developed by Schölkopf *et al.* [33]). Furthermore, the generalization bound of PPSVMs is analyzed and the result provides PPSVMs with a solid basis in the statistical learning theory. Note that when the posterior probability of each sample is either 0 or 1, the proposed PPSVMs coincide with regular SVMs. Intuitively, by weighing the samples, PPSVM is able to lessen the influence of the outliers with small weight. This is a natural way to make the algorithm more robust against outliers. Clearly, the potential advantage of PPSVMs lies in that it can be used to solve classification problems where the classes and samples are not equally distributed.

It is quite interesting to notice that the proposed approach for classification with posterior probabilities is very similar to what has been described in the Bayesian decision theory [7]. But it should be emphasized that our model is based on data rather than distributions.

However, in many real applications, there is no posterior probability information available for training samples. To make our PPSVMs applicable, we have proposed an empirical method for estimating the posterior probability of samples. The empirical method of calculating posterior probabilities also provides us a simply way to determine parameter ν in ν-SVMs, which is new and interesting. To verify the logical correctness of the proposed framework for weighing data, two classification examples are used to illustrate the relationship between our classifiers and Bayesian classifiers. These synthetic experiments demonstrate that the proposed PPSVMs are closer to Bayes optimal. Additional tests have also been conducted on several real data sets and the corresponding results demonstrate that the proposed PPSVMs are more robust than regular SVMs for some classification problems.

4.4.1 Unbalanced Classification Problems

In this section, the following unbalanced two-category classification problem is addressed:

$$(x_1, p(\omega|x_1)), (x_2, p(\omega|x_2)), \cdots, (x_l, p(\omega|x_l)) \in R^N \times I_0, I_0 = [0, 1], \quad (4.4.1)$$

where ω is one of the two categories, and $p(\omega|x_i)$ is a posterior probability. Note that the degree of unbalance in problem (4.4.1) is reflected by the distribution of posterior probabilities. Clearly, if the classification decision is based on the complete information of distribution function of posterior probabilities, the expected risk (or probability of error) can be minimized exactly using Bayesian decision theory [7]. Unfortunately, the Bayesian inference can not be applied here because the distribution function of all the samples is unknown and the only available information is the posterior probability of the training samples.

Using $\text{sign}(2p(\omega|x_i)-1)$ as the label of sample x_i, the information on posterior probabilities will be lost partly but regular SVMs can be employed to deal with the linear classification problem,

$$(x_1, \text{sign}(2p(\omega|x_1)-1)), (x_2, \text{sign}(2p(\omega|x_2)-1)), \cdots, (x_l, \text{sign}(2p(\omega|x_l)-1)).$$
$$(4.4.2)$$

In order to utilize fully the information provided by posterior probabilities in classification, a new continuous label is introduced to replace the original binary one for each sample. Define,

$$y_i = 2p(\omega|x_i) - 1, \quad 1 \le i \le l.$$

Obviously, $y_i = 1$ if $p(\omega|x_i) = 1$ and $y_i = -1$ if $p(\omega|x_i) = 0$. Now, a regular classification problem with +1 and -1 as labels can then be regarded as a special case of problem (1). For the sake of convenience, it is assumed that $p(\omega|x_i) \ne 0.5$ for all $1 \le i \le l$. In fact, if $p(\omega|x_i) = 0.5$, we can perturb $p(\omega|x_i)$ a little and force x_i to be in one of two categories.

To develop posterior probability support vector machines (PPSVMs) for solving unbalanced classification problems, we have to modify the loss function and risks first. Note that, in Bayesian decision theory, x_i is labeled 1 if $y_i > 0$ and x_i is labeled -1 if $y_i < 0$. Obviously, the modified loss function should coincide with Bayesian inference in this case. In the same time, the loss function here should be in agreement with the one in SVMs when the label of each sample is treated as $\text{sign}(2p(\omega|x_i) - 1)$. Based on these considerations, the loss function is redefined as follows.

Definition 4.4.1. Let $f(x, \alpha)\colon R^N \times \Lambda \to \{1, -1\}$ and let $H = \{f(x, \alpha)\colon \alpha \in \Lambda\}$ be a hypothesis space of problem (4.4.1), where Λ is a set of the parameters and a particular choice of α generates a so-called trained machine. Define L as follows:

$$L(y, f(x, \alpha)) = \begin{cases} 0, & \text{if } y > 0, \ f(x, \alpha) = 1; \\ 0, & \text{if } y < 0, \ f(x, \alpha) = -1; \\ 1, & \text{otherwise.} \end{cases}$$

L is called the loss function of problem (4.4.1). $\square$

Based on this loss function, the expected and empirical risks can be defined in terms of distributions and data, respectively [47, 48].

Now the linear separability for unbalanced classification can be easily defined as,

Definition 4.4.2. If there exists a pair $(w, b) \in R^N \times R$, such that

$$\begin{cases} w^T x_i + b > 0, & \text{if } y_i > 0; \\ w^T x_i + b < 0, & \text{if } y_i < 0; \end{cases} \tag{4.4.3}$$

Then the unbalanced classification problem (4.4.1) is called posterior probability linearly separable and

$$f(x) = w^T x + b$$

is called a posterior probability linear classifier. $\square$

The following remarks are useful in understanding the properties of linear separability:

- Obviously, the regular classification problem (4.4.2) is linearly separable if and only if problem (4.4.1) is posterior probability linearly separable.

- Since the norm of w is not restricted in Definition 4.4.2 and we can multiply (w, b) with a sufficiently large positive number, condition (4.4.3) is actually equivalent to: there exists a pair (w_1, b_1) such that

$$y_i(w_1^T x_i + b_1) \geq y_i^2, \quad 1 \leq i \leq l. \tag{4.4.4}$$

This property is important in specifying the constraint imposed by the linearly separability for the purpose of optimization. Specifically, if y_i is 1 or -1, equation (4.4.4) becomes

$$y_i(w_1^T x_i + b_1) \geq 1, \quad 1 \leq i \leq l.$$

This is just the canonical description of linear separability in regular SVMs.

- If $\|w\| = 1$, $|w^T x_i + b|$ is the distance of x_i to the hyperplane $w^T x + b = 0$. Therefore, for linear separable problems, $\dfrac{w^T x_i + b}{y_i}$ can be viewed as a weighted distance.

4.4.2 Weighted Optimal Hyperplane, Weighted Margin and Generalization

Generally, a large margin implies good generalization performance. Thus, to effectively deal with the unbalanced classification in problem (4.4.1), the concept of margin in SVMs has to be modified for the use by PPSVMs. Since the significance of a sample should be weakened while its margin be enlarged if its posterior probability is small, the weighted margin is defined as follows.

Definition 4.4.3. Let $w^T x + b$ be a linear classifier of posterior probability linearly separable problem (4.4.3) and $\|w\| = 1$,

$$\rho(w, b, x_i) = \frac{w^T x_i + b}{y_i}$$

is called the margin of x_i.

$$\rho(w, b) = \min\{\frac{w^T x_i + b}{y_i}, 1 \le i \le l\}$$

is called the margin of linear classifier $w^T x + b$. If there exists (w_0, b_0), such that,

$$\rho(w_0, b_0) = \max\{\rho(w, b) \colon (w, b) \in R^N \times R, \|w\| = 1\}.$$

The hyperplane $w_0^T x + b_0 = 0$ is called the optimal hyperplane for unbalanced classification problem (4.4.1), and $\rho(w_0, b_0)$ is called the maximal margin. □

As in SVMs, the maximal margin classifier is an important concept to PPSVMs, serving as a start point for analysis and construction of more sophisticated algorithms. Therefore, problems concerning the existence, uniqueness, implementation and other properties of the maximal classifier for PPSVMs must be investigated first.

Theorem 4.4.1. *There exists a unique optimal hyperplane for problem (4.4.1).*

Proof. From the linear separability, there exists $(w, b) \in R^N \times R$ and $\|w\| = 1$, such that

$$\frac{w^T x_i + b}{y_i} > 0, 1 \le i \le l.$$

The linear separability also implies

$$|b| \le \max\{\|x_i\|, 1 \le i \le l\} + 1.$$

Thus, we can define $\rho(w, b)$ in region $\{\|w\| \le 1\} \times \{|b| \le \max\{\|x_i\|\} + 1\}$. The existence of maximum of $\rho(w, b)$ follows directly from the continuity of $\rho(w, b)$ in the bounded region.

The maximum of $\rho(w, b)$ must be achieved at $\|w\| = 1$. Since $\|w\| > 0$, and if the maximum is achieved at some point (w, b) with $0 < \|w\| < 1$, then $\rho\left(\dfrac{w}{\|w\|}, \dfrac{b}{\|w\|}\right) = \dfrac{\rho(w, b)}{\|w\|}$ would achieve a larger value.

On the other hand, the maximum of $\rho(w, b)$ can not be achieved on two different points (w_1, b_1) and (w_2, b_2), where $\|w_1\| = 1$, $\|w_2\| = 1$ and $w_1 \ne w_2$. Otherwise, since function $\rho(w, b)$ is concave, the maximum would be achieved on the line that connects (w_1, b_1) and (w_2, b_2). Since there exists a pair (w, b) with $0 < \|w\| < 1$ on this line, it contradicts the fact that the maximum must be achieved on $\|w\| = 1$. Thus, theorem is proved.

Assume $\|w\| = 1$ and $f(x) = w^T x + b$ is a posterior probability linear classifier. Obviously, $\rho(w, b) = \min\left\{\dfrac{w^T x_i + b}{y_i}, 1 \le i \le l\right\}$ follows from the linear separability. Finding a maximal margin can now be formulated as solving the following optimization problem:

$$\begin{cases} \max\{\rho(w, b)\}, \\ w^T x_i + b \ge y_i \rho(w, b), & \text{if } y_i > 0, \\ w^T x_i + b \le y_i \rho(w, b), & \text{if } y_i < 0, \ \|w\| = 1. \end{cases}$$

As discussed in [48], the above optimization problem is equivalent to

$$\begin{cases} \min \dfrac{1}{2}\|w\|^2, \\ y_i(w^T x_i + b) \ge y_i^2, \quad 1 \le i \le l. \end{cases} \tag{4.4.5}$$

In linearly separable cases, condition (4.4.3) can be easily satisfied by multiplying w with a sufficiently large positive number. However, optimization problem (4.4.5) looks for w with the minimal norm. $\quad\square$

Note the generalization ability of the above maximal margin algorithm can be similarly analyzed as that in regular cases. Based on the direct proof in [5] (Theorem 4.3.2), this bound also holds for the expected risks of classification here. Moreover, the generalization ability of corresponding soft algorithms can be analyzed by employing the margin bound, as one can see in sequel.

4.4.3 Posterior Probability Support Vectors

Define Lagrangian function as,

$$L = \frac{1}{2}\|w\|^2 - \sum_{i=1}^{l} \alpha_i (y_i(w^T x_i + b) - y_i^2).$$

Then, the dual representation of optimization problem (4.4.5) is

$$\begin{cases} \min \frac{1}{2} \sum_{i,j=1}^{l} \alpha_i \alpha_j y_i y_j \langle x_i, x_j \rangle - \sum_{i=1}^{l} \alpha_i y_i^2, \\ \alpha_i \geq 0, \ 1 \leq i \leq l, \ \sum_{i=1}^{l} \alpha_i y_i = 0. \end{cases} \tag{4.4.6}$$

Let $\{\alpha_i^0, 1 \leq i \leq l\}$ be a solution of problem (4.4.6). Using the same inference as that in SVMs, optimal w_0 can be written as

$$w_0 = \sum_{i=1}^{l} \alpha_i^0 y_i x_i.$$

Definition 4.4.4. If $\alpha_i^0 \neq 0$, the corresponding x_i is called a posterior probability support vector. □

The following remarks are helpful to understand the nature of the proposed PPSVMs.

– Since $\|w\| \neq 0$, it is easy to see that there exists at least one $\alpha_i^0 \neq 0, 1 \leq i \leq l$, therefore, b_0 in the optimal hyperplane can be found by the Karush-Kuhn-Tucker complementarity conditions (cf. [17] and [9]) $\alpha_i^0(w^T x_i + b - y_i) = 0$. According to Theorem 4.4.1, we will obtain the same b_0 for all $\alpha_i^0 \neq 0$.

– If $p(\omega|x_i)$ is equal to 1 or 0, y_i will be 1 or -1 and all the above mentioned definitions and optimization problems coincide with those in regular SVMs. Thus, the proposed PPSVM is an extension of regular SVMs for unbalanced classification problems.

– For each posterior probability support vector, $w_0^T x_i + b_0 = y_i$. One of the most important differences between regular support vectors and posterior probability support vectors is that the former are the closest to the optimal hyperplane while the latter may not be.

– Note that optimization problem (4.4.6) uses inner products only, therefore, kernel techniques can be employed to solve the corresponding nonlinear problems.

– For linearly separable cases, it is interesting to notice that vector x on the optimal hyperplane satisfies $p(\omega|x) = 0.5$. This classification ruler is similar to the Bayesian decision [7] in principle. Particularly, it is commonly known that the Bayesian discriminant function of a normal density is linear in some special cases [7]. Does it coincide with the proposed PPSVM classifier? Theoretical analysis, unfortunately, is lacking at this point, but we will demonstrate this interesting fact in Section 4.4.6 through two designed examples.

– Apparently, the above classification idea looks as if we are searching a regression function with posterior probability $p(\omega|x) = 0.5$. However, in a regression problem, although a sample is correctly labeled in the classification, the distance from this sample to the boundary always has influence on deciding the regression function. Obviously, the main difference between classification and regression is in the objective function. From the definition of empirical and expected risks in Section 4.1, it can be found out that we are intrinsically solving a classification problem rather than a regression one. This is same with regular SVM classification and regression cases.

4.4.4 C-Soft Margin Algorithms and Fuzzy SVMs

Similarly as that in soft SVMs, for posterior probability linearly inseparable classification problems, the following optimization problem for soft margins can be formulated,

$$
\begin{cases}
\min \ \dfrac{1}{2}\|w\|^2 + C\sum_{i=1}^{l}\xi_i \\
\xi_i \geq 0, \ y_i(w^T x_i + b) \geq y_i^2 - y_i^2\xi_i, \ 1 \leq i \leq l.
\end{cases}
\tag{4.4.7}
$$

where C is a predefined positive real number and ξ_i's are slack variables.

In the constraint of equation (4.4.7), each slack variable ξ_i is multiplied with y_i^2. Obviously, other choices that could result in different implementations are also possible. Note that our hard margin of x_i is

$$
\frac{w^T x_i + b}{y_i}
$$

and it is relaxed to be

$$
\frac{w^T x_i + b}{y_i} + \xi_i.
$$

Therefore, the softness in equation (4.4.7) is in fact introduced through the relaxation of the margin, along the same line as that in soft SVMs in [5, 30]. As pointed out in [5, 30], the advantage of such relaxation is that each soft algorithm can be proved to be equivalent to a hard one in an augmented feature space and its generalization ability can thus be guaranteed.

To deal with the unbalanced classification problems, each x_i is weighted using a fuzzy membership s_i in [19]. Since the fuzzy membership is the attitude of the corresponding point toward one class and the punishment term ξ_i is a measure of errors in a SVM, $s_i\xi_i$ is a measure of errors with different weighting. The optimal hyperplane problem in [19] is then considered as the solution to

$$\begin{cases} \min \; \frac{1}{2}\|w\|^2 + C\sum_{i=1}^{l} s_i\xi_i \\ \xi_i \geq 0, \; y_i(w^T x_i + b) \geq 1 - \xi_i, \; 1 \leq i \leq l. \end{cases} \qquad (4.4.8)$$

Optimization problems (4.4.7) and (4.4.8) are very similar in format. Although it is claimed in [19] that small s_i can reduce the effect of ξ_i, the motivation is not clear from the point of view of statistical learning theory because the expected risk and the empirical risk are not defined. Consequently, optimization problem (4.4.8) can only be regarded simply as an extension of SVM-soft-margin optimization problem in its mathematical form. The difference between PPSVMs and Fuzzy SVMs is that the former is based on weighting the margin while the latter is based on weighting the penalty. It should also be pointed out that the optimization problem in [20] is almost identical to problem (4.4.8). Remarkably, it has the advantage of approaching the Bayesian rule of classification. Methods in [19, 20] can be considered as another two approaches to unbalanced problems. Unfortunately, the linearly separable problems can not be addressed if we use the ideas of reducing the effect of some samples as suggested in [19, 20]. Therefore, it is not possible to reformulate a complete SVM-like framework for unbalanced classification problems.

4.4.5 ν-SVM and Empirical Methods for Posterior Probability

Specifically, ν-SVMs are formulated as

$$\begin{cases} \min \; \frac{1}{2}\|w\|^2 - \nu\rho + \frac{1}{l}\sum_{i=1}^{l} \xi_i \\ \xi_i \geq 0, \; \dfrac{w^T x_i + b}{y_i} \geq \rho - \xi_i, \; 1 \leq i \leq l. \; \rho > 0 \end{cases}$$

where $\nu \geq 0$ is a constant and y_i is 1 or -1. To understand the role of ρ, note that for $\xi = 0$, the above constraints imply that the two classes are separated by the margin $2\rho/\|w\|$.

Based on the proposed margin of PPSVMs, a corresponding ν-SVM is formulated by the following modification,

$$
\begin{cases}
\min \ \dfrac{1}{2}\|w\|^2 - \nu\rho + \dfrac{1}{l}\sum_{i=1}^{l}\xi_i \\[2mm]
\xi_i \geq 0, \ y_i(w^T x_i + b) \geq y_i{}^2\rho - y_i{}^2\xi_i, \ 1 \leq i \leq l. \ \rho > 0.
\end{cases}
\tag{4.4.9}
$$

To obtain the dual, we consider the Lagrangian

$$
L(w, \xi, b, \alpha, \beta, \delta) = \frac{1}{2}\|w\|^2 - \nu\rho + \frac{1}{l}\sum_{i=1}^{l}\xi_i
$$

$$
- \sum_{i=1}^{l}\alpha_i(y_i(w^T x_i + b) - y_i{}^2\rho + y_i{}^2\xi_i) - \sum_{i=1}^{l}\beta_i\xi_i - \delta\rho
$$

and arrive at the following optimization problem

$$
\begin{cases}
\min \ \dfrac{1}{2}\sum_{i,j=1}^{l}\alpha_i\alpha_j y_i y_j \langle x_i, x_j \rangle \\[2mm]
\dfrac{1}{l} \geq \alpha_i y_i{}^2 \geq 0, \ 1 \leq i \leq l, \ \sum_{i=1}^{l}\alpha_i y_i = 0, \ \sum_{i=1}^{l}\alpha_i y_i{}^2 \geq \nu.
\end{cases}
\tag{4.4.10}
$$

It can be shown that the resulting decision function takes the form

$$
f^*(x) = \mathrm{sgn}(\sum_{i=1}^{l}\alpha_i y_i \langle x, x_i \rangle + b).
$$

To compute b, a point $x_+ \in \omega_1$ and $x_- \in \omega_2$ with $0 < \alpha_i y_+{}^2 < \dfrac{1}{l}$ and $0 < \alpha_i y_-{}^2 < \dfrac{1}{l}$ are selected. By the KKT conditions

$$
b = \frac{y_- w^T x_+ - y_+ w^T x_-}{y_+ - y_-}.
$$

Compared with regular ν-SVM, the only change in problem (4.4.10) is that each α_i is multiplied with $y_i{}^2$. Consequently, by the KKT conditions, the following theorem can be proved using the same method as that developed in Proposition 5 of [33]. This theorem indicates that the interpretation of ν in ν-PPSVMs is the same as that in regular ν-SVMs. However, it is more significant in the PPSVMs framework since a new approach to determine ν can be constructed based on this interpretation. The detailed discussion is presented at the end of this section.

Theorem 4.4.2. *Assume that the solution of (4.4.9) satisfies* $\rho > 0$, *then the following statements hold:*

   (i) ν *is an upper bound on the fraction of margin errors.*

   (ii) ν *is a lower bound on the fraction of support vectors.*   □

If the class prior probability and class-conditional probability of each sample are known, the posterior probability can be easily calculated using Bayesian formula. But in many real applications, the class-conditional probability is unknown. In order to improve the performance of SVMs using PPSVMs in applications, an empirical method for estimating the class-conditional probability is proposed here.

Let l_j be the number of the samples in class w_j ($j = 1, 2$). Obviously, $l = l_1 + l_2$.

Let $r > 0$ be a real number. The class-conditional probability of x_i is defined as

$$p(x_i|\omega_j) = \frac{l_{x_i}}{l_j} \; (j = 1, 2)$$

where l_{x_i} is the number of elements in $\{x \colon x \in w_j, \|x - x_i\| \leq r\}$.

Using the Bayesian formula, the posterior probability can be calculated as follows

$$
\begin{aligned}
p(\omega_1|x_i) &= \frac{p(x_i|\omega_1)p(\omega_1)}{p(x_i|\omega_1)p(\omega_1) + p(x_i|\omega_2)p(\omega_2)}, \\
p(\omega_2|x_i) &= \frac{p(x_i|\omega_2)p(\omega_2)}{p(x_i|\omega_1)p(\omega_1) + p(x_i|\omega_2)p(\omega_2)}.
\end{aligned}
\tag{4.4.11}
$$

From the above empirical posterior probability, if there are some training data in the other category around a sample, the influence of this sample on the classification is weakened and the degree of the weakening is decided by parameter r. Since we have only finite training samples in a classification problem, for each sample x_i, there must exist a sufficiently small $r > 0$ such that there are no other samples in $\{x \colon \|x - x_i\| \leq r\}$. In this case, $p(\omega_j|x_i) = 1$ if $x_i \in w_j$ ($j = 1, 2$). This indicates that PPSVMs will become SVMs if r is small enough. If the label of a sample does not coincide with its posterior probability, it will be regarded as an error. Let δ be a small positive real number such that $\delta < 0.5$. If $p(\omega_2|x_i) > 0.5$ for some $x_i \in \omega_1$, we force $p(\omega_2|x_i) = 0.5 - \delta$. If $p(\omega_1|x_i) > 0.5$ for some $x_i \in \omega_2$, we force $p(\omega_1|x_i) = 0.5 - \delta$.

Intuitively, a sample can be viewed as a margin error if the label determined by its posterior possibility contradicts with its actual label. By Theorem 4.4.2, ν can be taken to be the fraction of such errors. Clearly, this is an interesting method to determine ν. Hence, our empirical method for calculating the posterior probability also provides a new approach on determining the interpretable parameter in ν-SVMs. A PPSVM classifier determined by this approach will henceforth be referred to as an empirical PPSVM classifier. For example, we can let $\nu = \dfrac{l_0}{l}$ if the number of the wrong samples is l_0. Obviously, if an empirical PPSVM classifier coincides with its Bayesian classifier, the number of errors in PPSVM classification would almost equal to the number of the samples whose labels do not coincide with their posterior probabilities. It is for this reason that the method of determining ν is justified.

4.4.6 Experiments of PPSVMs

To show the relationship between Bayesian classifiers, SVM classifiers and PPSVM classifiers, the following two examples are designed.

Example 4.4.1. Let $p(\omega_1) = 0.7$ and $p(\omega_2) = 0.3$. Let

$$p(x|\omega_1) = \frac{1}{(2\pi)^{\frac{d}{2}}|\Sigma_1|^{\frac{1}{2}}} \exp\left\{ -\frac{1}{2}(x - \mu_1)^T \Sigma_1^{-1}(x - \mu_1) \right\},$$

$$p(x|\omega_2) = \frac{1}{(2\pi)^{\frac{d}{2}}|\Sigma_2|^{\frac{1}{2}}} \exp\left\{ -\frac{1}{2}(x - \mu_2)^T \Sigma_2^{-1}(x - \mu_2) \right\},$$

where $\mu_1 = (0,0)^T$, $\mu_2 = (2,2)^T$ and $\Sigma_1 = \Sigma_2 = \begin{bmatrix} 2 & 0 \\ 0 & 4 \end{bmatrix}$.

Total 100 samples are produced with labels randomly chosen according to the class prior probability. Obviously, if the label x_i is 1, we will calculate its class-conditional density according to $p(x|\omega_1)$, and if the label x_i is -1, we will calculate its class-conditional density according to $p(x|\omega_2)$. The posterior possibility of each sample can be easily derived by using Bayes formula. A linear kernel is used in this example. The relationship between an exact PPSVM classifier and the Bayesian classifier can be seen in Figure 4.4.1. In this case, results indicate that they almost coincide with each other. If the proposed empirical method is used to determine posterior probabilities and parameter ν, an empirical PPSVM classifier is constructed. The relationship between the empirical PPSVM and Bayesian classifiers is presented in Figure 4.4.2, while the relationship between a regular SVM and Bayesian classifier is presented in Figure 4.4.3.   □

Example 4.4.2. Let $p(\omega_1) = 0.7$ and $p(\omega_2) = 0.3$. Let

$$p(x|\omega_1) = \frac{1}{(2\pi)^{\frac{d}{2}}|\Sigma_1|^{\frac{1}{2}}} \exp\left\{ -\frac{1}{2}(x - \mu_1)^T \Sigma_1^{-1}(x - \mu_1) \right\},$$

$$p(x|\omega_2) = \frac{1}{(2\pi)^{\frac{d}{2}}|\Sigma_2|^{\frac{1}{2}}} \exp\left\{ -\frac{1}{2}(x - \mu_2)^T \Sigma_2^{-1}(x - \mu_2) \right\},$$

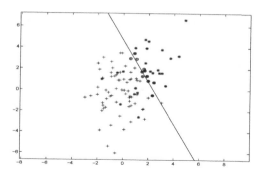

Figure 4.4.1: The relationship between an exact PPSVM classifier and the Bayesian classifier. The solid curve is for the exact PPSVM classifier ($\nu = 0.10$), while the dashed is for the Bayesian.

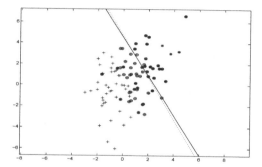

Figure 4.4.2: The relationship between an empirical PPSVM classifier and the Bayesian classifier. The solid curve is for the empirical PPSVM classifier ($\delta = 0.05$, $r = 0.4$), while the dashed is for the Bayesian.

where $\mu_1 = (0,0)^T$, $\mu_2 = (2,0)^T$, $\Sigma_1 = \begin{bmatrix} 0.25 & 0 \\ 0 & 1 \end{bmatrix}$ and $\Sigma_2 = \begin{bmatrix} 1 & 0 \\ 0 & 0.25 \end{bmatrix}$.

Total 200 samples are produced with labels randomly chosen according to the class prior probability. Obviously, if the label x_i is 1, we will calculate its class-conditional density according to $p(x|\omega_1)$, and if the label x_i is -1, we will calculate its class-conditional density according to $p(x|\omega_2)$. The posterior possibility of each sample can be easily derived by using Bayes formula. In this example, we use the standard quadratic polynomial kernels on both SVMs and PPSVMs. The posterior possibility of each sample can be easily derived by using class-conditional probability. The relationship between the exact PPSVM and Bayesian classifiers

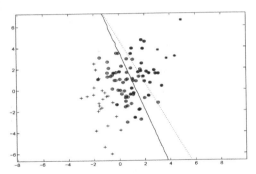

Figure 4.4.3: The relationship between a ν-SVM classifier and the Bayesian classifier. The solid curve is for the ν-SVM classifier ($\nu = 0.15$), while the dashed is for the Bayesian.

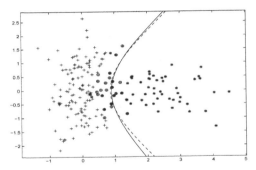

Figure 4.4.4: The relationship between an exact PPSVM classifier and the Bayesian classifier. The solid curve is for the exact PPSVM classifier ($\nu = 0.10$), while the dashed is for the Bayesian.

can be seen in Figure 4.4.4. The relationship between an empirical PPSVM classifier and the Bayesian classifier is presented in Figure 4.4.5, while the relationship between a regular SVM and Bayesian classifiers is presented in Figure 4.4.6. □

To some extent, the two examples illustrate that PPSVM formulation as well as the empirical estimation method is reasonable and closer to the Bayes optimal. According to the resulting relationship between PPSVM and Bayesian classifiers in the above examples, we can say that the proposed method of determining the parameter ν is reasonable.

To further demonstrate the performance of PPSVMs, a synthetic and several

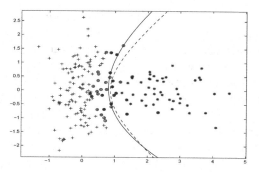

Figure 4.4.5: The relationship between an empirical PPSVM classifier and the Bayesian classifier. The solid curve is for the empirical PPSVM classifier ($\delta = 0.1, r = 0.2$), while the dashed is for the Bayesian.

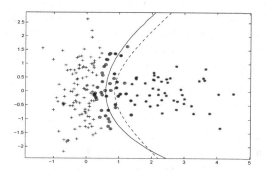

Figure 4.4.6: The relationship between an ν-SVM classifier and the Bayesian classifier. The solid curve is for the ν-SVM classifier ($\nu = 0.5$), while the dashed is for the Bayesian.

real classification problems taken from http://mlg.anu.edu.au/~raetsch/ are conducted. These as well as some discussions can be seen in our paper [42]. Obviously, we can use fuzzy membership to weight the margin, this work can be seen in [38].

4.5   The η-Unsupervised Learning Problems and Margin

In general discussions, we usually use the dichotomy of supervised and unsupervised learning to categorize learning methods. From a conceptual point of view, supervised learning is substantially simpler than unsupervised learning. In supervised learning, the task is to extract information from a set of training examples so that the associate labels of unseen input can be predicted. Since the loss function can be easily defined, the supervised learning problems are usually formulated as optimization problems of the expected risks. Unfortunately, this is not the case in unsupervised learning. There exist two competing objectives in unsupervised learning, which are called information preservation and dimension reduction. Similarly in supervised learning, the dimension reduction problem can usually be solved by defining a hypothesis space. Since the measure of loss depends on the particular problem, the information preservation problem can not easily defined. This results in that it is even not possible to find which one is really better in two admissible functions. This intrinsic ambiguity determines our philosophy in developing general purpose unsupervised methods. From the viewpoint of theory, without exactly defining the loss function, it is hard to obtain any meaningful theoretical results. From the viewpoint of application, every unsupervised learning algorithm must come with a set of parameters that can be used to adjust the algorithm to a particular problem. Desirably, the number of such parameters should be as small as possible, and their effect on the behavior of the algorithm should be as clear as possible.

A one-class problem is usually understood as computing a binary function that is supposed to capture regions in input space where the probability density lies, that is, a function such that most of the data will lie in the region where the function is nonzero [27, 28, 32]. So, one-class problem is a specific clustering problem in which there is only one class to be clustered. Over the last several years, SVM has already been generalized to solve one-class problems. Tax *et al.* presented a SVDD (support vector data description [43, 44]) method. The main idea in [43, 44] is to seek a minimal sphere (with minimal volume or minimal radius) containing all objects. By using kernel techniques, their SVDD can obtain more flexible and more accurate data descriptions. In real operations, they introduced the robust soft idea from SVM and allow some data points outside the sphere. The advantage of SVDD lies in that the algorithm is very intuitive and comprehensive. In 2001, Schölkopf *et al.* gave one-class problems a deep insight [32]. They presented a very simplified algorithm by changing it into a binary classification problem. The interesting fact is that their kernel algorithms coincide with the minimal sphere algorithms in feature spaces. But what still puzzles us now is that we cannot find out the exact definition of one-class problems and outliers in some important research papers, e.g., [32, 43]. Also, there seem to be no theoretical analysis about the relationship between the generalization ability and objective function of the learning algorithms in some important related papers.

As a specific η-unsupervised problem, our one-class and outlier problem is based on the new initial viewpoint that the number of outliers should be as few as possible. Such a viewpoint is different from those in previous papers. One of the main advantages of this new formulation is its complete statistical learning theory motivation. Besides, the important fact is that our maximum margin algorithms intrinsically coincide with that in [32, 43, 44] if the hypothesis space is the set of all constant functions. This illustrates that the algorithms in [32, 43] are theoretically motivated in the framework of our defined risks. It seems to be the first attempt to explain the intuitive minimal sphere algorithm in terms of margin. Furthermore, if the hypothesis space is the set of all hyper-planes, we can obtain a new algorithm [39].

As for clustering problems, the well-known k-means algorithm is a widely used dominant algorithm now. But in fact, k-means algorithm and almost all other clustering algorithms are only to minimize the empirical risk. This is reflected by the fact that the mean of samples is viewed as the expectation of corresponding distributions. On the other hand, some researchers have generalized some important ideas from SVM to solve clustering problems. In [12], Girolami presented a kernel-based clustering method by using k-means algorithms in feature spaces. In [1], the authors proposed a novel clustering method. They map data points by means of a Gaussian kernel to a high dimensional feature space, where they search for the minimal enclosing sphere. This sphere, when mapped back to data space, can separate into several components, each enclosing a separate cluster of points. In this chapter, we modify k-means algorithms by replacing the sample averages with the center of the corresponding minimal sphere. Then, the framework of SVMs, including soft margin and kernel techniques, can be employed to produce some robust and flexible clustering algorithms. This is another important contribution of this chapter. In our experiments, we find out that the width of the Gaussian kernel can control the shape of each cluster while the soft margin parameter in ν-SVM helps coping with outliers and overlapping clusters.

As far as the relationship between SVMs and PCA is concerned, a kernel method for directly performing nonlinear PCA was developed by Schölkopf *et al.* [31] in 1998. In 2003, a simple and straightforward primal-dual SVM formulation to PCA in dual variables, which is made in a similar fashion as least squares SVM (LS-SVM [36]) classifiers, was presented by Suykens *et al.* [37]. However, a complete SVM-like framework of PCA is not constructed, especially the important soft algorithms can not be obtained in terms of the formulations in [37]. The possibility that analyzing PCA in the framework of statistical learning theory is hardly considered likely. On the other hand, there are some papers about robust PCA, for example, see a recent paper [14]. However, since the authors in [14] discussed robustness problem using an ε-contamination model for a data distribution, their methods differ from the soft SVM algorithms in nature. In this chapter, we will establish a complete SVM-like framework for PCA and the robust PCA is achieved by using the soft idea in SVMs.

In this section, a complete SVM-like framework to learn the direction of linear PCA is presented, where new expected risk, linear separability and margin are defined. The concept of PCA margin leads to a new convex semi-definite optimization problem and a new definition of PCA support vectors. Since this framework is completely in the style of SVMs, we capture the nature of SVMs and will call this framework a linear Support Vector Principal Component Analysis (SVPCA). As a result, the generalization bound of SVPCA can be analyzed and SVPCA is then based on statistical learning theory. Simply in appearance, SVPCA is a weighted PCA. But the weight of each sample is determined by a semi-definite optimization problem. If all the samples are equally likely, i.e. they have the same weight, our SVPCA coincides with regular PCA. Furthermore, the robustness of SVPCA follows from the soft idea in SVMs. One potential advantage of SVPCA is that it can solve linear dimension reduction problems with good prediction ability. Since the proposed semi-definite optimization problem is convex, another important advantage is that our new learning algorithms have no local minima. It should be emphasized that SVPCA attempts to minimize the expected risk using PAC framework based on data rather than on distributions. It should also be emphasized that this chapter may be the first attempt to learn a linear PCA from the viewpoint of statistical learning theory. On the other hand, it should also be pointed out that there are still some unsolved problems in our framework, and these problems are discussed in the discussion section.

4.5.1 The Description of Unsupervised Learning Problems

Consider the following unsupervised problem

$$S = \{x_1, x_2, \cdots, x_l, \quad x_i \in R^N, i = 1, 2, \cdots, l\}. \tag{4.5.1}$$

where the samples are independently drawn and identified distributed according to a unknown density function $D(x)$. Let Z be an index set and $H = \{f : Z \longrightarrow R^n\}$ denote the hypothesis space. For $p \in [1, \infty)$ and $x \in R^N$, $\|x\|_p$ represents the l^p norm of x.

Definition 4.5.1 (Expected risk). Let $f \in H$. Let $\triangle(x, f(z))$ be a non-negative cost function to measure the closeness between x and $f(z)$. Then

$$err_D(f) = \int \triangle(x, f) D(x) dx \tag{4.5.2}$$

is called the expected risk of f about the unsupervised learning problem. □

Example 4.5.1 (Clustering and Optimal Vector Quantizer). Define $Z = \{1, 2, \cdots, k\}$, $f: i \rightarrow f(i)$ with $f(i) \in R^n$, and H to be set of all such functions. Set $\triangle(x_i, f(z)) = \min\{\|x_i - f(z)\|^2 : z = 1, 2, \cdots, k.\}$. A clustering problem is to solve the following optimal problem

$$\min\{\int \triangle(x, f(z))D(x)dx, \{f(1), f(2), \cdots, f(k)\} \subset H\}$$

and the solution $\{f^*(1), f^*(2), \cdots, f^*(k)\}$ is called the optimal vector quantizer.
$\square$

Example 4.5.2 (Principal Component Analysis). Define $f: R \rightarrow R^n, f(t) = x_0 + te$ where $t \in R$,

$$x_0 = \frac{1}{l}\sum_{i=1}^{l} x_i,$$

$e \in R^n$ and $\|e\| = 1$. Let the hypothesis space H be a set of all such functions and $\triangle(x, f) = \{\min\|x - x_0 - te\|^2 : t \in R\}$. A PCA problem is to solve the following optimal problem

$$\min\{\int \triangle(x, f)D(x)dx, f \in H\}$$

and the solution yields a line parallel to the direction of largest variance in $D(x)$.
$\square$

The above definition can be found in [16, 34].

4.5.2 η-Unsupervised Learning Problems and Margin

Let η be a threshold, which represents the scale of the desired region for a unsupervised learning problem.

Definition 4.5.2 (Loss function and the expected risk). Let $f \in H$. Let $\triangle(x, f(z))$ be a non-negative cost function to measure the closeness between x and $f(z)$. Let L be defined as follows:

$$L(\eta, x, f) = \begin{cases} 0, & \text{if } \min\{\triangle(x, f(z)): z \in Z\} \leq \eta; \\ 1, & \text{otherwise.} \end{cases}$$

Then L is called the loss function of f about an η-unsupervised learning problem, and

$$err_D(f, \eta) = \int L(\eta, x, f)D(x)dx \qquad (4.5.3)$$

is called the expected risk of f about the η-unsupervised learning problem. $\square$

From Definition 4.5.2, it is easy to know that the expected risk is the total probability that a point is beyond the admissible closeness. In fact, the expected risk reflects the averaged error of all the samples.

Definition 4.5.3 (Empirical risk). Let $f(x)\colon R^N \to R$ and $f \in H$. Then

$$err_{emp}(f, \eta) = \sum_{i=1}^{l} L(\eta, x_i, f)$$

is called the empirical risk of f about the η-unsupervised learning problem. □

Definition 4.5.4 (The optimal classifier and outlier of the η-unsupervised learning problem). Let $f_0 \in H$. If

$$err(f_0, \eta) = \min\{\int L(\eta, x, f)D(x)dx, f \in H\}$$

Then f_0 is called the optimal classifier. The outlier of f_0 in $\{x_1, x_2, \cdots, x_l,\ x_i \in R^N, i = 1, 2, \cdots, l\}$ is called an outlier of η-unsupervised learning problem. □

Note that an outlier should be defined in terms of distributions. But in the above definition, the outlier is only limited to the training samples. This is because the consideration of outliers is usually decided by the practical tasks.

Definition 4.5.5 (Margin). Let $f \in H$. The margin of a sample x_i ($i = 1, 2, \cdots,$ l) is defined as

$$m(f, x_i, \eta) = \eta - \min\{\triangle(x_i, f(z))\colon z \in Z\}.$$

The margin of f is defined as

$$m(f, S, \eta) = \min\{\eta - \min\{\triangle(x_i, f(z))\colon z \in Z\}, i = 1, 2, \cdots, l\ \}. \quad \square$$

Definition 4.5.6 (Separable). If there exists a $f \in H$ such that

$$\eta - \min\{\triangle(x_i, f(z))\colon z \in Z\} \geq 0$$

for $i = 1, 2, \cdots, l$, the η-unsupervised learning problem is called separable. □

Definition 4.5.7 (Maximum margin classifier). For a separable problem, if $f_0 \in H$ satisfies

$$m(f_0, S, \eta) = \max\{m(f, S, \eta), f \in H\}$$

Then f_0 is called the maximum margin classifier. □

Note that the unsupervised problem is formally described in [16, 34] and clearly stated in Section 4.5.1. The only change in our η-unsupervised learning problem is that a parameter η is introduced. The advantage is that the margin can be defined intuitively. This will enable us to investigate the η-unsupervised learning problems by using the nature in the framework of SVMs.

4.6 Some Marginal Algorithms for One-Class Problems

Obviously, an η-one-class problem can be naturally obtained from η-unsupervised learning problems by setting $Z = \{1\}$. In this section, we discuss η-one-class problems under two different hypothesis spaces.

4.6.1 The Minimal Sphere One-Class Problems

To find out the relationship between our marginal algorithms and the available one-class learning algorithms in [32, 43, 44], we first specify H to be the set of all constant functions and define $\triangle(x_i, f) = \|x_i - f\|$. Now, it is easy to know that the expected risk (4.5.3) is the total probability that a point is outside a ball.

Let the η-unsupervised learning problem under the above assumptions be separable, the margin of a "classifier" x_0 now is $\gamma = \min\{\eta - \|x_i - x_0\|, i = 1, 2, \cdots, l\}$. Then the maximum margin algorithm is formulated as the following optimization problem:

$$\begin{cases} \max_{\{x_0 \in R^n\}} \ \gamma \\ \eta - \|x_i - x_0\| \geq \gamma, \ \ i = 1, 2, \cdots, l. \end{cases} \tag{4.6.1}$$

Obviously, problem (4.6.1) is equivalent to

$$\begin{cases} \min_{\{R, x_0\}} \ R^2 \\ \|x_i - x_0\|^2 \leq R^2, \ \ i = 1, 2, \cdots, l. \end{cases} \tag{4.6.2}$$

Obviously, problem (4.6.2) is just the minimal sphere algorithm in [43, 44]. In many references, the terms smallest enclosing circle, smallest enclosing ball, smallest enclosing sphere, and minimum covering sphere are used in the same context. In this chapter, we use the term minimal sphere. This problem arises in applications such as location analysis and military operations and it is itself of interest as a problem in computational geometry; see [8, 11] for details.

As the maximum margin algorithm in SVMs, the minimal sphere optimization is an important concept, serving as a start point for the whole framework of one-class problems ([32, 43, 44]). Here, we only list the following ν-one-class optimization problems from [32]:

$$\begin{cases} \min_{\{R, x_0, \xi_i\}} \ R^2 + \dfrac{1}{\nu l} \sum_{i=1}^{l} \xi_i \\ \|x_i - x_0\|^2 \leq R^2 + \xi_i, \ \xi_i \geq 0, \ \ i = 1, 2, \cdots, l. \end{cases} \tag{4.6.3}$$

However, this chapter may be the first attempt to show that the minimal sphere optimization itself is a specific maximum margin algorithms.

4.6.2 The Minimal Slab One-Class Problems

To produce some new algorithms using our defined margins, a straightforward and the most natural way is to define some different hypothesis spaces. Motivated by the hypothesis spaces in SVMs, in this section, we set $H = \{w^T x + b: w \in R^n, b \in R^1, \|w\|_p = 1\}$ and define $\triangle(x_i, w^T x + b) = |w^T x_i + b|$. Now, it is easy to know that the expected risk (4.5.3) is the total probability that a point is outside a slab.

Let the one-class problem under the above assumptions be separable, the margin of x_i now is $\gamma = \eta - |w^T x_i + b|$. To interpret the geometric meaning of margin, let's introduce the following theorem from [23].

Theorem 4.6.1. *Let $z \in R^N$ be any point that is not on the plane $P = \{x, w^T x = 0\}$. Then for $p \in [1, \infty)$,*

$$\frac{|\langle w, z \rangle|}{\|w\|_p} = \|z - P\|_q$$

where $\|z - P\|_q$ denotes the distance of z to the plane P measured with respect to the dual norm l_q, i.e., with q satisfying $\frac{1}{p} + \frac{1}{q} = 1$. $\square$

If $p = 2$, $|w^T x_i + b|$ means the Euclidean distance of x_i to the hyperplane $|w^T x_i + b| = 0$. Note that η is a desired width, the margin of x_i in fact means the minimal Euclidean distance of x_i to two hyperplanes $w^T x_i + b + \eta = 0$ and $w^T x_i + b - \eta = 0$.

Clearly, the maximum margin algorithm for a separable problem is to solve the following optimization problem:

$$\begin{cases} \max_{\{w,b,\gamma\}} \; \gamma \\ \|w\|_p = 1, \; w^T x_i + b + \eta \geq \gamma, \; w^T x_i + b - \eta \leq -\gamma, \; i = 1, 2, \cdots, l. \end{cases}$$
$$(4.6.4)$$

Let $\rho = \eta - \gamma$, the optimization problem (4.6.4) becomes

$$\begin{cases} \min_{\{w,b,\rho\}} \; \rho \\ \|w\|_p = 1, \; w^T x_i + b \leq \rho, \; w^T x_i + b \geq -\rho, \; i = 1, 2, \cdots, l. \end{cases} \quad (4.6.5)$$

The interpretation of (4.6.5) is very clear, i.e., it is to obtain a minimal width zone that contains all the samples. Contrast to problem (4.6.2), the zone in problem (4.6.5) is a slab other than a ball and the center of the slab is a line other than a point.

Note that the optimization problem for a linearly separable one-class problem in [27, 32] is as follows

$$\begin{cases} \max_{\{w,b,\rho\}} \; \rho \\ \|w\|_p = 1, \; w^T x_i + b \geq \rho, \; i = 1, 2, \cdots, l. \end{cases} \quad (4.6.6)$$

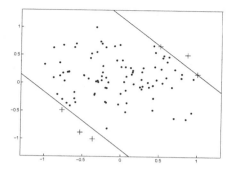

Figure 4.6.1: The minimal slab one-class linear classifier with $\nu = 5$.

Clearly, to estimate the desired slab for a one-class problem is more intuitive than to estimate a half space. Moreover, the complete formulation, which includes generalization error and learning algorithms, can be established. Furthermore, one-class problems and boosting algorithms are related.

If we let $p = 1$ in problem (4.6.2) and employ the idea in [28, 32], we can obtain the following LP problem

$$\begin{cases} \min_{\{w,b,\rho,\xi_i,\xi_i^*\}} \ \nu\rho + \sum_{i=1}^{l}(\xi_i + \xi_i^*), \\ w_1 + w_2 + \cdots + w_N = 1, w_i \geq 0, i = 1, 2, \cdots, N, \\ w^T x_i + b \leq \rho + \xi_i, \ w^T x_i + b \geq -\rho - \xi_i^*, \ i = 1, 2, \cdots, l, \\ \rho \geq 0, \ \xi_i \geq 0, \ \xi_i^* \geq 0, \ i = 1, 2, \cdots, l. \end{cases} \quad (4.6.7)$$

Problem (4.6.7) is a new algorithm for one-class problems. It is very much like the ν-LP problems in [28]. This similarity implies that problem (4.6.7) can be used to obtain a nonlinear solver by using boosting techniques and that the parameter ν has the same interpretation. The advantage of optimization problem (4.6.7) is that the parameter ν can efficiently control the number of support vectors and outliers. This parameterization has the additional benefit of enabling us to eliminate the regularization parameter C and the threshold η. From the viewpoint of unsupervised learning, parameters η and ν can satisfy the needs of different users. The details can be found in [39].

Problem (4.6.1) and (4.6.4) can be proved to be motivated in terms of statistical learning theory ([39]). From Section 4.6.1, it is easy to know that we in fact prove that the algorithms in [43], and [32, 44] are the statistical learning algorithms for our defined one-class problems. It may be the first time to endow the intuitive minimal sphere optimization problem in computational geometry with a interpre-

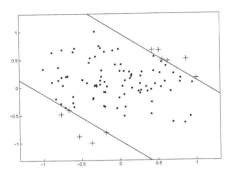

Figure 4.6.2: The minimal slab one-class linear classifier with $\nu = 10$.

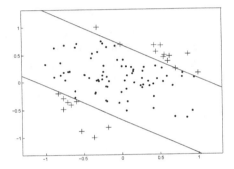

Figure 4.6.3: The minimal slab one-class linear classifier with $\nu = 20$.

tation of statistical learning theory. Moreover, from Section 6.2, it is easy to know that some new algorithms have been produced.

4.6.3 Experiments of the Minimal Slab One-Class Algorithms

Example 4.6.1. To illustrate the formulation in Section 4.6.2 intuitively, we design a synthetic example. The one-class linear classifiers are obtained by solving optimization problem (4.6.7). Please see Figure 4.6.1, Figure 4.6.2 and Figure 4.6.3.   □

From Figure 4.6.1, Figure 4.6.2 and Figure 4.6.3, it is easy to find that the parameter ν in optimization problem (4.6.7) is bigger than outliers and smaller than the sum of outliers and support vectors. They demonstrate to some extent the logical correctness of our theory in Section 4.6.2. The more detailed discussion of one-class problems can be seen in [39].

4.7 Some New Algorithms of Clustering Problems

Definition 4.7.1 (η-Clustering and η-Optimal Vector Quantizer). Define $Z = \{1, 2, \cdots, k\}$, $f: i \to f(i)$ with $f(i) \in R^n$, and H to be set of all such functions. Set

$$L(\eta, x, f(z)) = \begin{cases} 0, & \text{if } \triangle(x, f(z)) = \min\{\|x - f(z)\|: \\ & z = 1, 2, \cdots, k.\} \leq \eta; \\ 1, & \text{otherwise.} \end{cases}$$

An η-clustering problem is to solve the following optimal problem

$$\min\{\int L(\eta, x, f(z))D(x)dx, \{f(1), f(2), \cdots, f(k)\} \subset H\}$$

and the solution $\{f^*(1), f^*(2), \cdots, f^*(k)\}$ is called the η-optimal vector quantizer. $\square$

Obviously, the above η-clustering problem is a special case of η-unsupervised learning problems.

4.7.1 Some Available Clustering Algorithms

One of the most widely used algorithms for constructing locally optimal quantizers for distributions or data is the Generalized Lloyd (GL) algorithm [21] (also known as the k-means algorithm [22]). In this section, we denote $\triangle(x, y) = \|x - y\|^2$.

The well-known GL algorithm for a known distribution is described in the following

The GL Algorithm for Distributions.
Step 0. Set $j = 0$ and set $C^{(0)} = \{v_1^0, v_2^0, \cdots, v_k^0\}$ to an initial codebook.

Step 1 (Partition). Construct $V^j = \{V_1^j, V_2^j, \cdots, V_k^j\}$ by setting

$$V_l^{(j)} = \{x: \triangle(x, v_l^j) = \|x - v_l^j\|^2 \leq \triangle(x, v_m^j), m = 1, 2, \cdots, k\}$$

for $l = 1, 2, \cdots, k$.
Step 2 (Expectation). Construct $C^{(j+1)} = \{v_1^{j+1}, v_2^{j+1}, \cdots, v_k^{j+1}\}$ by setting

$$v_l^{(j+1)} = \text{argmin}_v E[\triangle(x, v)|x \in V_l^{(j)}] = E[x|x \in V_l^{(j)}]$$

for $l = 1, 2, \cdots, k$.

Step 3. Stop if $1 - \dfrac{\triangle(q^{j+1})}{\triangle(q^j)}$ is less than or equal to a certain threshold, where

$$\triangle(q^j) = \int \min\left\{ \|x - v\|^2 : v \in \{v_1^{j+1}, v_2^{j+1}, \cdots, v_k^{j+1}\} \right\} D(x)dx.$$

Otherwise, let $j = j + 1$ and go to Step 1. □

The GL algorithm can easily be adjusted to the case when the distribution $D(x)$ is unknown but a training set of independent observations of the underlying distribution is known instead. The modifications are straightforward replacements of the expectations by sample averages. The well-known GL algorithm for constructing a vector quantizer based on the training data is described in the following.

The GL Algorithm for Training Data.

Step 0. Set $j = 0$ and set $C^{(0)} = \{v_1^0, v_2^0, \cdots, v_k^0\}$ to an initial codebook.

Step 1 (Partition). Construct $V^j = \{V_1^j, V_2^j, \cdots, V_k^j\}$ by setting

$$V_l^{(j)} = \{x \colon \triangle(x, v_l^j) = \|x - v_l^j\|^2 \le \triangle(x, v_m^j), m = 1, 2, \cdots, k\}$$

for $l = 1, 2, \cdots, k$.

Step 2 (Expectation). Construct $C^{(j+1)} = \{v_1^{j+1}, v_2^{j+1}, \cdots, v_k^{j+1}\}$ by setting

$$v_l^{(j+1)} = \operatorname{argmin}_v \sum \{\triangle(x, v) \colon x \in V_l^{(j)} \cap S\} = \frac{1}{|V_l^{(j)}|} \sum \{x \colon x \in V_l^{(j)} \cap S\}$$

for $l = 1, 2, \cdots, k$.

Step 3. Stop if $1 - \dfrac{\triangle_l(q^{j+1})}{\triangle_l(q^j)}$ is less than or equal to a certain threshold, where

$$\triangle_l(q^j) = \frac{1}{l} \sum_{i=1}^{k} \sum \{\|v_i - x\|^2 \colon x \in V_i\}.$$

Otherwise, let $j = j + 1$ and go to Step 1. □

4.7.2 Some New Algorithms for η-Clustering Problems

In a learning problem, the distribution $D(x)$ is usually unknown. So in some applications, the density has to be estimated based on the available data. Clearly, we could solve whatever problems can be solved with the help of the density.

But in line with Vapnik's principle [47, 48], we avoid estimating the density when solving some special problems. Therefore, the objective of empirical quantizer design is to find a vector quantizer based on the training data. To design a quantizer with low distortion, most existing practical algorithms attempt to implement the empirical loss minimization principle. The performance of a vector quantizer is essentially measured by the empirical distortion.

Since the margin for clustering follows from that in η-unsupervised problems, it is easy to obtain some marginal algorithms. Unfortunately even in empirical loss cases, the global optimality is not feasible. So, we can only limit our discussion to a direct modification of k-means algorithms. In fact, our new clustering algorithms only replace the sample averages by the center of corresponding minimal sphere. We describe the details in the following.

The Minimal Sphere GL Algorithm.

Step 0. Set $j = 0$ and set $C^{(0)} = \{v_1^0, v_2^0, \cdots, v_k^0\}$ to an initial codebook.

Step 1 (Partition). Construct $V^j = \{V_1^j, V_2^j, \cdots, V_k^j\}$ by setting

$$V_l^{(j)} = \{x: \triangle(x, v_l^j) = \|x - v_l^j\|^2 \leq \triangle(x, v_m^j), m = 1, 2, \cdots, k\}$$

for $l = 1, 2, \cdots, k$.

Step 2 (Expectation). Construct $C^{(j+1)} = \{v_1^{j+1}, v_2^{j+1}, \cdots, v_k^{j+1}\}$. v_l^{j+1} is obtained by solving optimization problem (4.6.2) on the training set V_l^j for $l = 1, 2, \cdots, k$.

Step 3. Stop if $1 - \dfrac{\triangle_l(q^{j+1})}{\triangle_l(q^j)}$ is less than or equal to a certain threshold, where

$$\triangle_l(q^j) = \frac{1}{l} \sum_{i=1}^{k} \sum \{\|v_i - x\|^2 : x \in V_i\}.$$

Otherwise, let $j = j + 1$ and go to Step 1. □

Just like the GL algorithm, for a finite training set, the minimal sphere GL algorithm always converges in a finite number of iterations since the average distortion is non-increasing in both Step 1 and Step 2 and there is only a finite number of ways to partition the training set into k subsets. Since global optimality is not a feasible requirement, algorithms, even in theory, are usually designed to find locally optimal vector quantizers. A quantizer $\{f^*(1), f^*(2), \cdots, f^*(k)\}$ is said to be locally optimal if $min\{\int L(\eta, x, f(z))D(x)dx, \{f(1), f(2), \cdots, f(k)\} \subset H\}$ is only local, that is, slight disturbance of any of the code points will cause an increase in the distortion. It is known that GL algorithm [21] is a theoretical algorithm to find a locally optimal vector quantizer of a random variable. Since our minimal sphere GL algorithm essentially is a local margin algorithm, it can find a locally optimal vector quantizer.

Obviously, if $k(x, y)$ is used instead of $\|x - y\|^2$ in Step 1 and Step 3, and kernel-based form of optimization problem (4.6.3) is used instead of optimization problem (4.6.2) in Step 2, we can obtain kernel-based minimal sphere clustering algorithms. So far, we have in fact established a complete SVM formulation for clustering problems. Then, the robustness and flexibility can be achieved by using the soft idea and kernel techniques in SVMs. Furthermore, the margin idea in statistical learning theory can ensure the high performance of the minimal sphere GL algorithms.

We first independently present the Minimal Sphere GL Algorithm in [41], and we find that this similar clustering algorithm is also proposed in [10]. The difference is that we achieve this algorithm from the viewpoints of margin. There are a lot of experiments in [10]. Their experiments demonstrate that the Minimal Sphere GL Algorithm is powerful and practical.

4.8   New Marginal Algorithms for PCA

To discuss the marginal algorithms for PCA, define $f: R \to R^n$, $f(t) = x_0 + te$ where $t \in R$, $x_0 = \frac{1}{l} \sum_{i=1}^{l} x_i$, $e \in R^n$ and $\|e\| = 1$. Let the hypothesis space H be a set of all such functions and $\triangle(x, f) = \{\min \|x - x_0 - te\|^2 : t \in R\}$.

4.8.1 Maximum Margin Algorithms for PCA

In linearly separable cases, finding the maximal margin component can be described as solving the following optimization problem:

$$\begin{cases} \max_{\{e \in R^n\}} \; \gamma \\ e^T e = 1, \; \eta - \{\min \|x_i - x_0 - te\|^2 : t \in R\} \geq \gamma, \; i = 1, 2, \cdots, l. \end{cases}$$

It then can be reformulated as

$$\begin{cases} \min_{\{r^2, e\}} \; r^2 \\ e^T e = 1, \; e^T S_i e \geq r_i^2 - r^2, i = 1, 2, \cdots, l. \end{cases} \tag{4.8.1}$$

where $S_i = (x_i - x_0)(x_i - x_0)^T$, $r_i = \|x_i - x_0\|$. Note that in optimization problem (4.8.1), we are in fact looking for a specific e with the maximal margin. The maximum margin algorithm in SVMs can be formulated as a convex quadratic optimization problem by virtue of the KKT conditions. However, in PCA cases, the prime-dual formulation is more complex. Because problem (4.8.1) is not a convex, the powerful KKT conditions can not be applied. This may be the main obstacle to establish a SVM-like framework for solving PCA. Fortunately, we can employ some first order necessary conditions (Fritz John conditions and Farkas' Lemma [9] in optimization theory, which does not need the convexity assumption and is weaker than the KKT conditions. It should be pointed out that the regularity assumption, which guarantees that the desired Lagrangian multiplier is not zero, happens to hold for our optimization problem (4.8.1). And, it is just based on the the regularity assumption, problem (4.8.1) can be changed into a "dual" optimization problem. Here, we only give the following theorem. For the reader's convenience, its proof can be seen in [40].

Theorem 4.8.1. *Let e_0 be the solution of problem (4.8.1). Then e_0 must be an eigenvector of $\sum_{i=1}^{l} \alpha_i^0 S_i$ corresponding to the largest eigenvalue β, where $\{\alpha_i^0, 1 \leq i \leq l\}$ and β is a solution of the following problem*

$$
\begin{cases}
\max_{\{\alpha_1, \alpha_2, \cdots, \alpha_l, \beta\}} \sum_{i=1}^{l} \alpha_i r_i^2 - \beta \\
\sum_{i=1}^{l} \alpha_i S_i e_0 = \beta e_0 \text{ and } \beta \text{ is the largest eigenvalue of } \sum_{i=1}^{l} \alpha_i S_i. \\
\sum_{i=1}^{l} \alpha_i = 1, \ \alpha_i \geq 0, \ 1 \leq i \leq l.
\end{cases}
\tag{4.8.2}
$$

Definition 4.8.1. If $\alpha_i^0 \neq 0$, the corresponding vector x_i is called a support vector of PCA. Such a formulation to learn PCA is called SVPCA.   □

A few remarks need to be made on our SVPCA.

- If there exists a j ($1 \leq j \leq l$) such that $\alpha_j^0 \neq 0$,

$$
r_j^2 - e^T S_j e = \max\{r_i^2 - e^T S_i e, \ i = 1, 2, \cdots, l\}.
$$

This is completely analogous to the fact in SVMs, i.e., all the support vectors are equally closer to the optimal component and the distance from each support vector to the component is maximal among all the samples. Informally speaking, support vectors here are the patterns most informative for deciding the direction of data compression. In fact, one can further find out that only these vectors are involved in the expression of the component. It is for this reason they are called support vectors.

- If all the α_i are equal, β is the largest eigenvalue of

$$
\frac{1}{l} \sum_{i=1}^{l} S_i = M_l.
$$

This is just regular PCA. Hence, SVPCA is an extension of regular PCA. As can be seen from the appearance, SVPCA is a kind of weighted PCA. Each weight in fact reflects the extent of importance of the corresponding sample in a PCA problem. Intuitively, it is a natural way to make the algorithm more robust against outliers by weighting the samples according to the contribution of samples.

- The optimal solution of problem (4.8.2) is often called the first component. We will call an eigenvector with the second largest eigenvalue of $\sum_{i=1}^{l} \alpha_i^0 S_i$ the second component. Such a definition of the second component can ensure that the first and second components are orthogonal.

Until now, the above discussion looks nice and satisfactory. However, from the appearance of problem (4.8.2), it is easy to see that problem (4.8.2) can not be tackled using the regular methods in ordinary optimization theory. This may be the main obstacle to implement a SVM-like learning algorithm for solving PCA. Fortunately, problem (4.8.2) can be solved by using techniques from semi-definite programming (SDP), a branch of convex optimization that deals with the optimization of convex functions over the convex cone of positive semi-definite matrices or convex subsets. In order to change problem (4.8.2) into a SDP, the following lemma is introduced from [25].

Lemma 4.8.1. Let M be a symmetric real matrix and let λ be the largest eigenvalue of M. The following characterization holds

$$\lambda = \max\{x^T M x \colon \|x\| = 1\}. \qquad \square$$

According to Lemma 4.8.1,

$$x^T(\lambda I - M)x \geq 0, \forall x \in R^n.$$

So, problem (4.8.2) is equivalent to the following optimization problem with linear matrix inequalities (LMI) constraint:

$$\begin{cases} \max_{\{\alpha_1, \alpha_2, \cdots, \alpha_l, \beta\}} \sum_{i=1}^{l} \alpha_i r_i^2 - \beta \\ \sum_{i=1}^{l} \alpha_i S_i - \beta I \leq 0; \\ \sum_{i=1}^{l} \alpha_i = 1, \ \alpha_i \geq 0, \ 1 \leq i \leq l. \end{cases} \qquad (4.8.3)$$

where $\sum_{i=1}^{l} \alpha_i S_i - \beta I \leq 0$ represents that $\sum_{i=1}^{l} \alpha_i S_i - \beta I$ is negative semi-definite. Problem (4.8.3) is a standard SDP. Details of SDP can be found in [2, 46]. The solving of SDP leads directly to a learning algorithm for PCA. Importantly, this optimization problem is convex. So we obtain a method for learning PCA without local minima. Recently, the authors in [18] used SDP to learn a kernel matrix and a powerful transductive algorithm was presented. Moreover, their approach produce a convex method for learning 2-norm soft margin parameter in SVMs, solving an important open problem.

4.8.2 Robust SVPCA

If there is no f in H_0 such that $m(f, S) \leq \eta$, the η-PCA problem is not linearly separable. For problems which require classification violations to be allowed, Cortes and Vapnik proposed the soft SVM algorithms by introducing the slack variables in constraints [47, 48]. Similarly, for linearly inseparable PCA, the following optimization problem is formulated.

$$
\begin{cases}
\min_{\{r^2, e, \xi_1, \cdots, \xi_l\}} \quad r^2 + C \sum_{i=1}^{l} \xi_i \\
e^T e = 1, \quad e^T S_i e \geq r_i^2 - r^2 - \xi_i, \quad \xi_i \geq 0, \quad 1 \leq i \leq l.
\end{cases}
\tag{4.8.4}
$$

where C is a predefined positive real number controlling the trade-off between accuracy and margin.

As proved in the [40], we can obtain the following soft SVPCA with robustness.

Theorem 4.8.2. *Let e_0 be the solution of problem (4.8.4). Then e_0 must be an eigenvector of $\sum_{i=1}^{l} \alpha_i^0 S_i$ corresponding to the largest eigenvalue β, where $\{\alpha_i^0, 1 \leq i \leq l\}$ and β is a solution of the following problem*

$$
\begin{cases}
\max_{\{\alpha_1, \alpha_2, \cdots, \alpha_l, \beta\}} \sum_{i=1}^{l} \alpha_i r_i^2 - \beta, \\
\sum_{i=1}^{l} \alpha_i S_i e_0 = \beta e_0 \text{ and } \beta \text{ is the largest eigenvalue of } \sum_{i=1}^{l} \alpha_i S_i, \\
\sum_{i=1}^{l} \alpha_i = 1, \quad C \geq \alpha_i \geq 0, \; 1 \leq i \leq l.
\end{cases}
\tag{4.8.5}
$$

□

Note that the only difference from the maximum margin component case in optimization problem (4.8.2) is that α_i now has an upper bound of C.

4.8.3 Experiments and Discussion

To demonstrate the generalization ability of SVPCA, some comparison experiments are conducted on both synthetic and real data sets.

Table 4.8.1: Average values of regular PCA objective function on the testing sets

Data Sets	Regular PCA	Robust SVPCA	
mpg	1930.31 ± 1171.75	1928.82 ± 1160.90	$C = 0.0045$
housing	11196.33 ± 13528.43	9932.89 ± 6570.53	$C = 0.1$
uniform distribution	0.017488 ± 0.003161	0.016591 ± 0.001883	$C = 0.2$
normal distribution	0.091211 ± 0.049674	0.087080 ± 0.048731	$C = 0.2$

Example 4.8.1. There are four date sets in this example. The first two real data sets (mps and housing) are taken from

http://www.csie.ntu.edu.tw/~cjlin/libsvmtools/datasets/.

There are 50 samples in the third and forth data sets, generated according to the uniform distribution on $[-1, 1; -0.2, 0.2]$ and normal distribution with zero mean and covariance matrix $\begin{bmatrix} 4 & 0 \\ 0 & 0.1 \end{bmatrix}$ respectively. The standard 5-fold cross validation strategy has been employed for all data sets, i.e. all the samples in each data set are divided into five parts sequentially, and one part is used as testing set and the others are used as training set. The values of regular PCA objective function on the testing sets are listed in Table 4.8.1   □

To certain extent, this example shows that SVPCA sometimes has better generalization performance over regular PCA.

It is necessary to discuss some unsolved problems in our PCA framework.

It should be pointed out that we restrict our consideration to a simple hypothesis space $\{x_0 + te: x_0 = \frac{1}{l} \sum_{i=1}^{l} x_i, \|e\| = 1\}$ in this section. Since x_0 is fixed, our task in SVPCA is only to find a directional vector e. So the complete SVM framework for PCA and the robustness are only in terms of the direction vector. Obviously, if we add samples without changing the margin and x_0, the projection direction does not change. However, if a sample changes the center x_0 but does not destroy the margin, unfortunately, the component will change. Naturally, a direct idea is to seek both the optimal direction e and the optimal center x_0 simultaneously. Now, we have to say that this will be included in our future research. This is because there exist difficulties in both theory and implementation. In theory, we can not deduce that $x_0 = \sum_{i=1}^{l} \alpha_i x_i$ using the first order necessary conditions in optimization theory. In implementation, although we can force $x_0 = \sum_{i=1}^{l} \alpha_i x_i$ by the duality idea in SVMs, the constraint is no longer an LMI.

The introduction of kernel techniques significantly improve the flexibility of SVM. Similarly as that in [31], it is very desired to achieve a new kernel PCA in our framework.

To solve the above mentioned problems, it may need some more advanced techniques from SDP. One important aim of this chapter is that some researchers may be inspired and solve the problem of learning robust and flexible PCA systematically.

4.9 Conclusions

In this chapter, we first state SVMs in terms of the marginal generalization bound. Then we use margin method to investigate both supervised and unsupervised learning problems. As for supervised learning, a weighted margin idea is presented and some new marginal algorithms are obtained. As for unsupervised learning, a new concept of unsupervised learning problem is proposed. The great advantage is that the intuitive margin can be defined. As a result, the available one-class learning algorithms can be interpreted as a margin algorithm. Moreover, some new algorithms for one-class, clustering and PCA problems are produced.

Acknowledgements

This work has been supported by the National Science Foundation of China (60575001), the National Basic Research Program of China (2004CB318103), and the Excellent Youth Science and Technology Foundation of Anhui Province of China (04042069).

Bibliography

[1] A. Ben-Hur, D. Horn, H. T. Siegelmann, and V. Vapnik, "Support vector clustering," *Journal of Machine Learning Research*, vol. 2, pp. 135–137, 2001.

[2] S. Boyd and L. Vandenberghe, *Convex Optimization*, Lecture notes for EE364, Stanford University, available at http://www.stanford.edu/class/ee364, 2003.

[3] C. J. C. Burges, "A tutorial on support vector machines for pattern recognition," *Data Mining and Knowledge Discovery*, vol. 2, pp. 121–167, 1998.

[4] C. Cortes and V. Vapnik, "Support vector networks," *Machine Learning*, vol. 20, pp. 273–297, 1995.

[5] N. Cristianini and J. Schawe-Taylor, *An Introduction to Support Vector Machines*, Cambridge: Cambridge Univ Press, 2000.

[6] A. Demiriz, K. Bennett and J. Shawe-Taylor, "Linear programming boosting via column generation," *Machine Learning*, vol. 46, no. 1, pp. 225–254, 2002.

[7] R. O. Duda, P. E. Hart, and D. G. Stork, *Pattern Classification*, Second Ed., New York: John Wiley, 2001.

[8] J. Elzinga and D. Hearn, "The minimum covering sphere problem," *Management Sci.*, vol. 19, pp. 96–104, 1972.

[9] R. Fletcher, *Practical Methods of Optimization*, 2nd ed., New York: John Wiley, 1987.

[10] F. Camastra and A. Verri, "A novel kernel method for clustering," *IEEE Trans. Pattern Analysis and Machine Intelligence*, vol. 27, no. 5, pp. 801–805, 2005.

[11] B. Gärter, *Fast and robust smallest enclosing balls*, in *Algorithms-ESA99: 7th Annual European Symposium Proceedings*, J. Nestril, Ed., Lecture Notes in Computer Science, New York: Springer-Verlag, vol. 1643, pp. 325–338.

[12] M. Girolami, "Mercer kernel-based clustering in feature space," *IEEE Trans. Neural Networks*, vol. 13, no. 3, pp. 780–784, 2002.

[13] T. Hastie and W. Stuetzle, "Principal curves," *Journal of the American Statistical Association*, vol. 84, no. 406, pp. 502–516, 1989.

[14] I. Higuchi and S. Eguchi, "Robust principal component analysis with adaptive selection for tuning parameters," *Journal of Machine Learning Research*, vol. 5, pp. 453–471, 2004.

[15] I. T. Jolliffe, *Principal Component Analysis*, New York: Springer-Verlag, 1986.

[16] B. Kégl, *Principal Curves: Learning, Design, and Applications*, Ph. D. Dissertation, Concordia University, Canada, 1999.

[17] D. Kinderlerer and G. Stampcchia, *An Introduction to Variational Inequalities and Their Applications*, New York: Academic, 1980.

[18] G. R. G. Lanckriet, N. Cristianini, P. Bartlett, L. El Ghaoui, and M. I. Jordan, "Learning the kernel matrix with semi-definite programming," *Journal of Machine Learning Research*, vol. 5, pp. 27–72, Aug. 2004.

[19] C. Lin and S. Wang, "Fuzzy support vector machines," *IEEE Trans. Neural Networks*, vol. 13, no. 2, pp. 464–471, 2002.

[20] Y. Lin, Y. Lee, and G. Wahba, *Support vector machines for classification in nonstandard situations*, available at http://www.stat.wisc.edu/~yilin/papers/papers.html, 2000.

[21] Y. Linde, A. Buzo, and R. M. Gray, "An algorithm for vector quantizer design," *IEEE Trans. Communications*, vol. 28, no. 1, pp. 84–95, 1980.

[22] S. P. Lloyd, "Least squares quantization in PCM," *IEEE Trans. Information Theory*, vol. 28, pp. 129–137, 1982.

[23] O. L. Mangasarian, "Arbitrary-norm separating plane," *Operation Research Letters*, vol. 24, no. 1, pp. 15–23, 1999.

[24] K. R. Müller, S. Mika, G. Rätsch, K. Tsuda, and B. Schölkopf, "An introduction to kernel-based learning algorithms," *IEEE Trans. Neural Networks*, vol. 12, no. 2, pp. 181–201, 2001.

[25] M. Overton, "Large-scale optimization of eigenvalues," *SIAM J. Optimization*, 1991.

[26] J. C. Platt, *Probabilistic outputs for support vector machines and comparisons to regularized likelihood methods*, in *Advances in Large Margin Classifies*, A. J. Smola, P. L. Bartlett, B. Scholkopf, and D. Schuurmans, Eds., Cambridge, MA: MIT Press, 1999.

[27] G. Rätsch, *Robust Boosting via Convex Optimization*, Ph. D. Thesis, University of Posdam, 2001.

[28] G. Rätsch, S. Mika, B. Schölkopf, and K. R. Müller, "Constructing boosting algorithms from SVMs: An application to one-class classification," *IEEE Trans. Pattern Analysis and Machine Intelligence*, vol. 9, no. 4, pp. 1184–1199, 2002.

[29] R. Schapire, Y. Freund, P. Bartlett, and W. Sun Lee, "Boosting the margin: A new explanation for the effectiveness of voting methods," *Ann. Statist.*, vol. 26, no. 5, pp. 1651–1686, 1998.

[30] J. Schawe-Taylor, and N. Cristianini, "On the generalization of soft margin algorithms," *IEEE Trans. Information Theory*, vol. 48, no. 10, pp. 2721–2735, 2002.

[31] B. Schölkopf, A. J. Smolar, and K. R. Muller, "Nonlinear component analysis as a kernel eigenvalue problem," *Neural Computing*, vol. 10, pp. 1299–1319, 1998.

[32] B. Schölkopf, J. Platt, J. Shawe-Taylor, A. J. Smola, and R. C. Williamson, "Estimating the support of a high-dimensional distribution," *Neural Computation*, vol. 13, no. 7, pp. 1443–1471, 2001.

[33] B. Schölkopf, A. J. Smola, R. Williamson, and P. Bartlett, "New support vector algorithms," *Neural Computation*, vol. 12, pp. 1083–1121, 2000.

[34] A. J. Smola, S. Mika, B. Schölkopf, and R. C. Williamson, "Regularized principal manifolds," *Journal of Machine Learning Research*, vol. 1, pp. 179–200, 2001.

[35] P. Sollich, "Bayesian methods for support vector machines: Evidence and predictive class probabilities," *Machine Learning*, vol. 46, no. 1, pp. 21–52, 2002.

[36] J. A. K. Suykens and J. Vandewalle, "Least squares support vector machine classifiers," *Neural Processing Letters*, vol. 9, no. 3, pp. 293–300, 1999.

[37] J. A. K. Suykens, T. Van Gestel, J. Vandewalle, and B. De Moor, "A support vector machine formulation to PCA analysis and its kernel version," *IEEE Trans. Neural Networks*, vol. 14, no. 2, pp. 447–450, 2003.

[38] Q. Tao and J. Wang, "A new fuzzy support vector machine based on the weighted margin," *Neural Procession Letters*, vol. 20, no. 3, pp. 139–150, 2004.

[39] Q. Tao, G. Wu, and J. Wang, "A new maximum margin algorithm for one-class problems and its boosting implementation," *Pattern Recognition*, vol. 38, no. 7, pp. 1071–1077, 2005.

[40] Q. Tao, G. Wu, and J. Wang, *Learning robust and flexible PCA with convex semi-definite programming*, Technical Report, Institute of Automation, Chinese Academy of Sciences, 2004.

[41] Q. Tao, G. Wu, F. Y. Wang, and J. Wang, "Some marginal learning algorithms for unsupervised problems," *Lecture Notes in Computer Science*, New York: Springer, vol. 3495, 2005.

[42] Q. Tao, G. Wu, F. Y. Wang, and J. Wang, "Posterior probability support vector machines for unbalanced data," *IEEE Trans. Neural Networks*, vol. 16, no. 6, pp. 1561–1573. 2005.

[43] D. Tax and R. Duin, "Support vector domain description," *Pattern Recognition Letters*, vol. 20, pp. 1191–1199, 1999.

[44] D. Tax and R. Duin, "Support vector data description," *Machine Learning*, vol. 54, pp. 45–66, 2004.

[45] L. G. Valiant, "A theory of the learnable," *Communications of the ACM*, vol. 27, no. 11, pp. 1134–1142, 1984.

[46] L. Vandenberghe and S. Boyd, "Semi-definite programming," *SIAM Review*, vol. 38, no. 1, pp. 49–95, 1996.

[47] V. Vapnik, *The Nature of Statistical Learning Theory*, New York: Springer-Verlag, 1995.

[48] V. Vapnik, *Statistical Learning Theory*, Reading, MA: Addison-Wesley, 1998.

Chapter 5

Constraint Handling in Genetic Algorithm for Optimization

Gary G. Yen

Abstract: In this chapter we present a generic, two-phase framework for solving constrained optimization problems using genetic algorithms. In the first phase of the algorithm, the objective function is completely disregarded and the constrained optimization problem is treated as a constraint satisfaction problem. The genetic search is directed toward minimizing the constraint violation of the solutions and eventually finding a feasible solution. A linear rank based approach is used to assign fitness values to the individuals in phase one. The solution with the least constraint violation is archived as the elite solution in the population. In the second phase the simultaneous optimization of the objective function and the satisfaction of the constraints are treated as a bi-objective optimization problem. We elaborate on how the constrained optimization problem requires a balance of exploration and exploitation under different problem scenarios and come to the conclusion that a non-dominated ranking between the individuals will help the algorithm explore further while the elitist scheme will facilitate in exploitation. We analyze the proposed algorithm under different problem scenarios using Test Case Generator–2 and demonstrate the proposed algorithm's capability to perform well independent of various problem characteristics. In addition, the proposed algorithm performs competitively with the state-of-the-art constraint optimization algorithms on eleven test cases which were widely studied benchmark functions in the literature.

5.1 Introduction

Most real-world optimization problems involve constraints . Consider an optimization problem such as maximizing the profits of a particular production line.

145

The objective function to be maximized could be a function of various manipulating variables, including but not limited to the material consumption, the labor cost, the operating hours of the machines, and many additional factors. If the raw materials, manpower, and machines can be made available without limitation then there is no limit to the profit that can be achieved. However, in face of real-world complications they are most likely limited in the form of constraints imposed upon the optimization function. What constitute the difficulties of the constrained optimization problem are various limits on the decision variables, the constraints involved, the interference among constraints, and the inter-relationship between the constraints and the objective function. Taking a numerical example, suppose we want to maximize a function $f(X) = x_1 + x_2$ where the two variables are defined by $0 \leq x_1, x_2 \leq 1$. Under the presence of no additional constraint an optimum value of $f(X) = 2$ can be reached when $x_1 = 1$ and $x_2 = 1$. Assume that there is an equality constraint imposed upon these variables described by $g(X) \equiv x_1 - x_2 = 0.5$. Considering a resolution of up to two decimal places in the discrete search space, there are only 50 feasible solutions among 10,000 possible candidates. This implies that feasible space is only 0.5% of the actual parameter space. The best objective function value that can be attained is $f(X) = 1.5$ (for $x_1 = 1$ and $x_2 = 0.5$) and . The problem complexity can be greatly increased by the number of constraints or the types of constraints. The general constrained continuous-parameter optimization problem as succinctly defined in [33] is to search for X so as to

$$\text{Optimize } f(X), \quad X = (x_1, x_2, \cdots, x_n) \in \mathcal{R}^n \qquad (5.1.1)$$

where $X \in F \subseteq S$. The objective function f is defined on the search space $S \subseteq \mathcal{R}^n$, and the set $F \subseteq S$ defines the feasible region. Usually, the search space is an n-dimensional hyper box in $\mathcal{R}^n$. The domains of the variables are defined by their lower and upper bounds as,

$$l(i) \leq x_i \leq u(i), \quad 1 \leq i \leq n, \qquad (5.1.2)$$

whereas the feasible region F is restricted by a set of m additional constraints $(m \geq 0)$,

$$\text{with } q \text{ inequality constraints } g_j(X) \leq 0 \ (j = 1, \cdots, q), \qquad (5.1.3)$$

and

$$\text{with } m - q \text{ inequality constraints } h_j(X) = 0 \ (j = q + 1, \cdots, m). \qquad (5.1.4)$$

The inequality constraints that take the value of 0, i.e., $g_j(X) = 0$ at the global optimum to the problem are called the *active constraints*. In the following discussion and in the remaining of this chapter without loss of generality we shall consider the minimization of the objective function unless specified otherwise. In addressing the constrained optimization problem in the real world scenario we can arguably say that obtaining a feasible solution (one that is usable under the problem formulation) takes precedence over optimizing the objective function (which minimizes the cost involved). There are also problems with higher complexity in which finding a single feasible solution itself can be a monumental task. These problems are treated as constraint satisfaction problems and various evolutionary algorithms have been proposed to solve them effectively, e.g., [11]. The main challenge in constrained optimization is simultaneously handling the constraints as well as optimization of the objective function. The constraint handling methods have primarily focused on various designs of fitness formulation for each individual in the population depending on its objective function and constraint satisfaction. Over the past decade various constraint-handling techniques using genetic algorithms [9, 11, 13] and benchmark test functions [26] for constrained optimization problems have been proposed.

We summarize below our motivations to design a new constraint handling scheme to complement the existing methods in literature. (i) Reliability: From a practical point of view, it is essential that the GA used for constrained optimization produces feasible optimal solutions for every run. While this may be too much to ask for we would certainly hope that at least the feasibility criteria can be met for every run and that adequate optimization can be achieved. (ii) A Generic Framework: There are various types of constrained optimization problems that we may encounter and it would be impractical if the GA used has to be tuned to fit for specific problem or if it uses special operators that cannot be implemented in all problem domains. While a generic framework may not result into the most efficient design for each problem setting, it is the most advisable at the algorithm design stage, while necessary modifications can be made to tailor for particular problems (such as exploiting carefully crafted genetic operators).

The chapter is organized as follows. In Section 5.2 a thorough review of the essential features of the best known constraint handling schemes pertinent to this research is presented. In Section 5.3 we introduce the proposed constraint handling scheme and elaborate it in detail. In Section 5.4 we analyze the distinctions between the proposed constraint-handling scheme and another scheme that favors domination of feasible solutions in tackling problems with different characteristics. Using the Test Case Generator-2 (TCG-2) proposed in [38], we demonstrate the effectiveness and efficiency of the proposed constraint handling method in Section 5.5. In Section 5.6, we evaluate the performance of the proposed scheme on the eleven test cases proposed in [26] and provide a fair comparison with the best results from the state-of-the-art in literature. Finally we conclude in Section 5.7 with some observations about the proposed algorithm.

5.2   Literature Survey

In this section we will review some of the relevant methods proposed for constraint handling using genetic algorithms. We have broadly categorized these methods into i) methods based on penalty functions, ii) methods based on preference of feasible solutions over infeasible ones, and iii) methods based on multiobjective optimization. Techniques involving special operators [31], decoders [26], repair mechanisms [32], and hybrid approaches [24] are considered irrelevant to the algorithm proposed herein. We intend to present the basic ideas underlying the design of each of the above constraint-handling methods and provide the rationale to justify our proposed design framework.

i) *Methods based on penalty functions*

Penalty functions were popularly used in the conventional methods for constrained optimization [17] and were amongst the first methods used to handle constraints with evolutionary algorithms. In these methods the individuals are penalized based on their constraint violations. The penalty imposed on infeasible individuals can range from completely rejecting the individual to decreasing its fitness based on the degree of violation. There are different types of penalty functions based on this principle and some of them are discussed below.

In the *death penalty method* the infeasible solutions are not considered for selection for the next generation and this is the greatest penalty that can be imposed on an infeasible solution. Among all penalty methods, the death penalty is the simplest to implement. This approach has the drawback of not exploiting any information from the infeasible individuals to guide the search [9]. Also when the initial population consists of no feasible solution, the whole population has to be rejected and a new one is randomly generated. This technique may work well in problems where the feasible space is convex and covers a large part of the search space but is not generally used otherwise.

In the *static penalty method* the penalty is a weighted sum of the constraint violations. The objective function is modified as,

$$\text{obj}(X) = f(X) + \sum_{j=1}^{m} r_j c_j(X) \tag{5.2.1}$$

where $f(X)$ is the actual objective function value, r_j is the penalty coefficient for constraint j, $c_j(X)$ is the degree of violation of constraint j corresponding to the individual X, and $\text{obj}(X)$ is the modified objective function value after adding penalty.

The success of the static penalty method depends on the proper penalty coefficients chosen for constraints. This has to be determined carefully based on the difficulties of these constraints.

In [23] a *dynamic penalty method* was proposed where the penalty assigned to each individual depends on the generation number and a scaling constant C in addition to its constraint violation. The authors of [23] claim that this as an important distinction which applies more selective pressure on nearly feasible solution thus making them feasible. However the difficulty involved in tuning many parameters for dynamic penalty method has significantly limited its applicability.

While the penalty function methods discussed so far are easy to implement they require some degree of parameter tuning to tailor for each problem. From [36] we summarize some of the guidelines for penalty function methods: (i) penalties which are functions of the distance from feasibility perform better than those which are merely functions of the number of violated constraints; (ii) for a problem having few constraints and few feasible solutions, penalties which are solely functions of the number of violated constraints are not likely to find solutions; and (iii) the more accurate the penalty is estimated, the better quality of solution can be found.

A method with a *self-adaptive penalty* function to alleviate the difficulty of choosing penalty coefficients was proposed in [16]. The authors proposed a two-stage penalty function that requires no explicit definition of any parameters. The method was formulated to ensure that slightly infeasible solutions with a low objective function value remain fit. The first stage ensures that the worst of the infeasible solutions has a penalized objective function that is higher than or equal to that of the best solution in the population (all other solutions are penalized by a lesser amount depending on their feasibility). The second penalty increases the penalized objective value of the worst of the infeasible solutions to twice the objective value of the best solution. The method however, required the definition of a scaling factor.

In [3] another parameter less adaptive penalty method is proposed. Information from the population such as the average of the objective function and the degree of violations of each constraint during the evolution is used in order to define different penalties for different constraints. The basic idea is to allow those constraints which are more difficult to be satisfied to have relatively higher penalty coefficients. However in this method the penalty coefficients do not depend on the number of generations so that no pressure is applied to infeasible solutions as the evolution process continues.

Because penalty functions combine the objective function value and the constraint violation value to decide the fitness of each individual, there is a domination relationship between the constraint violation and the objective function in deciding the fitness of the individual. In [37], the authors characterize the problem of choosing the appropriate penalty coefficient (r_j) for each constraint and describe how it affects the domination between the constraint violation and the objective function in deciding the rank of each individual. To overcome the burden of choosing an optimal r_j the authors propose a probability factor P_f which denotes the probability of the objective function used to allocate rank to the individual. The ranking method incorporated assures that feasible solutions are ranked based only on their objective function while the probability factor P_f determines whether objective function or constraint violation should be used to rank infeasible individuals. A P_f value of 0.45 was found to produce very good results in [37]. This implies that infeasible solutions would be ranked less often based on their objective function value (45%) and more often based on their constraint violation value (55%). While this method produced the best results for all of the problems evaluated, there was one significant drawback with the results obtained. For a particular test problem the method could only produce feasible solutions 6 out of 30 runs. This can be attributed to the selection scheme in which constraint violation does not dominate the objective function even when ranking infeasible solutions.

ii) *Methods based on preference of feasible solutions over infeasible solutions*

In [35], the authors suggested a penalty function method in which feasible solutions would always have higher fitness than infeasible ones. A rank-based selection scheme was used and the rank was assigned based on the objective function values mapped into $(-\infty, 1)$ for feasible solutions and the constraint violation mapped into $(1, +\infty)$ for infeasible solutions. Hence in this technique all feasible solutions dominate the infeasible ones. Infeasible solutions will be compared based on their constraint violation while feasible solutions will be compared based on their objective function value only. This method presents some interesting properties: (i) as long as no feasible solution is found, the objective function will produce no effect on the rank of the individual; (ii) once there is a combination of feasible and infeasible solutions in the population then feasible solutions will be ranked ahead of all infeasible solutions; and (iii) feasible solutions will be ranked based on their objective function values. The major drawback which we could experience in this method is a lack of diversity either explicitly defined or as part of the selection scheme. This deficiency will occur in problems with disconnected feasible components in which cases the GA may be stuck within one of the feasible components and never get to explore.

In [34] the authors suggested a method based on infeasibility degree (IFD) selection. IFD of a solution is defined as the sum of the square value of all the constraint violations of that solution. The whole procedure is divided into two stages: an initial infeasibility selection procedure and infeasibility degree selection genetic algorithm. An infeasible solution is accepted if its IFD value is less than or equal to a threshold value. Otherwise it is rejected and is replaced by the most feasible solution in the current population. The threshold IFD value is reduced in each iteration in order to minimize the search space. The problem with this method is that users should supply a parameter, called annealing coefficient, to calculate the threshold value.

The same idea as described above formed the basis of [13] where selection was based upon the following underlying principles: (i) a feasible solution wins over any infeasible solutions; (ii) two feasible solutions are compared only based on their objective function values; (iii) two infeasible solutions are compared based on the amount of their constraint violations; and (iv) two feasible solutions i and j are compared only if they are within a critical distance $\bar{d}$, otherwise another solution j is checked n_f times before i is chosen as the winner. The authors of [13] also argued that real coded representation was better suited for constrained optimization problems as it affords a greater chance of maintaining feasibility. In addition the authors used a niching scheme to maintain diversity among feasible solutions and binary tournament selection to make pair wise comparisons. The penalty approach was different in the sense that the coefficient r_j was unity for all constraints and all the constraints were normalized to allot equal importance to each constraint. This method performed very well on a variety of benchmark test problems and niching operator was incorporated to overcome the problem of stagnation discussed above. However, this method requires heuristically chosen parameters on critical distance $\bar{d}$ and n_f.

iii) *Methods based on multiobjective optimization techniques*

Constrained Optimization by Multiobjective Genetic Algorithm (COMOGA) was proposed in [40] where the solutions are first ranked based on non-domination of their constraint violations and then also ranked based on their objective function. A P_{cost} factor selects solutions based on objective function while the others are selected based on constraint violation. The P_{cost} is adjusted depending on the target proportion of feasible solutions in the population. The obvious drawback of this technique is the high computational complexity especially as the number of constraints increases. Also there is not enough evidence to suggest that treating each constraint independently and ranking them based on Pareto domination is an efficient design.

In [44] a technique based on Pareto strength and Minimal Generation Gap (MGG) is proposed. The constraint optimization problem is converted into a bi-objective optimization problem. One is the original objective function and the other is the degree of constraint violation. In each generation two offspring are selected; one with the highest Pareto strength and another with the lowest degree of violating constraints. If the individuals with the highest Pareto strength are not unique, then the one with the lowest degree of violating constraints is selected. The process is repeated until the whole population is replaced by a new population.

In [1] the authors suggested a multiobjective method inspired by penalty approach. The approach is based on the following criteria: 1) if no feasible individual exists in the current population, the search should be directed toward the feasible region; 2) if a majority of the individuals in the current population are feasible, the search should be directed toward the unconstrained optimum; 3) A feasible individual closer to the optimum is always better than a feasible individual further away from the optimum; 4) An infeasible individual might be a better individual than a feasible individual if the number of feasible individuals is high. To fulfill the above criteria two ranking schemes are used. *Rank1* is based on the objective function and *rank2* based on non dominated ranking with respect to the constraints. A new objective function is then formulated based on the two ranks and individuals are ranked according to the new objective function. The main problem with this method is that it did not perform well for problems containing equality constraints.

In [9] the author proposed a subpopulation based approach similar to VEGA by using $m+1$ subpopulations where m denotes the number of constraints and the first subpopulation is devoted to optimizing the objective function. The method differs from [40] in that non-dominated ranking is never employed but the fitness function for each problem is changed so that initially the fitness function for each subpopulation (except the first one which is based on the objective function) depends on the violation of its constraint. If the solution evaluated does not violate the constraint corresponding to the subpopulation but is infeasible, then the subpopulation will minimize the total number of violations. Finally once the solution becomes feasible it will be merged with the first subpopulation and look to minimize the objective function. While the results produced were satisfactory, the choice of the size of each subpopulation remained an open question.

From our analyses of the algorithms previously proposed to solve the constrained optimization problem, we notice three common features: (i) lack of elitism; (ii) choices of parameters that require a priori knowledge about the problem characteristics; and (iii) lack of an assurance of producing feasible solutions. Elitism can be very effective in maintaining the best feasible solution in the population and can guide the genetic search to concentrate around this solution thus providing an exhaustive search. At the same time elitism can lead to poor diversity in the population. The algorithm proposed herein blends the elitist approach with a non-dominated ranking approach thus providing a delicate balance between exploration and exploitation. We believe that a possible rationale that many methods fail to produce feasible solutions for every run is the simultaneous handling both the objective function and the constraints in all stages of the algorithm. This prevents the GA from moving toward the feasible regions without being distracted by the objective function landscape. Our proposed method provides a greater assurance of producing feasible solutions because the search is directed only based on the constraint violation value till feasible solutions are found.

While it is commonly accepted that no algorithm will be effective in solving all types of problems as documented in the No Free Lunch Theorem [43], we have proposed a generic framework that requires absolutely no definition of any problem dependent parameter. This is in keeping up with the development of algorithms with an increasing level of generality by the usage of hyper-heuristics. Hyper-heuristics are heuristics that choose between the lower-level heuristics. Hence with respect to genetic algorithms, hyper-heuristics define an approach where the higher level is a genetic algorithm that decides which heuristic to call next. That is, a genetic algorithm is coded such that it represents a sequence of heuristic calls, rather than a representation of the problem itself. The lower level of the hyper-heuristic is the set of heuristics which operate directly on the solution. The interested reader is referred to [4, 5] for more reading. Our proposed algorithm is an effort to define a hyper-heuristic to solve constrained optimization problems. Throughout this research we have not made an effort to experiment with lower-level heuristics. In the following, Section 5.3 describes the proposed constraint handling scheme and how it can satisfy the design requirements of reliability and problem independence.

5.3 Proposed Constraint Handling Scheme

GA's being a stochastic search technique can offer no guarantee of producing feasible solutions. To address this concern, we have formulated the GA in such a way that finding feasible solutions is the prioritized objective. Once a feasible solution is found, then the best one is kept in the population using the elitist scheme thus assuring that the found feasible solution is never lost. However preferring feasible solutions over infeasible ones could cause the GA to be stuck in one particular feasible component where there are disconnected feasible components and the GA may never get to explore the other feasible components containing the global optimum. So exploring the search space guided by both the constraint satisfaction and the objective function optimization will be the secondary objective. The proposed constraint handling scheme consists of two phases and the algorithm switches smoothly from the first phase to the second based on a simple conditional statement.

i) Phase I (Constraint Satisfaction Algorithm): In the first phase of the algorithm the objective function is completely disregarded and the entire search effort is directed toward finding a single feasible solution. Each individual of the population is ranked based on its constraint violation only and fitness is assigned to each individual based on its rank. The elitist strategy is used and the solution with the least constraint violation is copied to the next generation. This phase takes care

of the feasibility criteria and provides a usable solution (one that satisfies all constraints). We find this technique to be especially suitable for highly constrained
problems wherein finding a feasible solution may be extremely difficult. In such
problems it would be worthwhile and efficient to explore the search space based
on the constraints alone without taking the objective function into consideration.

ii) Phase II (Constrained Optimization Algorithm): The algorithm switches to
this phase once at least one feasible solution has been identified. This phase is
treated as a bi-objective optimization problem where the constraint violations and
the objective functions have to be minimized simultaneously in a modified objective space that we call the "objective function-constraint violation space," or f-v
space for short. We have used a non-dominated sorting like in [13] to rank the
individuals. We save the feasible individual with the best objective function in the
population as the elitist solution. We also use a niching scheme in the f-v space
so that sufficient diversity is maintained and the GA will continue to explore. We
believe that this multi-objective evolutionary algorithm (MOEA) based approach
will search to minimize both the objective function and constraint violation simultaneously and guide the algorithm in exploring the region between the constrained
and unconstrained optima and the feasible and infeasible parts of the search space.
The details of implementation of the algorithm are given below.

5.3.1 Scalar Constraint Violation

From the problem formulation we have m constraints and the constraint violation for an individual is a vector of m dimensions. Using a tolerance (δ) of
0.001 for equality constraints, the constraint violation of individual X on the jth
constraint is calculated by,

$$c_j = \begin{cases} \max(0, g_j(X)), & j = 1, \cdots, q, \\ \max(0, |h_j, (X)| - \delta), & j = q+1, \cdots, m, \end{cases} \quad (5.3.1)$$

where $|\cdot|$ denotes the absolute operator.

Each constraint violation is then normalized by dividing it by the largest violation of that constraint in the population. We use normalized constraint violations
to treat each constraint equally. First we find the maximum violation of each constraint in the population by using (5.3.2),

$$\mathrm{cmax}(j) = \max_X c_j(X). \quad (5.3.2)$$

These maximum constraint violation values are used to normalize each constraint violation. The normalized constraint violations are added together to produce a scalar constraint violation $v(X)$ for that individual which takes a value
between 0 and 1.

$$v(X) = \frac{\sum_{j=1}^m c_j(X)/\mathrm{cmax}(j)}{m}. \quad (5.3.3)$$

5.3.2 Rank Based Fitness Allocation

In both phases of the proposed algorithm we allocate a fitness to each individual based on its rank in the population. In the first phase all the individuals are ranked based on their scalar constraint violation. The rank based fitness function is implemented from [6]. In the second phase the individuals are sorted into different fronts based on non-domination and rank is assigned to each individual based on the front it belongs to.

Find ϕ - number of feasible solutions in the population
  if ($\phi == 0$)
    // PHASE I
    objective $\Rightarrow$ minimize $v(X)$
    elite solution $\Rightarrow$ solution with least $v(X)$
    $r(X) \Rightarrow$ rank based fitness of individual based on violation $v(X)$
    fitness$(X) = r(X)$
  else
    // PHASE II
    $f(X) \Rightarrow$ given objective function
    objective $\Rightarrow$ minimize $(f(X), v(X))$
    elite solution $\Rightarrow$ feasible solution with least $f(X)$
    $r(X) \Rightarrow$ nondominated rank based fitness of individual
    $d(X) \Rightarrow$ crowding-distance assignment of individual (0–1)
    fitness$(X) = r(X) + d(X)$
  end
  Apply genetic operators on current population
  generation = generation + 1
end

Figure 5.3.1: Pseudo code of the proposed constraint handling algorithm

5.3.3 Crowding-Distance Assignment

It is desirable to have a diverse set of solutions in the f-v space to maintain the exploratory power of the algorithm and hence a niching scheme based on the distance of the nearest neighbors to each solution is applied. To get an estimate of the density of solutions surrounding a particular individual in the second phase we calculate the normalized average distance of two points on either side of this point along each one of the dimensions. This quantity $d(X)$ takes a value between 0 (the individual has multiple copies in the population) to 1 (the individual is not crowded). The fitness of the individual based on its rank and crowding-distance is given by,

$$\text{fitness}(X) = r(X) + d(X). \tag{5.3.4}$$

Note here that the elitist individual is chosen irrespective of its *fitness* but based only on the conditions for each of the two phases. The pseudo code of the algorithm is given in Figure 5.3.1.

In the following section we discuss the algorithm design and how we come up with it based on the difficulties associated with different problem scenarios.

5.4   Constrained Optimization – Algorithm Design

As discussed before, one of the major challenges for constrained optimization is to search for optimal solutions that are feasible with respect to the constraints. One of the approaches for effectively solving the constrained optimization problem is to treat the constraints as "objectives with goals" and define preference among individuals as described in [18]. However this can lead to an extremely high dimensional objective space as the number of constraints grows. The computational complexity will become unmanageable. Hence we have used a single parameter, the *scalar constraint violation* (SCV), representing normalized net violation of constraints by an individual. To analyze the proposed algorithm further let us consider each of the two phases individually. Let us define the usage of the following terms,

F – feasible region, i.e., the domain of the search space S that is feasible;

C_i – the ith disconnected feasible component in the search space S,

   $i = 1, \cdots, k$;

$C_i \cap C_j = \phi, \quad i \neq j, \quad i, j = 1, \cdots, k$; and

$C_1 \cup C_2 \cup \cdots \cup C_k = F$.

Phase I - Constrained Satisfaction Problem:

Goal: To find a feasible solution from a random initialization.

If there are m constraints and k feasible components then a GA with selection based only on constraint violation will find a feasible solution with probability one as $t \to \infty$. This is true because in this case the scalar constraint violation is only a measure of distance from the feasible region. As selection favors minimizing this distance, a feasible solution will be eventually reached. Since there are k feasible components, the probability of the first feasible solution being found in any one of the feasible components is approximately $1/k$.

Next we begin our analysis of the second phase of the algorithm where the actual optimization takes place. During the design of our fitness scheme we could have chosen either one of the following two schemes: the *preference scheme* or the *non-dominated scheme*. The *preference scheme* based on [30] is defined by,

(i) Any feasible solution is better than any infeasible solution; and

(ii) Among two feasible solutions i and j, assign greater probability of selection to the solution with the better objective function.

We shall compare this with the *non-dominated scheme* in which solutions are ranked based on the non-domination of their constraint violations and objective function values.

In analyzing these two selection schemes we try to draw meaningful conclusions about the design of the selection scheme under different problem scenarios.

Phase II - Constrained Optimization Problem:

Goal: To search for the feasible global optimum after a single feasible solution is found.

We define the efficiency of a search technique by its speed (with respect to the number of function evaluations) at which it can get to the global optimum as opposed to an exhaustive brute-force search.

A major issue in solving the constrained optimization problem is the balance between the exploration and exploitation. Let us consider the f-v space. In our algorithm we maintain the feasible solution with the best objective function unchanged in our population and this can be regarded as an artificial way of creating a genetic drift phenomenon which helps in exploitation. At the same time we apply a niching scheme in the f-v space looking for a well extended and uniform Pareto front thus helping the algorithm explore even when it is converging. The following cases illustrate why and when this property of the algorithm is essential.

Case 1: There is only one feasible component ($k = 1$) – a need for exploitation

In this case our initial feasible solution will belong to this feasible component and the global optimum is also located within this component. Selection based on the *preference scheme* will be more efficient in converging to the global optimum than the *non-dominated scheme* as (i) there is no need to explore; (ii) the infeasible solutions (which carry no useful genetic information in this case) are not encouraged in the population and this technique can lead to the global optimum in a less number of evaluations.

Case 2: If there are k (> 1) disconnected feasible components – a need for exploration.

In this case the *preference scheme* may not be efficient in converging to the global optimum. This is because the chances of the feasible initial solution being located in the feasible component with the global optimum is $1/k$ and becomes less as the number of disconnected components increase so there may be a need for the GA to search for solutions in the other components. This presents a need

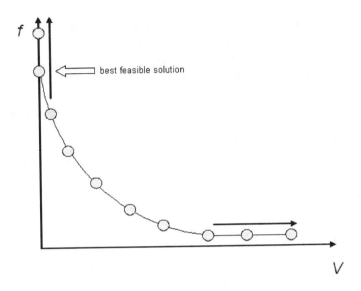

Figure 5.4.1: Schematic of the non-dominated ranking used in the GA

for exploration to find feasible solutions. We analyze the two methods by taking 2 solutions i and j from the population and evaluate the selection scheme that will increase the probability of converging to the global optimum. We also assume in our discussion that the feasible solution with the best objective function is saved as the elitist solution in the population.

(i) If solution i is feasible and j is infeasible, we could

(a) assign a greater probability of selection to i irrespective of the objective function values of i and j;

(b) check if j has a better objective function value than i and consider both i and j non-dominated if it does.

Otherwise assign a greater probability of selection to i.

In designing our algorithm, we chose option (b). This is because our elitist scheme already saves the best solution in the population. This elitist solution is obviously feasible and has an objective function value that is just as good or better then i. So irrespective of whether i or j is chosen, the elitist scheme assures that a part of the genetic search proceeds along the direction of the feasible solution with the best objective function. Hence by giving j an equal probability of selection, we are also favoring genetic search in the infeasible regions that may have good objective function values.

(ii) Among two feasible solutions i and j, consider i and j non-dominated irrespective of the objective function values. This helps the algorithm explore more as the best feasible solution is already stored as the elitist solution. Hence by giving both i and j an equal probability of selection we are giving the algorithm a better chance to explore.

Figure 5.4.1 shows a non-dominated set of solutions in the f-v space and all these individuals are ranked one. The niching scheme assigns different fitness to these solutions based on how crowded they are. So there is a greater selective pressure on solutions at the two corners. The niching scheme tries to extend the Pareto front along the directions of the two solid (black) arrows. Since the elitist solution is saved unchanged in the population, there is a greater probability of solutions around it. Hence we have indicated this solution in the figure using the hollowed (white) arrow.

In the next section, we introduce the Test Case Generator-2 (TCG-2) [38] and perform actual experiments using the two selection schemes under the problem scenarios discussed above in Cases 1 and 2.

5.5   Selection Scheme Comparison Using TCG-2

In this section, we use the Test Case Generator-2 (TCG-2) to simulate different problem scenarios and evaluate the performance of the two selection schemes proposed. The TCG-2 is an enhanced version of the Test Case Generator (TCG) proposed in [30]. The nine different tunable features of the TCG-2 are:

n – dimensionality of the problem,

m – number of feasible components,

ρ – feasibility ratio of the search space,

c – complexity of the feasible search space,

a – number of active constraints,

p – number of peaks of the objective function,

σ – width of the peaks,

α – decay of height of the peaks, and

d – distance between the different feasible components.

The search space is composed of an n-dimensional hypercube with each dimension ranging in the closed interval of [0,1]. The feasible regions of the search space are determined by m, ρ, c, and d. The general idea behind the TCG-2 is to randomly create m non-overlapping boxes (or rectangular areas) in the search space. The total occupancy of the m feasible components put together is $\rho \times \mid S \mid \times (1 - c)$. Considering a two-dimensional search space, if the complexity c is zero then there are m feasible components and each one of them is a perfect rectangle. New boxes are attached to the existing ones maintaining a minimum distance d between the feasible components for the remaining $\rho \times \mid S \mid \times c$ part of the search

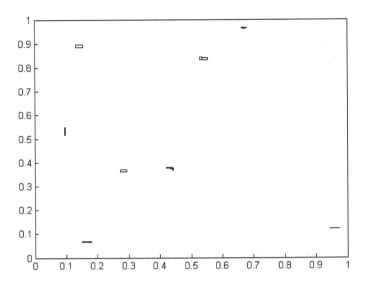

Figure 5.5.1: Feasible components for $n = 2$, $\rho = 0.001$, $c = 0.2$, $m = 8$, and $d = 0.1$.

space. Figure 5.5.1 shows the feasible components in a two-dimensional search space for $n = 2$, $\rho = 0.001$, $c = 0.2$, $m = 8$, and $d = 0.1$.

Based on the created feasible components, the constraint violation function is defined. The constraint violation value is zero inside the feasible components, while outside the feasible components the constraint violation value is the distance to the closest center of all the feasible components. The constraint violation is defined by,

$$cv(X) = \begin{cases} 0, & \text{if } X \text{ is inside a feasible component} \\ |cc_{\text{feasible}} - X|, & \text{otherwise} \end{cases}$$

where cc_{feasible} is the closest center to any feasible component.

The objective function is defined using a set of p randomly placed Gaussian $g_k(X)$, where h_x is the height of the peak k and $\bar{c}_k$ is the center of the peak k. In order to evaluate the objective function $f(X)$, the closest center $\bar{c}_i$ of the solution vector is found and then the Gaussian function $g_k(X)$ is evaluated.

All centers $\bar{c}_k$ are placed randomly in the search space with the exception of the global optimum that is placed such that there are exactly a active constraints at the global optimum. All peaks heights are evenly distributed between $[\alpha, 1]$ such that the global optimum has the highest peak $h_k = 1$ while the lowest peak has $h_k = \alpha$. The global optimum is placed either inside the feasible regions (if

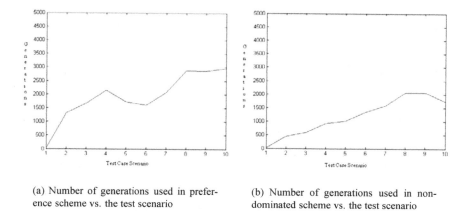

(a) Number of generations used in prefer- (b) Number of generations used in non-
ence scheme vs. the test scenario   dominated scheme vs. the test scenario

Figure 5.5.2: Number of generations used in preference/non-dominated scheme vs. the test scenario.

$a = 0$) or at the borders (if $a > 0$). Hence the global optimum always satisfies the constraints and has a value of 1. The test scenario chosen is specified by a function TCG-2 $(n, m, \rho, c, a, p, \sigma, \alpha, d)$. We have implemented two tests to verify the results from our previous discussion regarding selection schemes for different types of problems. In each of these tests we have defined 10 levels of difficulties and the performance of each algorithm is plotted based on how it treats problems of increasing difficulty.

To perform this test we increase our problem complexity from a problem with 1 feasible component and one optimum to a problem with 10 disconnected feasible components and 10 peaks. All the other characteristics of the problem such as the feasibility ratio and number of active constraints are kept the same. In short, we basically have 10 test scenarios defined by TCG-2 $(2, m, 0.005, 0.2, 1, p, 0.2, 0.5, 0.1)$ where m and p are varied from 1 to 10. We allow the algorithm to run a maximum of 5000 generations. In addition, since we know that the global optimum is 1, we stopped the algorithm if the best feasible objective function value crosses 0.999 borderline. When implementing both the *preference scheme* and the *non-dominated* scheme, we use niching in the second phase of the both algorithms to maintain diversity. The same elitist scheme is also employed in both designs. The following plots, Figures 5.5.2(a) and 5.5.2(b), compare the results from the *preference scheme* method and the *non-dominated scheme*.

Starting with the same number of generations required for m and p value of 1, the non-dominated scheme shows a much better performance when the num-

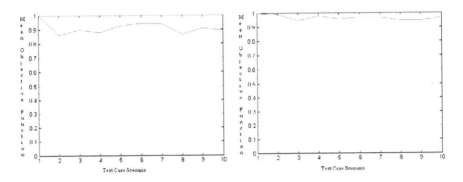

(a) Mean objective function with a maximum of 5,000 generations using preference scheme vs. the test scenario number

(b) Mean objective function with a maximum of 5,000 generations using non-dominated scheme vs. the test scenario number

Figure 5.5.3: Mean objective function with a maximum of 5,000 generations using preference/non-dominated scheme vs. the test scenario number.

ber of disconnected components and the number of peaks increase. Even though the increase in the number of generations required to solve the problem does not increase linearly with m and p, we can clearly see that under all scenarios the non-dominated scheme performs better. Figures 5.5.3(a) and 5.5.3(b) show the mean objective function values obtained under various problem scenarios.

Again we notice that better performance is obtained by the *non-dominated scheme* for all choices of m and p greater than 1. By acknowledging that the *non-dominated scheme* achieved these results with a less number of generations as shown in Figures 5.5.2(a) and 5.5.2(b), we can argue with confidence that the non-dominated scheme performs more efficiently in the current design scheme. In the next section we extend the tests to the eleven test problems from [33] used frequently in literature and provide a fair comparison with some state-of-the-art approaches.

5.6   Test Results

We apply the proposed constraint handling scheme to eleven test cases from [26] as shown in Table 5.6.1 using real-coded individuals with a probability of mutation $1/n$, where n is the number of decision variables involved. The crossover is implemented by using the binary representation of the chromosomes [6] with

Table 5.6.1: Summary of test cases

Function	n	Type of optimization	ρ	LI	NE	NI	a
Min.G1	13	Quadratic	0.0111 %	9	0	0	6
Max.G2	k	Nonlinear	99.8474 %	0	0	2	1
Max.G3	k	Polynomial	0.0000 %	0	1	0	1
Max.G4	5	Quadratic	52.1230 %	0	0	6	2
Min.G5	4	Cubic	0.0000 %	2	3	0	3
Max.G6	2	Cubic	0.0066 %	0	0	2	1
Max.G7	10	Quadratic	0.0003 %	3	0	5	6
Max.G8	2	Nonlinear	0.8560 %	0	0	2	0
Max.G9	7	Polynomial	0.5152 %	0	0	4	2
Max.G10	8	Linear	0.0010 %	3	0	3	6
Max.G11	2	Quadratic	0.0000 %	0	1	0	1

LI - linear inequalities, NE - nonlinear equalities, NI- nonlinear inequalities, a - active constraints, and ρ - the feasibility ratio $\rho = |F \cap S|/|S|$.

a probability of 0.9. For all of the eleven problems we use a population size of only 10 individuals and 10% elitism. The linear rank based fitness assignment is adopted from [6]. We run the proposed algorithm for 5,000 generations in each of 50 runs. The characteristics of these test problems are given in Table 5.6.1 reproduced from [26].

The feasibility ratio $\rho = |F \cap S|/S$ is determined experimentally in [32] by calculating the percentage of feasible solutions among 1,000,000 randomly generated individuals. For G2 and G3 a value of $k = 50$ was used in [32] which is different from 20 used in our experiments. Also because we treat our equality constraints by relaxing them using a threshold value, the values of ρ would be slightly different for G3, G5 and G11. From Table 5.6.1 we can clearly see that we have a variety of test functions involving both maximization (Max) and minimization (Min) problems with different types of objective functions (e.g., quadratic, nonlinear or polynomial) and constraints (e.g., linear equality or nonlinear equality). We have used the proposed constraint handling scheme *without* any modifications (on the scheme and on the parameter setting) for solving all of these 11 problems. The results are shown in Table 5.6.2.

One of the first observations is that all of the 50 runs produce feasible solutions for all the test problems and this assures that in a "real-world scenario" we produce usable solutions in every run of the algorithm. This is largely credited to

Table 5.6.2: Summary of test cases

Function	Optimum value	Worst	Best	Median	Standard deviation	MFG	Infeasible Runs
Min.G1	-15	-11.9999	-14.9999	-14.9997	0.8514	11.24	0
Max.G2	0.803553	0.672169	0.803190	0.755332	0.0327	0	0
Max.G3	1.0	0.7855820	1.00009	0.94899	0.0489	31.68	0
Max.G4	-30665.5	-30651.96	-30665.53	-30663.36	3.3103	0	0
Min.G5	5126.498	6112.2231	5126.5096	5170.5294	341.225	1807.82	0
Max.G6	-6961.8	-6954.3186	-6959.568	-6959.568	1.2691	289.52	0
Max.G7	24.306	35.881930	24.410977	26.735666	2.6139	53.22	0
Max.G8	0.095825	0.095825	0.095825	0.095825	0	9.28	0
Max.G9	680.63	684.131429	680.76	681.706	0.7443	5.84	0
Max.G10	7049.33	12097.4078	7060.55	7723.167	798.68	99.86	0
Max.G11	0.75	0.8094	0.7490	0.7493	0.0093	13.32	0

MFG - Mean generation number when the first feasible solution is found.

the first phase of our algorithm which treats the constrained optimization problem as a constraint satisfaction problem. In problems G2 and G4 a feasible solution is always found in the random initial population itself and this is because the feasible space occupy a large portion of the search space in these problems. Even though G1 contains 9 constraints, they are all linear and hence the feasible space is convex. Thus by minimizing the distance to the feasible regions the constraints can be satisfied effectively and a feasible solution is found early in the search process. G3 has only one constraint and it was relaxed slightly by using a threshold of 0.001 to help find feasible solutions. Finding feasible solutions presents the greatest challenge in G5 where the combination of nonlinear equalities and linear inequalities causes some complications in locating feasible solutions. Nonlinear inequalities again cause some appreciable delay in finding feasible solutions for G8 while a combination of linear inequalities and nonlinear inequalities introduces some delay in finding feasible solutions for G7 and G10.

Since finding feasible solutions is independent of the objective function in the first phase of our algorithm we can draw some conclusions about how constraints affect the GA's ability in finding feasible solutions.

(i) Nonlinear constraints in general introduce more difficulty in finding feasible solutions than linear constraints. This can be understood by acknowledging that GA's are stochastic search techniques that work on reinforcement learning based on the fitness values. This fitness value is a trustable indicator of how far each solution is from the feasible region when there is a linear mapping between the decision variables and the constraints. However, when the mapping becomes nonlinear then the distance between slightly infeasible solutions and completely feasible solutions in the decision space may be disproportionate to the differences in the constraint violation values. Hence one step toward feasibility in the constraint space may involve an exhaustive search in the decision space.

(ii) The feasibility ratio and the type of constraints combine together to define the degrees of difficulty in the constraint satisfaction problem. We know from Table 5.6.1 that G2 and G4 have a large ρ value and a feasible solution was found even in the random initial population on all 50 runs in spite of the fact that nonlinear constraints were present in both problems. At the same time even for problems with low ρ like G1, feasible solutions could be easily found because the constraints are all linear. G5 involves a combination of nonlinear constraints and very low ρ and this probably caused the difficulty in finding feasible solutions. Also G6 and G10 which required relatively more generations to find feasible solutions involve a combination of low ρ and nonlinear constraints.

In the results for the best values found by the algorithm from Table 5.6.3, we see that the algorithm has produced results extremely close to the optimum value known for all of the 11 test problems. For G1 even though the value of -15.0 could not be reached accurately, the algorithm consistently produced -14.9999 as the optimal value. In G2 again the optimal value of 0.8031 found is fairly close to the optimum value of 0.8035. In addition to reaching the optimum in G3, the algorithm produces a result closed to the optimum in G4. In G5, G6, G7 and G9 the best results produced by the algorithm differ in decimal places from the optimum value. G8 was a very easy problem and the optimum results were obtained for all 50 runs. In G10 again the algorithm having produced 7060.55 was quite close to the 7049.33 optimum which apparently no GA has reached as shown in Table 5.6.3. The best value for G11 is only better than the 0.75 optimum because of the tolerance in the equality constraint used. Also note that the standard deviation over 50 runs for all the problems other than G5 and G10 is extremely small and the median is very near the best values obtained. This implies that the algorithm is robust in obtaining consistent results. Table 5.6.3 reproduced from [16] compares the best results obtained from the other algorithms in literature to those obtained with the proposed constraint handling scheme in the last column.

We can see that the algorithm has performed very well for all of the test problems reaching or obtaining values extremely near the global optimum. In fact from the table it is obvious that stochastic ranking scheme proposed in [37] has produced the best results known so far for all the test problems. But a downside to that approach as pointed out in [16] is that for G10 only 6 out of 30 runs produced feasible solutions. In [16], 17 runs out of 20 produced feasible solutions while our algorithm produced feasible solutions in all of the 50 runs. We in a way have tackled the same problem of domination between the objective function and constraint violation in assigning the fitness to the individual brought out in [16], but have solved it using non-dominated ranking as opposed to using a probability

Table 5.6.3: Comparison of best results

Function	Optimum	Koziel and Michalevicz	Runarsson and Yao	Deb 2000	Farmani and Wright 2003	Proposed Constraint Handling Scheme
Min.G1	-15	-14.7864	-15.0000	-15.0000	-15.0000	-14.9999
Max.G2	0.803553	0.799530	0.803515		0.802970	0.803190
Max.G3	1.0	0.9997	1.0000		1.0000	1.0000
Max.G4	-30665.5	-30664.900	-30665.539	-30665.537	-30665.500	-30665.5312
Min.G5	5126.4981		5126.4970		5126.9890	5126.63094
Max.G6	-6961.8	-6952.100	-6961.814		-6961.800	-6961.17856
Max.G7	24.306	24.620	24.307	24.373	24.480	24.410977
Max.G8	0.095825	0.095825	0.095825		0.095825	0.095825
Max.G9	680.63	680.91	680.63	680.63	680.64	680.7622
Max.G10	7049.33	7147.90	7045.32	7060.22	7061.34	7060.5528
Max.G11	0.75	0.75	0.75		0.75	0.7490 *

* – a value lower than the global minimum has been obtained due to relaxation of the equality constraints.

factor P_f [37].

The efficiency of each of the above algorithms can be measured by comparing the number of function evaluations used by each of the algorithms. The number of function evaluations in general is equal to the (population size)×(number of generations) as each solution is evaluated once in every generation. The algorithms presented in [16, 26] used 1,400,000 function evaluations while the algorithm in [37] used 350,000 function evaluations. The algorithm in [13] used different number of function evaluations based upon the difficulty of the problem ranging from 50,000 to 350,000. In comparison, the proposed algorithm herein used 50,000 function evaluations to produce the results shown in Table 5.6.2 for all the problems.

5.7 Conclusions

We have implemented a two-phase genetic algorithm to solve the constrained optimization problem. This algorithm has the advantage of being problem independent and does not rely on any parameter tuning. The proposed constraint handling scheme was tested on TCG-2. We assigned various levels of difficulty based on the number of disconnected components and peaks in the decision space. We provided the rationale behind using a non-dominated ranking scheme for selecting individuals in a modified objective space of the objective function plotted versus the constraint violation. This in addition to the elitist scheme helps to provide a delicate balance between exploration and exploitation. The experiment verified that the proposed algorithm is an efficient design as the number of disconnected feasible components and the number of peaks increase. We then extended the tests to the eleven test problems commonly used in literature. Apart from finding optimal solutions that are extremely close to the optimum values, the proposed algorithm also found feasible solutions in every run. We attribute this to the first phase of the algorithm where the complete search effort is devoted to finding feasible solutions.

Bibliography

[1] A. Angantyr, J. Andersson and J. Aidanpaa, "Constrained optimization based on a multiobjective evolutionary algorithm," *Proc. Cong. Evol. Comput.*, Canberra, Australia, Dec. 2003, pp. 1560–1567.

[2] T. Bäck,"Selective pressure in evolutionary algorithms: A characterization of selection mechanisms," *Proc. 1st IEEE Conf. Evol. Comput.*, Orlando, FL, June 1994, pp. 57–62.

[3] H. J. Barbosa and A. C. Lemonge, "A new adaptive penalty scheme for genetic algorithms," *Information Sci.*, vol. 156, pp. 215–251, 2003.

[4] E. Burke, E. Hart, G. Kendall, J. Newall, P. Ross and S. Schulenburg, "Hyper-heuristics: An emerging direction in modern search technology," in *Handbook of Meta-Heuristics*, F. Glover and G. Kochenberger, Editors, Berlin: Kluwer, 2003, pp. 457–474.

[5] E. Burke, G. Kendall and E. Soubeiga, "A tabu-search hyper-heuristic for timetabling and rostering," *Journal of Heuristics*, vol. 9, pp. 451–470, 2003.

[6] A. Chipperfield, P. Fleming, H. Pohlheim and C. Fonseca, "Genetic algorithm toolbox for use with Matlab," http://www.shef.ac.uk/~gaipp/ga-toolbox, 2004.

[7] C. Coello, "Use of a self-adaptive penalty approach for engineering optimization problems," *Comput. Ind.*, vol. 40, pp. 113–127, 2000.

[8] C. Coello, "Treating constraints as objectives for single-objective evolutionary computation," *Eng. Opt.*, vol. 32, pp. 275–308, 2000.

[9] C. Coello, "Theoretical and numerical constraint-handling techniques used with evolutionary algorithms: A survey of the state of the art," *Comput. Methods Appl. Mech. and Eng.*, vol. 191, pp. 1245–1287, 2002.

[10] D. Coit, A. Smith and D. Tate, "Adaptive penalty methods for genetic optimization of constrained combinatorial problems," *INFORMS J. Comput.*, vol. 8, pp. 173–182, 1996.

[11] B. Craenen, A. Eiben and J. Van Hemert, "Comparing evolutionary algorithms on binary constraint satisfaction problems," *IEEE Trans. Evol. Comput.*, vol. 7, pp. 424–444, 2003.

[12] J. Culberson, "On the futility of blind search: An algorithmic view of No Free Lunch," *Evol. Comput.*, vol. 6, pp. 234–242, 1998.

[13] K. Deb, "An efficient constraint handling method for genetic algorithms," *Comput. Methods Appl. Mech. and Eng.*, vol. 186, pp. 311–338, 2000.

[14] K. Deb and S. Agrawal, "A niched-penalty approach for constraint handling in genetic algorithms," *Proc. Int. Conf. on Artificial Neural Network and Genetic Algorithms*, Portoroz, Slovenia, Apr. 1999, pp. 235–243.

[15] K. Deb, A. Pratap, S. Agarwal and T. Meyarivan, "A fast and elitist multiobjective genetic algorithm: NSGA-II," *IEEE Trans. Evol. Comput.*, vol. 6, pp. 182–197, 2002.

[16] R. Farmani and J. Wright, "Self-adaptive fitness formulation for constrained optimization," *IEEE Trans. Evol. Comput.*, vol. 7, pp. 445–455, 2003.

[17] R. Fletcher, "Practical methods of optimization," 2nd Edition, New York: John Wiley, 1990.

[18] C. Fonseca and P. Fleming, "Genetic algorithms for multiobjective optimization: formulation, discussion and generalization," *Proc. 5th Int. Conf. Genetic Algorithms*, Urbana-Champaign, IL, June 1993, pp. 416–423.

[19] C. Fonseca and P. Fleming, "Multiobjective optimization and multiple constraint handling with evolutionary algorithm– Part I: A unified formulation," *IEEE Trans. Syst., Man and Cybern.-A*, vol. 28, pp. 26–37, 1998.

[20] M. Gen and R. Cheng, "A survey of penalty techniques in genetic algorithms," *Proc. Cong. Evol. Comput.*, Nagoya, Japan, May 1996, pp. 804–809.

[21] L. Han and G. Kendall, "Investigation of a tabu assisted hyper-heuristic genetic algorithm," *Proc. Cong. Evol. Comput.*, Newport Beach, CA, June 2003, pp. 2230–2237.

[22] E. Hart, P. Ross and J. Nelson, "Solving a real-world problem using an evolving heuristically driven schedule builder," *Evol. Comput.*, vol. 6, pp. 61–80, 1998.

[23] J. Joines and C. Houck, "On the use of non-stationary penalty functions to solve nonlinear constrained optimization problems with GA's," *Proc. Cong. Evol. Comput.*, Orlando, FL, June 1994, pp. 579–584.

[24] J. Kim and H. Myung, "Evolutionary programming techniques for constrained optimization problems," *IEEE Trans. Evol. Comput.*, vol. 1, pp. 129–140, 1997.

[25] T. Kiyota, Y. Tsuji and E. Kondo, "Unsatisfying functions and multiobjective fuzzy satisfying design using genetic algorithms," *IEEE Trans. Syst., Man, Cybern.-B*, vol. 33, pp. 889–897, 2003.

[26] S. Koziel and Z. Michalewicz, "Evolutionary algorithms, homomorphous mappings, and constrained parameter optimization," *Evol. Comput.*, vol. 7, pp. 19–44, 1999.

[27] A. Kuri and J. Gutierrez, "Penalty function methods for constrained optimization with genetic algorithms: A statistical analysis," *Proc. of 2nd Mexican Int. Conf. Artificial Intelligence*, Merida, Mexico, Apr. 2002, pp. 108–117.

[28] Z. Michalewicz, "Genetic algorithms, numerical optimization and constraints." *Proc. Int. Conf. Genetic Algorithms*, Pittsburgh, PA, Jul. 1995, pp. 151–158.

[29] Z. Michalewicz and N. Attia, "Evolutionary optimization of constrained problems," *Proc. 3rd Annual Conf. Evol. Programming*, San Diego, CA, Feb. 1994, pp. 98–108.

[30] Z. Michalewicz, K. Deb, M. Schmidt and T. Stidsen, "Test-case generator for nonlinear continuous parameter optimization techniques," *IEEE Trans. Evol. Comput.*, vol. 4, pp. 197–215, 2000.

[31] Z. Michalewicz and C. Janikow, "GENOCOP: a genetic algorithm for numerical optimization problems with linear constraints," *Commun. ACM*, pp. 122–133, 1996.

[32] Z. Michalewicz and G. Nazhiyath, "GENOCOP III: a co-evolutionary algorithm for numerical optimization problems with nonlinear constraints," *Proc. Cong. Evol. Comput.*, Perth, Australia, Nov. 1995, pp. 647–651.

[33] Z. Michalewicz and M. Schoenauer, "Evolutionary algorithms for constrained parameter optimization problems," *Evol. Comput.*, vol. 4, pp. 1–32, 1996.

[34] S. Mu, H. Su, J. Chu and Y. Wang, "An infeasibility degree selection based genetic algorithms for Constrained Optimization Problems," *Proc. Cong. Evol. Comput.*, Canberra, Australia, Dec. 2003, pp.1950–1954.

[35] D. Powell and M. Skolnick, "Using genetic algorithms in engineering design optimization with nonlinear constraints," *Proc. 5th Int. Conf. Genetic Algorithms*, Fairfax, VA, June 1989, pp. 424–431.

[36] J. Richardson, M. Palmar, G. Liepus, and M. Hillard, "Some guidelines for genetic algorithms with penalty functions," *Proc. Int. Conf. Genetic Algorithms*, Fairfax, VA, June 1989, pp. 191–197.

[37] T. Runarsson and X. Yao, "Stochastic ranking for constrained evolutionary optimization," *IEEE Trans. Evol. Comput.*, vol. 4, pp. 344–354, 2000.

[38] M. Schmidt and Z. Michalewicz, "Test-case generator *TCG-2* for nonlinear parameter optimization," *Proc. Parallel Problem Solving from Nature*, Paris, France, Sept. 2000, pp. 539–548.

[39] M. Schoenauer and S. Xanthakis, "Constrained GA optimization," *Proc. Int. Conf. Genetic Algorithms*, Urbana-Champaign, IL, June 1993, pp. 573–580.

[40] P. Surry, N. Radcliffe and I. Boyd, "A multi-objective approach to constrained optimization of gas supply networks: The COMOGA method," *Proc. AISB Workshop on Evolutionary Computing*, Sheffield, UK, Apr. 1995, pp. 166–180.

[41] J. van Hemert and T. Bäck, "Measuring the searched space to guide efficiency: The principle and evidence on constraint satisfaction," *Proc. Int. Conf. Parallel Problem Solving from Nature*, Granada, Spain, Sept. 2002, pp. 23–32.

[42] D. Whitley, "The GENITOR algorithm and selection pressure: Why rank-based allocation of reproductive trials is best," *Proc. Int. Conf. Genetic Algorithms*, Fairfax, VA, June 1989, pp. 116–123.

[43] D. Wolpert and W. Macready, "No free lunch theorems for optimization," *IEEE Trans. Evol. Comput.*, vol. 1, pp. 67–82, 1997.

[44] Y. Zhon, Y. Li, J. He, and L. Kang, "Multiobjective and MGG evolutionary algorithm for constrained optimization," *Proc. Cong. Evol. Comput.*, Canberra, Australia, Dec. 2003, pp.1–5.

Chapter 6

Hybrid PSO-EA Algorithm for Training Feedforward and Recurrent Neural Networks for Challenging Problems

Xindi Cai, Ganesh K. Venayagamoorthy, and Donald C. Wunsch II

Abstract: Artificial neural network applications are numerous and rapidly growing over time for solving nonlinear complex problems. This chapter presents a hybrid training algorithm based on particle swarm optimization (PSO) and evolutionary algorithm (EA) for feedforward and recurrent neural networks. The application of the hybrid PSO-EA algorithm for training of a feedforward neural network as a board evaluator for the game Go and training of a recurrent neural network to predict the missing values from a time series are presented to show the potential of the algorithm. Results show the hybrid algorithm performs better than when the PSO or the EA algorithm is used individually as a training algorithm.

6.1 Introduction

Artificial neural networks have been applied to many world applications due to their nonlinear system dynamics and strong capabilities to carry out function approximation. Fast and efficient training algorithms are being developed as the problems become more and more complex. In general, backpropagation, with its numerous variants such as quasi-Newton and Levenberg-Marquardt, is a popular method used for training neural networks [4, 28, 61]. Although there are successful applications in various fields, conventional gradient decent algorithms

171

have drawbacks in one way or another, from the born-with local minima, non-differentiable error function to the exponentially increased calculation of derivative matrices and approximate error covariance matrices. Alternative approaches for neural network training are required to overcome the above mentioned and many other shortcomings with the existing traditional techniques.

Stochastic search techniques such as evolutionary approaches can perform better on global search problems that are complex, non-differentiable and high dimensional [63]. Particle swarm optimization (PSO) and evolutionary algorithm (EA) are two successful evolutionary algorithms among many variants for optimization.

PSO is based on the simulation of the behavior of a flock of birds or a school of fish. The main concept here is to utilize the communication involved in such swarms or schools. The PSO particles or individuals in a swarm are accelerated towards their best positions and swarm best position from iteration to iteration. Thus, the particles have a tendency to fly towards the better and better regions of the search space over time, which results in a fast convergence of searching. Previous work by one of the authors [27] illustrated that PSO requires less number of iterations and training samples, in achieving same error goal in comparison with the backpropagation algorithm for training feedforward neural networks.

Evolutionary algorithms (EA) with selection and mutation operators are other popular search algorithms. The selection operator forces the individual to find better fitness in order to survive to the next generation. The mutation operator brings about diversity in the population to avoid pre-mature convergence, the key issue to avoid of trapping in local minima. The main concept in EA is to keep the competition in the population.

The innovative hybrid PSO-EA algorithm presented in this chapter is based on the inheritance of the advantages of both PSO and EA, i.e., the fast convergence and good diversity. The hybrid algorithm integrates evolutionary operators, such as selection and mutation, into the standard PSO algorithm. By applying the selection operation to PSO, the algorithm saves the limited computational source by eliminating the unpromising particles, and keeps the best performing particles as well [2]. The mutation operation prevents the particles being too close to each other under the influence of the global best particle. This is more likely to occur after during the first half of the search process and can be potential risk of getting trapped in a local minimum [6].

This chapter presents the work of the authors [7–10] in training feedforward and recurrent neural networks using the hybrid PSO-EA algorithm for solving some challenging problems. Section 6.2 of this chapter describes the PSO, EA and the hybrid algorithms. Section 6.3 demonstrates the training of a multilayer perceptron (MLP) feedforward neural network as a board evaluator for the game of Go. Section 6.4 illustrates the training of an Elman recurrent neural network to predict the missing values of a time series. Finally, the conclusions and future work are given in Section 6.5.

6.2   PSO, EA and the Hybrid Algorithm

6.2.1 Particle Swarm Optimization

Particle swarm optimization (PSO) is a form of evolutionary computation technique developed by Kennedy and Eberhart [18, 31, 32]. Similar to EA, PSO algorithm is a population based optimization technique, where the system is initialized with a population of random potential solutions and the algorithm searches for optima satisfying some performance index over generations. It is unlike an EA, however, in that each potential solution is also assigned a randomized velocity, and the potential solutions, called *particles*, are then "flown" through the m-dimensional problem space.

Each particle i has a position represented by a position vector X_i. A swarm of particles moves through an m-dimensional problem space, with the velocity of each particle represented by a vector V_i. At each time step, a function f_i representing a quality measure is calculated by using X_i as input. Each particle keeps track of its own best position, which is associated with the best fitness it has achieved so far in a vector P_b of dimension m referred to as the *pbest* of the particle. The best of the Pbest of all the particles obtained so far in the population is referred to as the *gbest* and is stored in a vector P_g of dimension m.

At each time step t, by using the individual best position, $P_b(t)$ and global best position, $P_g(t)$ (the particle in the population with the best fitness), a new velocity for particle i is calculated using:

$$V_i(t+1) = w \times V_i(t) + c_1\varphi_1(P_b(t) - X_i(t)) + c_2\varphi_2(P_g(t) - X_i(t)) \quad (6.2.1)$$

where c_1 and c_2 are positive acceleration constants, φ_1 and φ_2 are uniformly distributed random numbers in the range [0, 1], and w is the inertia weight. The term V_i is limited to the range $\pm V_{\max}$. Changing the velocity this way enables the particle i to search around its individual best position P_b and the global best position P_g. Based on the updated velocities, each particle changes its position according to:

$$X_i(t+1) = X_i(t) + V_i(t+1) \quad (6.2.2)$$

Based on (6.2.1) and (6.2.2), the population of particles tends to cluster together with each particle moving in a random direction (Figure 6.2.1).

The pseudocode for PSO is given as follows:

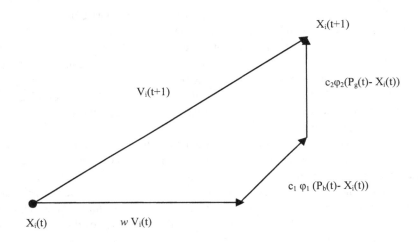

Figure 6.2.1: PSO concept of a particle's position and velocity update illustrated in two dimensions. $X_i(t)$ and $V_i(t)$ denote the particle's position and the associated velocity vector in the searching space at generation t, respectively. Vector $c_1\varphi_1(P_b(t) - X_i(t))$ and $c_2\varphi_2(P_g(t) - X_i(t))$ describe the particle's "cognitive" and "social" activities, respectively. The new velocity $V_i(t + 1)$, given by (6.2.1), is determined by the momentum/inertia weight part, "cognitive" part, and "social" part. The particle's position, given by (6.2.2), at generation $t + 1$ is updated with $X_i(t)$ and $V_i(t + 1)$.

Initialize a population of particles with random positions and velocities in the m-dimensional problem space.
Do
    Evaluate the fitness according to some given fitness function.
    Update the P_b if the current particle's fitness value is better than that of P_b.
    Determine P_g: choose best of the P_b.
    For each particle:
        Calculate particle's new velocity according to (6.2.1);
        Calculate particle's new position according to (6.2.2).
While a sufficiently good fitness or a maximum number of iteration is not yet attained.

The velocity update of a PSO particle given in (6.2.1), consists of three parts. The first part is the momentum part, preventing abrupt changes. The second part represents learning from its own flying experience. The third part represents the collaboration among particles–learning from group best's flying experience. The balance among these three parts determines the balance of the global and local search ability, therefore the performance of a PSO [56].

The inertia weight w controls the balance of global and local search ability. A large w facilitates the global search while a small one enhances local search. The introduction of an inertia weight also frees the selection of maximum velocity $V_{\max}$. The $V_{\max}$ can be reduced to $X_{\max}$, the dynamic range of each variable which is easier to learn; and the PSO performance is as good or better [53].

There always exist cases, perhaps pathological ones, where heuristics, such as presented here, will fail [62]. With that caveat, our observations are that the static parameter settings of PSO, if well selected, can do a good job, but much better performance can be achieved if a dynamically changing scheme for the parameters is well designed. Examples include a linearly decreasing inertia weight [53], a nonlinearly fuzzy adaptation [54], or involving a random component rather than time-decreasing [55]. All intuitively assume that the PSO should favor global search ability at the beginning and local search at the end.

6.2.2 Evolutionary Algorithm

The evolutionary algorithm (also called evolution strategy in [51]) begins with a random population of n neural networks; P_i, $i = 1, \cdots, n$, defined with weights and bias for each network. Weights and biases are generated by random sampling from a uniform distribution. Each neural network in the population has an associated self-adaptive parameter σ and this controls the step size of the search in the next generation. To be consistent with the range of initialization, the self-adaptive parameters for weights and biases are initially set to a constant.

The best half of the population is used as parents to create offspring for next generation. Each parent generates an offspring by varying all of the associated weights and biases. Specifically, for each parent P_i, $i = 1, \cdots, n$, an offspring P'_i, is created by

$$\sigma'_i(j) = \sigma_i(j) \exp(\tau N_j(0, 1)), \qquad i = 1, \cdots, n, \ j = 1, \cdots, N_w \qquad (6.2.3)$$

$$w'_i(j) = w_i(j) + \sigma'_i N_j(0, 1), \qquad i = 1, \cdots, n, \ j = 1, \cdots, N_w \qquad (6.2.4)$$

where N_w is the number of weights and biases in the neural network, τ is the size of the dimensions (m) of the individuals/particles given by $\tau = 1 / \sqrt{2\sqrt{N_w}}$, and $N_j(0, 1)$ is a standard Gaussian random variable resampled for every j [11].

The pseudocode for EA is:

Initialize a population of individuals with random weights, bias and self-adaptive parameters in the m-dimensional problem space.
Do
    Evaluate the fitness according to some given fitness function in m variables.
    Compare the fitness values to find the individuals to produce offspring.
    Keep the winners and replace losers with offspring for the next generation.
    For each offspring:
        Calculate the self-adaptive parameter according to (6.2.3);
        Calculate the weights according to (6.2.4).
While a sufficiently good fitness or a maximum number of generations is not attained.

6.2.3 Hybrid of PSO and EA

The social and cognitive adaptation makes PSO focus more on the co-operation among the particles. With memory, each particle tracks the best performance in its own history, and its neighborhood throughout the entire evolution when sharing the memory. However, particles of PSO are not eliminated even if they are ranked to have the worst fitness in the population, which may waste the limited computational resources. On the other hand, individuals in EA compete for survival, but the "winning" survivors do not all retain sufficient history. Clearly, the advantage of one algorithm can complement for the other's shortcoming. Thus, the motivation to develop a hybrid technique based learning algorithm. The computations involved in the PSO algorithm are easy, and adds only a slight computation load when it is incorporated into EA. Furthermore, the flexibility of PSO, to control the balance between local and global exploration of the problem space, helps to overcome premature convergence of an elite strategy in EA, and also enhances searching ability.

Based on the complementary properties of PSO and EA, a novel hybrid algorithm is designed to combines the co-operative and competitive characteristics of both PSO and EA. In other words, the PSO is applied to improve the survival individuals, and maintain the properties of competition and diversity in EA. In each generation, the hybrid algorithm selects half of the population as the winners (elites) according to the fitness, and discards the rest half as losers. These elites are enhanced, sharing the information in the community and benefiting from their learning history, using the standard PSO procedure. The enhanced elites then serve as parents for an EA mutation operation to produce same amount of offspring to fill up the vacuum that the discarded individuals left in the population size. The offspring also inherit the social and cognitive information from the corresponding parents and carry this to the next generation in the case that they become winners. Figure 6.2.2 illustrates this hybrid PSO-EA algorithm concept.

The pseudocode for hybrid PSO-EA is summarized as follows:

Initialize a population of individuals with random positions and velocities in the m-dimensional problem space.
Do
    Evaluate the fitness according to some given fitness function.
    Compare the fitness values to find the winners.
    Enhance the winner with PSO.
    For each elite:
        Update the P_b if the current particle's fitness value is better than the P_b;
        Determine P_g : choose the particle with the best fitness value of winners;
        Calculate particle's new velocity according to (6.2.1);
        Calculate particle's new position according to (6.2.2).
    Use the enhanced elites as parents to produce offspring with EA to replace losers for the next generation.
    For each offspring:
        Save parent's P_i as current P_i for further comparison;
        Use parent's velocity as self-adaptive parameters;
        Calculate the self-adaptive parameter according to (6.2.3);
        Calculate the position according to (6.2.4).
While a sufficiently good fitness or a maximum number of iterations/generations are not attained.

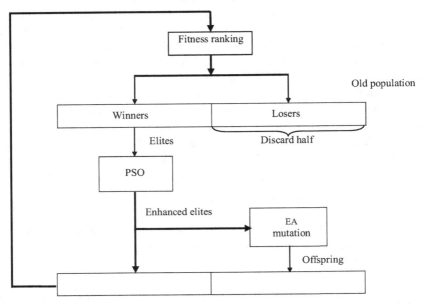

Figure 6.2.2: Flow chart of the hybrid PSO-EA algorithm. The winners made of half the population, are enhanced by PSO and kept in the population for the next generation. The PSO enhanced winners called the "Enhanced elites" are the parents in EA operation to produce offspring. The offspring replace the discarded losers in the old population to maintain a constant population size for every generation.

The following two sections describe the application of the hybrid PSO-EA algorithm to train feedforward and recurrent neural networks for real world applications.

6.3 Feedforward Neural Networks as Board Evaluator for the Game Capture Go

6.3.1 Background

The foundation of game theory was first laid down by von Neumann in 1944 [42], and further developed by [3, 37]. Claude Shannon [49] proposed that a mechanical algorithm could play a game if that algorithm contained two ingredients: an evaluation function - a mathematical formula that assigned credits according to different board positions - and a rationale, which he called "minimax," that sought to minimize the maximum damage that the opponent could do in any circumstance [40]. Guided by pioneering works, computer game engines have beaten their designers in varieties of games, from Tic-Tac-Toe to Chess. The brute-force search algorithm, combined with an expert database, reached its peak, given the computational power of current machines, when Chinook won the world checkers title in 1994 [49] and IBM's deep blue beat the World Chess Champion Garry Kasparov in 1997 [29].

Unfortunately, traditional game solutions are ineffective in the game of Go, due to the tremendous size of its game tree, dynamic changes of the crucial building blocks, and vague rank information of its stones/strings/groups. Unlike most other games of strategy, Go has remained elusive for computers, and it is increasingly recognized as a "grand challenge" of artificial intelligence, which attracts researchers from a diversity of domains, such as game theory [5], pattern recognition, reinforcement learning [64], and even cognitive psychology [14].

Even though it beats professional masters at the late end-game stage [5], computer Go still plays at amateur level for the rest of the game (especially the middle game), despite continued, albeit slow, progress [24, 41]. In particular, neural networks and evolutionary computation provide useful means for evaluating a board position [22, 46]. Unlike a rigid look-up table in a knowledge database, the neural evaluator adapts to the nonlinear stimulus-response mappings of the game. Evolutionary algorithms have shown to be a promising approach to solve complex constrained optimization problems. The method of two populations of evolution strategies is among many effective algorithms in solving the min-max problem [58]. In this approach, the fitness of an individual depends on not only the individual itself but also the individuals of the other population. Chellapilla and Fogel succeeded in evolving an expert-level neural board position evaluator for checkers without any domain expertise [12, 13, 23]. Their work concludes that computer game engines can learn, without any expert knowledge, to play a game at an expert level, using a co-evolutionary approach.

The trend of [11] is followed and PSO is applied in combination with an EA to develop a neural evaluator for the game of "Capture Go." As a simplified version of Go, Capture Go has the same rules as Go, but has a different goal—whoever captures first, wins. The system for this game, with minor modifications, should be a useful subsystem for an overall Go player.

Previous work [34] on Capture Go showed that the simplified game is a suitable test bench to analyze typical problems, such as lifetime learning, incremental evolution [26], open ended evolution and scalability, involved in evolution-based algorithms. Growing from zero knowledge, the game engine presented in this chapter extends our work [7] and demonstrates its potential in learning strategies. The large-scale game engine, a neural network with more than 6000 parameters, is trained by a PSO-enhanced evolutionary algorithm.

6.3.2 Go and Capture Go

The Game of Go

Go is a deterministic, perfect information, zero-sum game of strategy between two players. More details on the game of Go can be found in [57]. Players take turns to place black and white pieces (called *stones*) on the intersections of the lines in a 19×19 grid, called the Go board. Once played, a stone cannot be removed, unless captured by the other player. To win the game, each player seeks to surround more territory (empty grids) by one's own stones than is surrounded by the opponent's stones.

Adjacent stones of the same color form *strings*, and hence *groups*; an empty intersection adjacent to a stone, a string, etc., is called its *liberty*. A group is *captured* when its last liberty is occupied by the opponent's stone. A player cannot make a *suicidal move* by placing a stone on an intersection with no liberty. An *eye* is a formation of stones of special significance in Go. When an empty intersection is completely surrounded by stones of same color, it is known as an eye. An opponent cannot place a stone on that intersection unless it is a capturing move, i.e., unless placing the stone causes one or more of the stones that surround the intersection to be captured. A string with two eyes cannot be captured because filling one of the eyes would be a suicidal move and is therefore illegal. Having two eyes is the line between "alive" strings and "dead" ones. The potential of making two eyes for a string, e.g., whether surrounding enough large eye space, capturing enemy stones in neighborhood or connecting to "alive" friend strings,

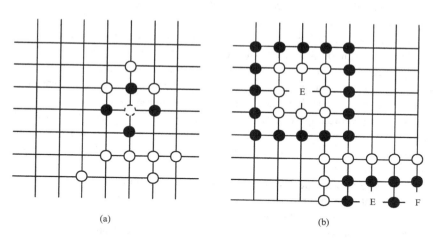

(a)   (b)

Figure 6.3.1: Go board, terms and rules. Only a portion of the 19 × 19 Go board is shown. In (a), the lower 5 T-like white stones form a string, and hence a group with the white stone to their left. In the middle, black captures a white stone, the dashed one, and removes it from the board, by placing the top black stone on the board. However, white cannot immediately put a stone onto the dashed-stone intersection to capture the top black stone because such a move would repeat the board position when. That is the rule of "Ko." In the middle of (b), white has an eye at "E." White cannot put his stone at "E," which results in no liberty of the whole string. That is a suicide move and forbidden by the rules. Black can put his stone at "E," even though there is no liberty of that stone either, but it is allowed because it captures the white string. So the white string with only one eye is dead and will be removed from the board at the end of the game. The black string at the corner has two eyes, i.e., "E" and "F," and hence is alive since the white cannot seize both of the eyes simultaneously.

will greatly affect the final fate of the string. Those strings that are incapable of forming two eyes will be considered as captured and hence removed when calculating territories at the end of the game. Evaluating whether stones can be formed into strings, and furthermore into groups, and whether strings are capable of forming two eyes is the fundamental skill in Go because it represents the game strategies of "attack," making oneself strong and being aggressive, and "defense," avoiding vulnerability and surviving (see Figure 6.3.1).

To prevent loops, it is illegal to make moves that recreate prior board positions (rule of *Ko*). The rule for Go is simple: one can place a stone on any empty intersection unless it is a suicidal or Ko move. A player can pass any time. The game ends when both players pass in consecutive turns.

Go starts with an empty board and fills with stones as the game progresses. The branching factor, the legal moves on the board at any time, is around 200 on average, more at the beginning, and less at the end of the game. The game length varies from 150 to 300 moves. All these factors result in a game tree varying in size from 10^{360} to 10^{690} nodes. In general, a game tree of approximately 10^{575} nodes [1] is accepted by most researchers. A modest 7-ply ("ply" means a round of moves by two players) look-ahead search needs to handle ten thousand trillion positions, which makes a brute force search approach infeasible even with the most powerful computer.

The vague and frequently changing rank information of strings/groups make board evaluation extremely difficult. A single stone has no rank at all. The importance of a stone, a string, and a group changes, affected by numbers of factors such as connectivity, liberty, position on board, correlation with neighboring friend and/or foe, and even the player aesthetics, as the game carries on. Additionally, tactical skills are not as important as in chess. Winning a tactical struggle over a group may not lead clearly to winning the game. Therefore, creating an evaluation function by employing pattern recognition, tactical search and rule-based reasoning, mainly based on matching the expert knowledge, is not promising. It is true that such techniques are predominant in the top computer Go engines. But, all of these engines have remained weak. When dealing with complex, subtle, and unsettled board situations, such as the middle-game, learning is crucial [25, 39].

In order to reduce the tremendous size of the game tree and approximate the dynamic nonlinear evaluation function of Go, researchers have explored the use of neural networks with machine learning methods such as supervised learning (training on moves from professional games) [21] and reinforcement learning [46, 50]. A hybrid PSO-EA is applied for those tasks on the simplified version of Go described below.

Capture Go

Capture Go (also known as Atari Go) inherits all rules from Go, with a modified winning criteria, i.e., whoever captures first wins the game. In case of no capture, the player who has occupied more territory wins. Moreover, there is no Ko move in Capture Go because repeated board positions never occur. In this chapter, Capture Go games are played on a 9×9 board.

Capture Go, though simple, maintains most characteristics of Go. This simplified version emphasizes the most important strategies, i.e., how to attack and defend. It also addresses the strategic balance between capture and territory expansion, in the sense that both players must still attend to fortifying their own territories. In addition, compared to the high branching factor and long-range spatiotemporal interaction in Go, Capture Go has a typically shorter game length, with a maximum number of intersections on the board, and clear game ending criteria, i.e., first capture wins. Also, the candidate moves often fall into a small neighborhood of the last move, which greatly reduces the game tree branches. A well-developed Capture Go evaluator can be plugged directly into an overall computer Go engine. All these make Capture Go a perfect starting point in searching for effective multi-objective mechanisms for Go.

6.3.3 Architecture of Neuro-Game Engine

The idea from [11] is used in developing a feedforward neural network (multilayer perceptron MLP) to carry out the board evaluation function, assigning credits for leaves in the game search tree of Capture Go. The best candidate move is then chosen according to alpha-beta minimax search from the game tree. The board information is represented by a vector of length 81, with each element corresponding to an intersection on the board. Elements in the vector are from $[-1, 0, +1]$, where "-1" denotes that the position on board is occupied by a black stone, "0" denotes an empty position, and "1" denotes a white stone. The feedforward neural evaluator consists of three hidden layers and one output node. The second, third, and output layer are fully connected. The first hidden layer is designed specially, following [11], to process spatial information from the board. Each neuron has a bipolar sigmoid activation function, i.e.,

$$\tanh(\lambda x) = \frac{e^{\lambda x} - e^{-\lambda x}}{e^{\lambda x} + e^{-\lambda x}}$$

with a variable bias term. Figure 6.3.2 shows the general structure of the swarm neuro-game engine. The number of weights and biases, i.e., N_w, in such a MLP (size: $140 \times 40 \times 10 \times 1$) was 6216, and hence τ was 0.0796 (Eq. (6.2.3)).

The one-dimensional input vector destroys the spatial characteristics such as neighborhood and distance of the board. The first hidden layer is implemented to remove the handicap. Each neuron in the layer covers an $n \times n$, $n = 3, \cdots, 9$, square overlapping a subsection of the board. Figure 6.3.3 illustrates some samples of the overlapping subsections. All 49 possible 3×3 square subsections are assigned to the first 49 neurons in the first hidden layer. The following 36 4×4 squares are allocated to the next 36 neurons until the entire board is allocated to the last neuron in the layer [11]. Therefore, 140 neurons are set in the first hidden layer to enable the capture of possible features that are processed in subsequent hidden layers (of 40 and 10 hidden nodes). The connecting weights between input layer and first hidden layer are designed specially to reflect the symmetric

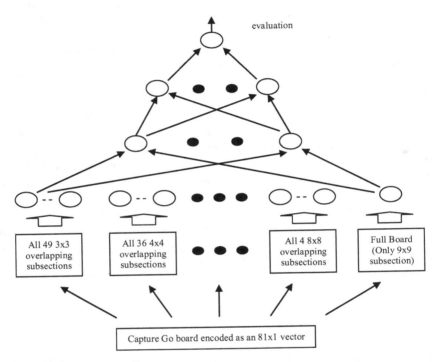

Figure 6.3.2: Architecture of the swarm neuro-game engine. This feedforward neural network evaluates the given board. Different sets of parameters of the neuro-game engine lead to different credit assignments for the given board, and hence represent different strategies. The board pattern is interpreted by 140 sub-squares. Each of these subsquares is assigned to a node in the first layer of the network for spatial preprocessing purpose. The outputs are then passed through two hidden layers of 40 and 10 nodes, respectively. The output node of the entire network is scaled between $[-1, 1]$ with "-1" in favor for the white and "1" for black.

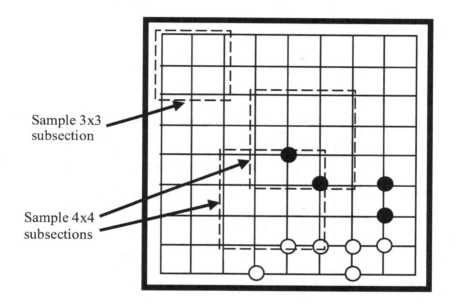

Figure 6.3.3: Game board preprocess in first hidden layer of neuro-game engine. The number of $n \times n$ subsections is $(9 - n + 1) \times (9 - n + 1)$ for $n = 3, \cdots, 9$. The neuro-player was able to learn the string/group patterns of Capture Go based on these subsections.

property of the board (see Figure 6.3.4). Each intersection on the 9×9 board is assigned a weight. Only 15 of them, in the dashed triangle area in Figure 6.3.4, are allowed to be different. The rest of the weights are duplicated by those 15, flipping along horizontal, vertical, and diagonal axes, the dashed lines in Figure 6.3.4. The board information of each square subsection, i.e., black, white, and empty, is weighted, summed up, and passed to the first hidden layer for further processing. This weight assignment guarantees that patterns shown on symmetric square subsections will contribute equally to the network. Also, those weights, when trained, will teach the neuro-engine to play at favorable subareas, e.g., corners, of the board.

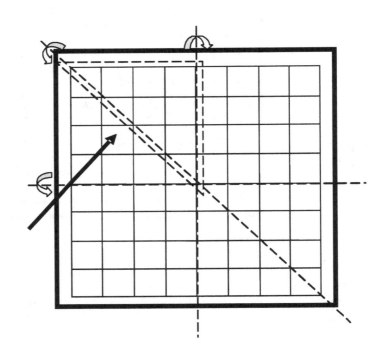

Figure 6.3.4: Share-weight designed to represent the symmetric nature of the game board. Weights in the dashed triangle area are duplicated along three axes to reflect the board symmetry. This share-weight scheme guarantees that patterns shown on symmetric square subsections will contribute equally to the network.

6.3.4 Simulation Results

In the self-play stage, a population of 40 individuals, each representing a game engine, is evolved by playing games of Capture Go. Each individual earns credits based on its game results. Each player, always black, plays one game against each of eight randomly selected opponents, always white, from the population. The game is scored for each player as -2, 0, or $+1$ points depending on the results of loss, draw, or win. In total, there are 320 games per generation, with each engine participating in average of 16 games. After all games are complete, each individual accumulates the scores it earned as its fitness value and updates according to the algorithms employed, i.e., PSO, EA, or the hybrid.

In the population evolved by PSO, particles compare their current scores to their records to determine the local best. They are also sorted to verify if a new global best is found. The positions of particles are then updated according to (6.2.1) and (6.2.2). In the population evolved by the EA, 20 strategies that receive the greatest total scores are retained as parents and generate offspring based on (6.2.3) and (6.2.4) to replace the losing 20 strategies. In the population evolved by the hybrid algorithm, the 20 particles with the highest total scores are enhanced by standard PSO, i.e., (6.2.1) and (6.2.2). Using an EA, the enhanced particles are also used as parents to produce offspring, based on (6.2.3) and (6.2.4), to replace the 20 discarded particles as shown in Figure 6.2.2.

Each game is played using an alpha-beta minimax search of the associated game tree for each board position, looking a pre-defined number of moves into the future. The depth of the search is set to six, ensuring the board is evaluated after the opponent has an opportunity to respond to the player's move. In addition, some forced moves, such as immediate capture attempt/save, will not be counted as one of the six depths in the alpha-beta search procedure.

The weights of each swarm neuro-engine are generated randomly from a uniform distribution over $[-0.2, 0.2]$. The self-adaptive parameters for the EA are initially set to 0.05. The PSO parameters are set to

$$w = 0.4667, \; c_1 = 1.7971, \; c_2 = 2.4878,$$

respectively. The value of $v_{\max}$ for PSO is set to 2.0. The whole evolutionary process is iterated for 100 generations. After every 20 generations of training, a tournament among the current best game engines of PSO, EA, and the hybrid is held to show the fitness evolving process (see Figure 6.3.5). At last, the best neuro-engine (at the generation 100) of each category, i.e., PSO, EA, and the hybrid, is then used to play against each other and a random player. Table 6.3.1 summarizes their performance in 100 games in the tournament. All players illustrate success in learning of strategies in Capture Go game because they overwhelm the random player with only 28 moves per game on average. The hybrid PSO-EA player in

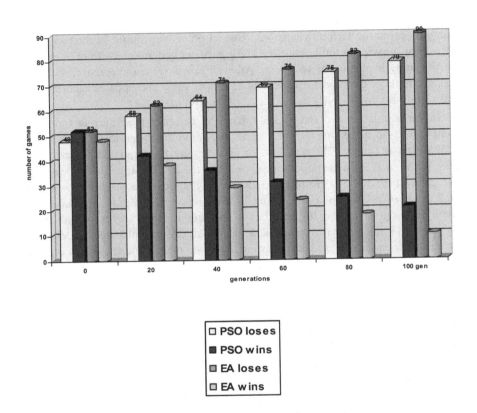

Figure 6.3.5: Tournament results among hybrid PSO-EA, PSO and EA after every 20 generations of training. A tournament of 100 games are played between the hybrid and PSO (indicated by the cubic), the hybrid and EA (indicated by the cylinder), respectively. The hybrid always plays black. In each tournament, the first columns of the cubic and the cylinder show the number of games won by the hybrid when playing against PSO and EA, respectively.

Table 6.3.1: Tournament results among hybrid PSO-EA, PSO, EA, HC and random players in 100 games. Black players are listed in row and white players in column. For example, result in row 2 column 4 means that hybrid PSO-EA player in black wins 90 to 10 against EA player in white.

	Hybrid PSO-EA	PSO	EA	HC	Random
Hybrid PSO-EA	/	79/21	90/10	76/24	100/0
PSO	62/38	/	76/24	70/30	100/0
EA	53/47	68/32	/	70/30	100/0

black dominates both PSO and EA players. Considering the advantage that black plays first, the hybrid in white is roughly equivalent to the EA and slightly weaker than the PSO. The PSO and EA players are at the same level. Another game engine is trained by a simple learning method, hill-climbing (HC), for Capture Go, to verify if the dynamics of the game and the co-evolutionary setup of the training are the key issues of the game engine learning [43]. The tournament results show that PSO, EA, and the hybrid players outperform the HC player, which indicates that the improving of game engines comes mainly from the learning algorithms. Finally, the best hybrid player is compared with a web Capture Go player [48], and wins by 6/4 in 10 games.

In addition to self-play, a defensive player of Capture Go is hand-coded. This player takes defensive strategies with the following priorities: 1) connect all its stones into one string; 2) choose a move that maximizes its liberties (the liberty count saturates when it makes two eyes); 3) surround more empty intersections with a wall; and 4) attack weak stone(s) of its opponent. The player is hard to capture because it is difficult to seize all its liberties before it makes two eyes or captures an opponent's stone(s) instead (see Figure 6.3.6). The defensive player also beats the web player, with an 8/2 in 10 games. The game result indicates that it is more likely to defeat this defensive player by occupying more territories, rather than capturing its stones. Competing with this player teaches our hybrid engine to manage the balance between seizing its own territories and capturing enemy stones.

Tables 6.3.2 and 6.3.3 contain the complete sequence of moves of two selected games between the best-evolved hybrid PSO-EA and the defensive player. The notation for each move is given in the form of "n. $A: (a, b)$," where n is the nth move, A denotes the black/white player and (a, b) is the board position of row a and column b, of the move. Table 6.3.2 illustrates the process of the first game where the hybrid engine manages to expand its own territory and restrict its opponent's. The hybrid engine wins the game by seizing 19 more intersections

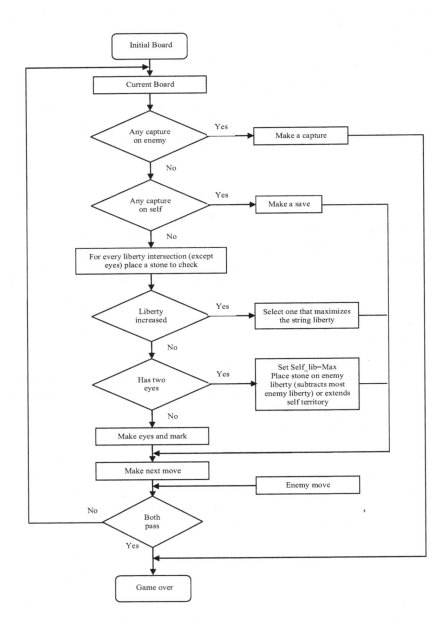

Figure 6.3.6: Flowchart of the defensive player.

Table 6.3.2: Game one between the hybrid PSO-EA and the defensive player. The hybrid engine plays black, the defensive player plays white; comments on moves are offered in brackets.

Hybrid PSO-EA engine	Defensive expert	Comments
1. B: (4, 4)	1. W: (5, 6)	
2. B: (6, 8)	2. W: (5, 5)	
3. B: (7, 7)	3. W: (5, 7)	[good move of hybrid engine, trying to make an eye at (7, 8) or (8, 8)]
4. B: (6, 7)	4. W: (4, 7)	
5. B: (6, 6)	5. W: (3, 7)	
6. B: (5, 8)	6. W: (3, 6)	
7. B: (5, 4)	7. W: (3, 5)	
8. B: (5, 3)	8. W: (2, 7)	
9. B: (4, 8)	9. W: (1, 7)	[Defensive expert tries to extends its territory in all directions in the past 8 moves]
10. B: (3, 8)	10. W: (1, 6)	[Hybrid engine seized opponent's liberties and potential eye position]
11. B: (3, 4)	11. W: (1, 5)	
12. B: (3, 3)	12. W: (1, 4)	
13. B: (3, 2)	13. W: (1, 3)	
14. B: (3, 1)	14. W: (1, 2)	
15. B: (3, 0)	15. W: (4, 6)	[Both sides set up the territory boundaries in the past consecutive moves]
16. B: (6, 5)	16. W: (1, 1)	[Hybrid engine blocks the entrance of possible invasion to its territory]
17. B: (2, 3)	17. W: (1, 0)	
18. B: (2, 2)	18. W: (1, 8)	
19. B: (2, 8)	19. W: (0, 7)	
20. B: PASS	20. W: (0, 1)	
21. B: (2, 1)	21. W: (0, 5)	[The hybrid seizes the rival's liberties in the past 5 moves]
22. B: PASS	22. W: (0, 3)	[The defensive is making eyes in the past 4 moves]

Table continued on the next page.

Table 6.3.2: Continued.

Hybrid PSO-EA engine	Defensive expert	Comments
23. B: (2, 5)	23. W: (2, 0)	[The hybrid make a vulnerable move]
24. B: (2, 4)	24. W: PASS	[The hybrid saves the vulnerable move]
25. B: (7, 2)	25. W: PASS	
26. B: (7, 1)	26. W: PASS	
27. B: (7, 0)	27. W: PASS	[The hybrid occupies possible invasion positions in its territory in the past 3 moves]
28. B: PASS	28. W: PASS	[Both sides pass and game ends. The hybrid wins by seizing 19 more intersections]

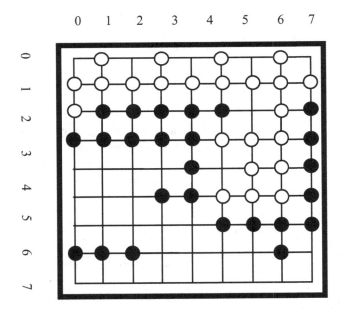

Figure 6.3.7: Final board of game one between the hybrid PSO-EA and the defensive player.

Table 6.3.3: Game two between the hybrid PSO-EA and the defensive player. The hybrid engine plays black, the defensive player plays white; comments on moves are offered in brackets.

Hybrid PSO-EA engine	Defensive expert	Comments
1. B: (4, 4)	1. W: (6, 6)	
2. B: (2, 7)	2. W: (6, 5)	
3. B: (6, 7)	3. W: (6, 4)	
4. B: (7, 2)	4. W: (6, 3)	
5. B: (6, 1)	5. W: (5, 6)	
6. B: (4, 5)	6. W: (5, 3)	
7. B: (3, 3)	7. W: (5, 2)	
8. B: (5, 1)	8. W: (7, 6)	
9. B: (5, 5)	9. W: (7, 4)	[The defensive player tries to make an eye]
10. B: (6, 2)	10. W: (4, 2)	[The hybrid engine enhances its territory at the bottom-left corner]
11. B: (4, 0)	11. W: (3, 2)	
12. B: (3, 4)	12. W: (3, 1)	
13. B: (2, 3)	13. W: (2, 1)	
14. B: (4, 3)	14. W: (1, 1)	
15. B: (1, 2)	15. W: (1, 0)	
16. B: (0, 2)	16. W: (5, 7)	[The hybrid draws the top-left boundary of its top territory in the past 5 moves]
17. B: (7, 7)	17. W: (4, 7)	[The hybrid protects its eye space at the bottom-right corner]
18. B: (2, 6)	18. W: (5, 8)	[The hybrid tries to close the top-right boundary of its top territory]
19. B: (3, 6)	19. W: (8, 6)	
20. B: (6, 8)	20. W: (8, 5)	[The hybrid is making an eye at the bottom-right corner to save its group there]
21. B: (8, 7)	21. W: (5, 4)	[The hybrid makes an eye at the bottom-right corner and saves its group there]
22. B: (3, 5)	22. W: (0, 1)	[The defensive makes an eyes at top-left corner]
23. B: (1, 3)	23. W: (3, 0)	
24. B: (4, 1)	24. W: (3, 7)	
25. B: (4, 6)	25. W: (3, 8)	

Table continued on the next page.

Table 6.3.3: Continued.

Hybrid PSO-EA engine	Defensive expert	Comments
26. B: (1, 8)	26. W: (7, 3)	[Both sides try to settle down their territory boundary at the top-right corner]
27. B: (7, 1)	27. W: (8, 3)	
28. B: (8, 1)	28. W: (8, 2)	
29. B: (8, 0)	29. W: (2, 8)	[Both sides finalize their territory boundary at the bottom-left corner]
30. B: (1, 6)	30. W: PASS	[Both sides finalize their territory boundary at the top-right corner]
31. B: (2, 2)	31. W: PASS	
32. B: PASS		[Both sides pass and game ends. The hybrid wins by seizing 11 more intersections]

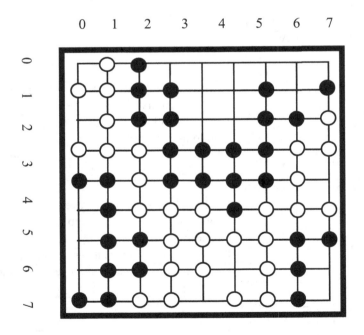

Figure 6.3.8: Final board of game two between the hybrid PSO-EA and the defensive player.

than the defensive player (see Figure 6.3.7). Table 6.3.3 illustrates the process of the second game where the hybrid demonstrates its capability in making eyes and hence maintaining safety for multiple groups. The hybrid engine wins the game by seizing 11 more intersections than the defensive player (see Figure 6.3.8).

6.4 Recurrent Neural Networks for Time Series Prediction

Time series analysis in various fields, such as finance, biology, physics, require models and techniques for chaotic systems [17, 45, 47]. Numerous time series prediction competitions bring researchers from diverse disciplines together to quantitatively compare different approaches to a common problem [38, 60]. The ensemble approach consists of models taken from a single class, e.g. neural networks, support vector machine or regression tree since the model diversity of an ensemble leads to a better performance than that of a single predictor [35]. Kalman smoother and extended Kalman filter, either applied directly to the time series or used to train neural networks to predict, also work for the time series analysis [30].

Recurrent neural networks (RNN), due to the feedback connections in their topologies, are ideal for such temporal information processing problems [47]. Gradient-based training algorithms, such as backpropagation through time (BPTT) [61] and extended Kalman filter (EKF) [44], have been known to suffer from local minima and have heavy computational load for obtaining the derivative information. As the network architecture grows, the dimension of derivative matrices and approximate error covariance matrices in those algorithms increases exponentially, which makes them unfeasible for large scale recurrent networks [33].

For the IJCNN 2004 CATS time series prediction competition, a series with 5000 data is given, where 100 values are missing, as shown in Figure 6.4.1. An Elman RNN, trained by a novel non-gradient algorithm, is selected to predict the missing data [8].

6.4.1 Architecture of Recurrent Neural Networks

The recurrent architecture allows the identification of dynamic systems with an explicit model of time and memory. An obvious generalization of the recurrent network is the extension of the system outputs in the future time steps. If this so called overshooting leads to a good prediction, a whole series of forecasting is available as an output [65]. In practice, the recurrence of the system is approximated with a finite unfolding which truncates after a certain number of time steps.

The Elman's architecture [19, 20] is chosen for the time series prediction, which consists of a context layer, an input layer, two hidden layers, and an output

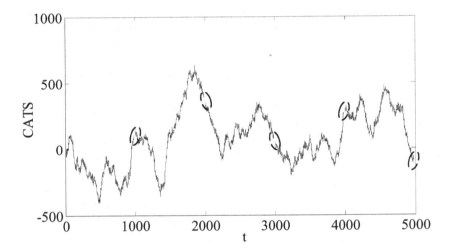

Figure 6.4.1: Plot of the CATS time series consisting of 5000 given data points with 100 missing data, indicated by the dashed ovals.

layer, as shown in Figure 6.4.2.

The context layer inputs in this network are obtained from the outputs of the first hidden layer units. The hidden unit activation pattern is copied verbatim through weight connections set equal to 1 and stored in the context units. The context layer has 40 neurons, the input layer has 100 neurons, the first hidden layer has 40 neurons, the second hidden layer has 20 neurons, and the output layer has 5 neurons. Neurons between adjacent layers are fully connected, as indicated by the bold arrows in Figure 6.2.2. The transfer functions used for the neurons in the two hidden layers and the output layer are

$$\text{tansig}(x) = \frac{e^x - e^{-x}}{e^x + e^{-x}}.$$

The number of weights and biases, i.e., N_w, in such a RNN (size: $100 \times 40 \times 20 \times 5$) was 6605, and hence τ was 0.0784 (cf. Eq. (6.2.3)).

The dynamics of the system is determined partly by the autonomous development and partly by external system environment inputs. In real-world applications, like time series prediction, the information of external environment is usually incomplete, or even unavailable. In literature, different architectures are proposed to address this problem [66]. Some approach utilizes error between system output and observation as an indicator of external influence. Others separate the state equation of recurrent network into a past and a future part. Under the assumption of a constant environment, the observation and expectation data are presented, respectively, to compensate the partially known external drivers. To be consistent in both training and predicting, the recurrent network used in this chapter employs its own predictions as future inputs to replace the unknown future observations. In addition, intermediate states, where no external input is presented to the system between two time steps, are implemented to improve the autonomous development of the dynamic, and furthermore the system stability.

6.4.2 Simulation Result

Training

Different numbers of neurons in the input/hidden layers have been tried to determine the best structure of the RNN for the time series prediction. Regarding the input layer, it is feasible that the number of original data is larger than that of the predicted data. The dominance of the original data will dilute the error introduced by the predicted data. Increasing the number of neurons in the hidden layers may benefit the system dynamic, but in return deal with more computational burden. The model shown in Figure 6.4.2 is selected as the baseline structure. The number of neurons in one layer is tuned with the neurons in rest of the layers frozen.

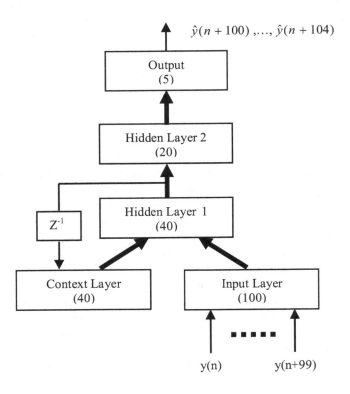

Figure 6.4.2: Elman's recurrent neural network architecture for CATS. The number of neurons in each layer is listed in parentheses. Bold arrows indicate fully connections between layers. The inputs $y(n)$ may contain previous prediction $\hat{y}(n)$.

Table 6.4.1 summaries a few of the trials carried out in determining the best RNN structure.

Five identical RNNs (size: $100 \times 40 \times 20 \times 5$) are used for prediction on the 5 data sets, each containing 980 samples. The input vector is composed of original samples and the network's previous predictions. At the first round, 100 consecutive original training samples are fed to the network to predict the next 5 points. When the prediction is available, the first 5 data points in the input vector are discarded, and RNN outputs, i.e., the predictions, are concatenated to the end of the input vector. This queue-like operation is continued until there are 20 predictions. The next round of training starts with another 100 consecutive original samples, whose starting point is 20 samples away from that of the previous 100, supplied to the RNN. The last 120 samples, 100 as inputs and 20 as targets, are allocated for testing purpose. When the RNN is well trained and tested, the last 100 samples are presented to predict the missing samples (see Figure 6.4.3). Batch training method is used here, and the weights are updated based on a cumulative error. The process is repeated over a number of epochs.

For the convenience, the original data are normalized. The mean value of the original data is subtracted first. Such a zero mean sequence is then divided by its maximum absolute value to fit between -1 and 1.

Based on previous research in literature [27, 53, 56], the PSO parameters are set to $w = 0.8$, $c_1 = 2.0$, $c_2 = 2.0$, respectively. The value of $v_{\max}$ for PSO is set to 2.0. 40 particles are used for the time series prediction.

Besides training the RNNs with the original time series, the RNNs are also trained with the series of sequence differences, defined as

$$y^*(n) = y(n) - y(n-1),$$

where the $y(n)$ and $y(n-1)$ are obtained from the given time series data. This small scale dynamic signal is fed into RNNs the same way as the original data. A process of smoothing is applied for estimation of the main trend of the CATS. The smoothed series consists of the mean value of data and its nearest neighbors to the left and to the right. The deviation prediction by sequence difference is added to the main trend for the final prediction.

Results

Due to the assumption that missing data are only affected by recent samples, only a portion of 980 samples in each set is used for training. To find the best size of the training set, segments with different length are tested, shown in Figure 6.4.4. The simulation indicates that the employing 240 samples, i.e., 601st - 940th, are enough for maximizing accuracy and minimizing the computational burden. The

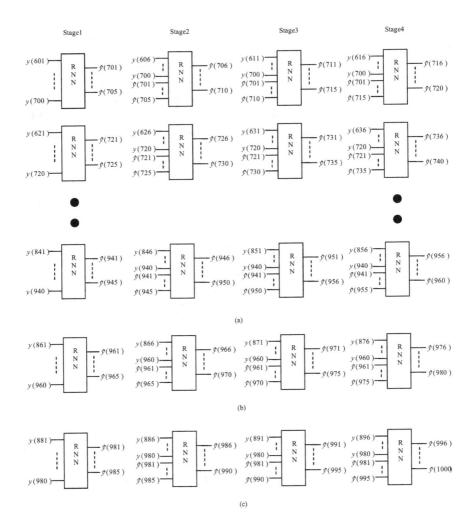

Figure 6.4.3: Illustration of RNN training, testing and predicting. (a) The RNN training procedure on one data set. (b) The RNN testing procedure on the same data set. (c) The prediction of the 20 missing samples for the same data set in (a) and (b). $y(n)$ and $\hat{y}(n)$ denote the original value and the RNN predicted values, respectively.

Table 6.4.1: System performance for different RNN structures based on the fifth training segment, i.e., 4601st–4860th, of original sequence data.

Number of neurons in the input layer	60	80	100	120
MSE after 5000 generations	3140	1741	393	746
Number of neurons in the hidden layer #1	20	30	40	50
MSE after 5000 generations	1057	777	393	652
Number of neurons in the hidden layer #2	10	15	20	25
MSE after 5000 generations	1803	746	393	497

Table 6.4.2: The MSE and standard deviation (STD) of prediction for both original sequence and sequence difference on training and validation sets

	Training set, i.e., 4601st–4860th		Validation set, i.e., 4861st–4880th	
	MSE	STD	MSE	STD
Prediction with original sequence data set	393	40	484	22
Prediction with sequence difference data set	102	10	157	12

corresponding predictions on the training set are 701st - 960th. The last 100 samples, i.e., 861st–960th, are set side for verifying the network prediction on 961st–980th. Figure 6.4.5 shows the training errors on one data set, i.e., 4601st–4860th, for both the standard PSO and the hybrid PSO-EA. It's clear that the hybrid algorithm outperforms the standard PSO. Therefore, we employ the hybrid PSO-EA for the rest of the training and prediction. The MSE of the prediction is 351 on 100 missing points. The predictions can be further improved by optimizing the PSO parameters as explained in [16, 59].

The training of sequence difference samples is organized the same way as above. In Figure 6.4.6, the training performance on the two different data sets is compared. The sequence difference training sample provides a better estimation for the time series (see Table 6.4.2). Thus, the sequence difference data sets are used for training the RNNs and prediction. Figure 6.4.7 clearly demonstrates the improvement of prediction based on sequence difference. The MSE and standard deviation of the prediction on the five training sets and five validation sets are listed in Table 6.4.3. Finally, the predictions and the errors on the five missing

Table 6.4.3: The MSE and standard deviation of prediction on five training sets and five validation sets, base on sequence difference

	Training data sets				
	601st–860th	1601st–1860th	2601st–2860th	3601st–3860th	4601st–4860th
MSE	123	94	79	70	102
Standard deviation	11	10	9	8	10
	Validation sets				
	861st–880th	1861st–1880th	2861st–2880th	3861st–3880th	4861st–4880th
MSE	135	169	194	242	157
Standard deviation	12	13	14	16	12

Table 6.4.4: The MSE of prediction on 100 missing data base on sequence difference

Missing data	981st–1000th	1981st–2000th	2981st–3000th	3981st–4000th	4981st–5000th	Total
MSE	139	173	675	371	395	351

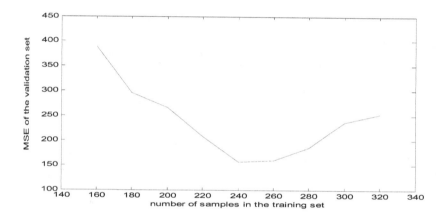

Figure 6.4.4: System performance on different length of training segments. The RNN is trained with different number of sequence difference samples in the fifth segment of the time series. The MSEs are all calculated on the validation set, i.e., 4861st–4880th, with respective to different training length.

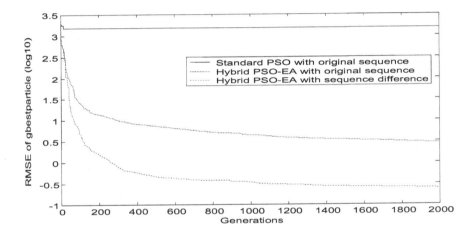

Figure 6.4.5: Training error for the hybrid PSO-EA and the standard PSO algorithms. The errors reflect the performance of the best particle, i.e., the P_g, at each generation for the fifth training data set. For hybrid, both original sequence and sequence difference data are used.

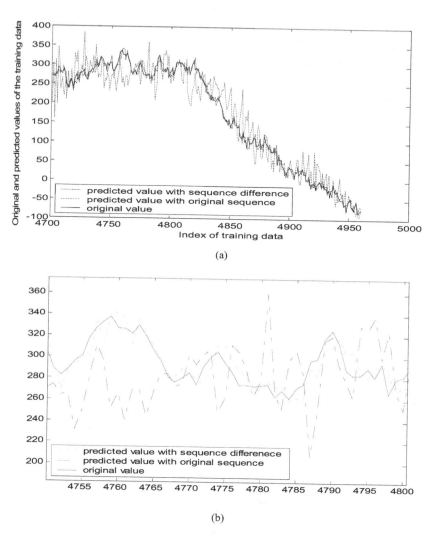

(a)

(b)

Figure 6.4.6: Hybrid PSO-EA predictions by the *gbest* particles from both the original sequence training set and the corresponding sequence difference one after 5000 generations. (a) shows the prediction on the whole data set, and (b) is a shorter segment of (a), which clearly illustrates the curves for original data, predicted values from original sequence and predicted values from sequence difference, and their tendencies.

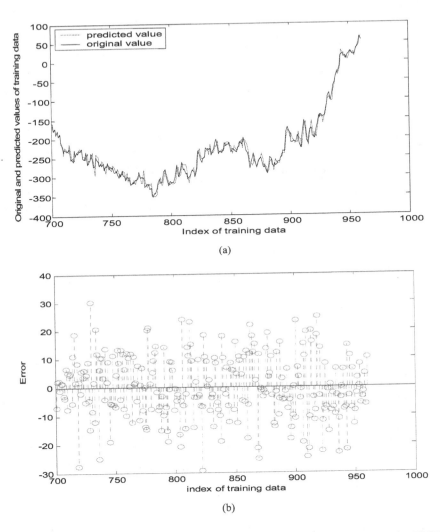

(a)

(b)

Figure 6.4.7: Prediction on the first training data set, i.e., 701st - 960th. The RNN is trained by hybrid PSO-EA with sequence difference data. The prediction is given by the gbest particle after 5000 generations. (a) shows the predicted values on the whole data set, and (b) is the corresponding errors.

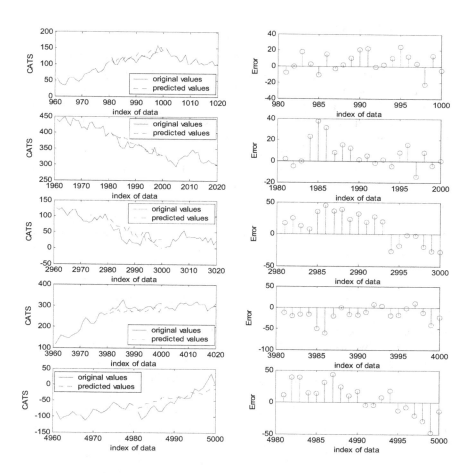

Figure 6.4.8: Missing values, the approximations and the corresponding errors. The first column in the figure demonstrates the prediction by best trained RNN with hybrid PSO-EA on the five gaps of missing data. The original values for the missing data are also presented for comparison. The errors of prediction are illustrated on the second column, in the corresponding rows, in the figure.

gaps are presented in Figure 6.4.8, along with the original values [36]. The MSE of the prediction is presented in Table 6.4.4.

6.5 Conclusions

An innovative algorithm, based on a hybrid of PSO and EA, has been presented and discussed in this chapter for training two popular neural network architectures, the feedforward and recurrent networks, for challenging problems. The combination of the search capabilities of these two global optimization methods has been explored. The hybrid algorithm inherits both cooperative and competitive characteristics from PSO and EA methods. It shares searching information within the promising individuals, which leads to faster convergence, and simultaneously replaces the losers with the offspring of elites, who head to new territories in the search space. The purpose of applying EA mutation to PSO is to increase the diversity of the population, and enable the PSO to escape local minima.

The hybrid PSO-EA algorithm is used to train a swarm neuro-engine using a feedforward neural network as the evaluation function of the leaf nodes in a game tree, with zero expertise involved. After generations of self-play training, results of tournaments among hybrid PSO-EA, PSO, and EA show that the hybrid PSO-EA approach performs better than both PSO and EA. The hybrid engine is also trained and tested to play with a defensive knowledge-based Capture Go player. The hybrid engine manages to expand and fortify its own territory, instead of capturing the opponent's stone, to countervail the defensive strategies of the hand-coded player. Two winning games by the swarm neuro-game engine demonstrate that it balances the goals of occupying territory and capturing stones, as well.

The hybrid PSO-EA learning algorithm is successful in training RNN for the time series prediction. Given a low correlation, the difference of the sequence may better reflect the properties of the series than the original sequence at short term. We pre-process the time series to obtain the sequence difference samples and save them for training, which works better in the simulations. Experimental results show that recurrent neural networks, trained by the hybrid algorithm, are able to predict the missing values in the time series with minimum error, in comparison with those trained with standard EA and PSO algorithms.

Further investigation on the hybrid PSO-EA learning algorithm will focus on the parameter setting of (w, c_1, c_2) and communication topology of the particles. Instead of keeping them fixed, other dynamically changing scheme for the parameters [16, 59], if well designed, can lead to promising performance. One example is to employ hierarchical PSO [15], using the outer PSO for parameter optimization and the inner one for problem optimization. Setting a proper propagation delay for the global best position through different communication topologies, such as tours and cluster, may help to avoid the local minima with the price of convergence speed. Benchmark problems will be employed to verify the significance of the algorithm.

Acknowledgements

The authors gratefully acknowledge the support from the National Science Foundation under grants ECS # 0300526 and ECS # 0348221, the M. K. Finley Missouri endowment and the University of Missouri-Research Board.

Bibliography

[1] L. V. Allis, H. J. van den Herik, and I. S. Herschberg, "Which games will survive?" in *Heuristic Programming in Artificial Intelligence 2-The Second Computer Olympiad*, D. N. L. Levy and D. F. Beal, Editors, Ellis Horwood, 1991, pp. 232–243.

[2] P. J. Angeline, "Using selection to improve particle swarm optimization," *Proc. IEEE Congress on Evolutionary Computation*, 1998, pp. 84–89.

[3] R. Axelrod, *The Evolution of Cooperation*, New York: Basic Books, 1984.

[4] R. Battiti, "First and second order methods of learning: between the steepest decent and Newton's method," *Neural Computation*, vol. 4, no. 2, pp. 141–166, 1992.

[5] E. Berlekamp and D. Wolfe, *Mathematical Go: Chilling Gets the Last Point*. A .K. Peters, Ltd, 1994.

[6] T. Blackwell and P. Bentley, "Don't push me! Collision-avoiding swarms," *Proc. 2002 Congress on Evolutionary Computation*, 2002, vol. 2, pp. 1691–1696.

[7] X. Cai and D. Wunsch II, "Evolutionary computation in playing Capture Go game," *Proc. of ICNNS'04*, Boston, 2004.

[8] X. Cai, N. Zhang, G. K. Venayagamoorthy, and D. C. Wunsch II, "Time series prediction with recurrent neural networks using hybrid PSO-EA algorithm," *Proc. INNS-IEEE International Joint Conference on Neural Networks (IJCNN)*, Budapest, Hungary, July 2004, vol. 2, pp. 1647–1652,

[9] X. Cai, G. K. Venayagamoorthy, and D. C. Wunsch II, "Evolutionary swarm neural network game engine for Capture Go," *IEEE Trans. on Evolutionary Computation*, under revision.

[10] X. Cai, N. Zhang, G. K. Venayagamoorthy, and D. C. Wunsch II, "Time series prediction with recurrent neural networks using hybrid PSO-EA algorithm," *Neurocomputing*, accepted for publication, 2005.

[11] K. Chellapilla and D. B. Fogel, "Evolution, neural networks, games, and intelligence," *Proc. IEEE*, Special Issue on Computational Intelligence, vol. 87, no. 9, pp. 1471–1496, 1999.

[12] K. Chellapilla and D. Fogel, "Evolving neural networks to play checkers without relying on expert knowledge," *IEEE Trans. on Neural Networks*, vol. 10, no. 6, pp. 1382–1391, Nov. 1999.

[13] K. Chellapilla and D. Fogel, "Evolving an expert checkers playing programs without using human expertise," *IEEE Trans. on Evolutionary Computation*, vol. 5, no. 4, pp. 422–428, Aug. 2001.

[14] X. Chen, D. Zhang, X. Zhang, Z. Li, X. Meng, S. He, and X. Hu, "A functional MRI study of high-level cognition II. The game of Go," *Cognitive Brain Research*, vol. 16, no. 1, pp. 32–37, 2003.

[15] S. Doctor, G. K. Venayagamoorthy, and V. G. Gudise, "Optimal PSO for collective robotic search applications," *Proc. IEEE Congress on Evolutionary Computation*, 2004, vol. 2, pp. 1390–1395.

[16] S. Doctor and G. K. Venayagamoorthy, "Improving the performance of particle swarm optimization using adaptive critic design," *Proc. IEEE Swarm Intelligence Symposium*, 2005, pp. 393–396.

[17] R. Durbin, S. Eddy, A. Krogh, and G. Mitchison, *Biological Sequence Analysis: Probabilistic Models of Proteins and Nucleic Acids*, Cambridge: The Cambridge University Press, 1998.

[18] R. Eberhart and Y. Shi, "Particle swarm optimization: developments, applications and resources," *Proc. 2001 Congress on Evolutionary Computation*, 2001, vol. 1, pp. 81–86.

[19] J. L. Elman, "Finding structure in time," *Cognitive Science*, vol. 14, pp. 179–211, 1990.

[20] J. L. Elman, "Distributed representations, simple recurrent networks, and grammatical structure," *Machine Learning*, vol. 7, pp. 195–225, 1991.

[21] H. D. Enderton, "The Golem Go program," Tech. Rep. CMU-CS-92-101, Carnegie Mellon University, 1991.

[22] M. Enzenberger, "The integration of a priori knowledge into a Go playing neural network," 1996. Available: http://www.markus-enzenberger.de/neurogo.ps.gz

[23] D. Fogel, *Blondie24: Playing at the Edge of AI*, San Francisco, CA: Morgan Kaufmann, 2002.

[24] D. Fotland, "The 1999 FOST (fusion of science and technology) cup world open computer championship," Tokyo, 1999. Available: http://www.britgo.org/results/computer/fost99htm

[25] J. Fürnkranz. "Machine learning in games: A survey," in *Machines that Learn to Play Games*, J. Fürnkranz and M. Kubat, Eds., Huntington, NY: Nova Scientific Publishers, 2001, Chapter 2, pp. 11–59.

[26] F. Gomez and R. Miikkulainen, "Incremental evolution of complex general behavior," *Adaptive Behavior*, vol. 5, pp. 317–342, 1997.

[27] V. G. Gudise and G. K. Venayagamoorthy, "Comparison of particle swarm optimization and backpropagation as training algorithms for neural networks," *Proc. IEEE Swarm Intelligence Symposium*, 2003, pp. 110–117.

[28] S. Haykin, *Neural Networks: A Comprehensive Foundation*, 2nd Ed., New Jersey: Prentice Hall, 1999.

[29] F. Hsu, *Behind Deep Blue*, Princeton, NJ: Princeton Univ. Press, 2002.

[30] A. Jazwinski, *Stochastic Processes and Filtering Theory*, Academic Press, 1970.

[31] J. Kennedy and R. Eberhart, "Particle swarm optimization," *Proc. IEEE International Conference on Neural Networks*, 1995, vol. 4, pp. 1942–1948.

[32] J. Kennedy, R. Eberhart and Y. Shi, *Swarm Intelligence*, San Meteo: Morgan Kaufmann, 2001.

[33] J. F. Kolen and S. C. Kremer, Eds., *A Field Guide to Dynamic Recurrent Networks*, New York: IEEE Press, 2001.

[34] G. Konidaris, D. Shell, and N. Oren, "Evolving neural networks for the capture game," *Proc. SAICSIT Postgraduate Symposium*, Port Elizabeth, South Africa, 2002. Available from: http://www-robotics.usc.edu/~dshell/res/evneurocapt.pdf

[35] A. Krogh and P. Sollich, "Statistical mechanics of ensemble learning," *Physical Review E*, vol. 55, no. 1, pp. 811–825, 1997.

[36] A. Lendasse, "Time series prediction competition: The CATS benchmark," available at http://www.cis.hut.fi/~lendasse/competition.html

[37] J. M. Smith, *Evolution and the Theory of Games*. Cambridge, UK: Cambridge University Press, 1982.

[38] J. McNames, J. Suykens, and J. Vandewalle, "Winning entry of the K. U. Leuven time-series prediction competition," *International Journal of Bifurcation and Chaos*, vol. 9, no. 8, pp. 1485–1500, 1999.

[39] D. Mechner, "All systems Go," *Science*, vol. 38, no. 1, 1998.

[40] P. Morris, *Introduction to Game Theory*, New York: Springer-Verlag, 1994.

[41] M. Muller, "Decomposition search: A combinatorial games approach to game tree search, with applications to solving Go endgames," *Proc. of IJ-CAI*, vol. 1, pp. 578–583, 1999.

[42] J. von Neumann and O. Morgenstern, *Theory of Games and Economic Behavior*, Princeton, NJ: Princeton Univ. Press, 1944.

[43] J. B. Pollack and A. D. Blair, "Co-evolution in the successful learning of Backgammon strategy," *Machine Learning*, vol. 32, pp. 226–240, 1998.

[44] G. V. Puskorius and L. A. Feldkamp, "Neurocontrol of nonlinear dynamical systems with Kalman filter trained recurrent networks," *IEEE Trans. Neural Networks*, vol. 5, no. 2, pp. 279–297, 1994.

[45] A. P. Refenes, A. Zapranis, and G. Francis, "Stock performance modeling using neural networks: A comparative study with regression models," *Neural Networks*, vol. 7, no. 2, pp. 375–388, 1994.

[46] N. Richards, D. Moriarty, P. Mc Questen, and R. Miikkulainen, "Evolving neural networks to play Go," *Applied Intelligence*, vol. 8, pp. 85–96, 1998.

[47] E. W. Saad, D. V. Prokhorov, and D. C. Wunsch II, "Comparative study of stock trend prediction using time delay, recurrent and probabilistic neural networks," *IEEE Transactions on Neural Networks*, vol. 9, no. 6, pp. 1456–1470, 1998.

[48] http://www.schachverein-goerlitz.de/Foren/Fun/Go/go.htm

[49] J. Schaeffer, *One Jump Ahead: Challenging Human Supremacy in Checkers*, New York: Springer, 1996.

[50] N. Schraudolph, P. Dayan, and T. Sejnowski, "Temporal difference learning of position evaluation in the game of Go," *Advances in Neural Information Processing*, vol. 6, pp. 817–824, 1994.

[51] H. P. Schwefel, *Evolution and Optimum Seeking*, New York: Wiley, 1995.

[52] C. E. Shannon, "Automatic chess player," *Scientific American*, vol. 182, no. 48, 1950.

[53] Y. Shi and R. C. Eberhart, "Parameter selection in particle swarm optimization," *Proc. 1998 Annual Conference on Evolutionary Computation*, Mar. 1998.

[54] Y. Shi and R. C. Eberhart, "Fuzzy adaptive particle swarm optimization," *Proc. 2001 Congress on Evolutionary Computation*, 2001.

[55] Y. Shi and R. C. Eberhart, "Particle swarm optimization with fuzzy adaptive inertia weight," *Proc. Workshop on Particle Swarm Optimization*, 2001.

[56] Y. Shi, "Particle swarm optimization," *IEEE Connections*, the newsletter of the IEEE Computational Intelligence Society, vol. 2, no. 1, pp. 8–13, 2004.

[57] A. Smith, *The Game of Go*, Tokyo, Japan: Charles Tuttle Co., 1956.

[58] M. Tahk and B. Sun, "Coevolutionary augmented Lagrangian methods for constrained optimization," *IEEE Trans. on Evolutionary Computation*, vol. 4, no. 2, pp. 114–124, 2000.

[59] G. K. Venayagamoorthy, "Adaptive critics for dynamic particle swarm optimization," *Proc. IEEE International Symposium on Intelligent Control*, pp. 380–384, 2004.

[60] A. Weigend and N. Gershenfeld, Eds., *Time Series Prediction: Forecasting the Future and Understanding the Past*, Reading, MA: Addison-Wesley, 1994.

[61] P. J. Werbos, "Backpropagation through time: What it does and how to do it," *Proc. IEEE*, vol. 78, no. 10, pp. 1550–1560, 1990.

[62] D. H. Wolpert and W. G. Macready, "No free lunch theorems for optimization," *IEEE Trans. on Evolutionary Computation*, vol. 1, no. 1, pp. 67–82, Apr. 2001.

[63] X. Yao, "Evolving artificial neural networks," *Proc. IEEE*, vol. 87, no. 9, pp. 1423–1447, 1999.

[64] R. Zaman, D. V. Prokhorov, and D. C. Wunsch II, "Adaptive critic design in learning to play the game of Go," *Proc. International Joint Conference on Neural Networks*, Houston, TX, 1997, vol. 1, pp. 1–4.

[65] H. G. Zimmermann and R. Neuneier, "Neural network architectures for the modeling of dynamical systems," *A Field Guide to Dynamic Recurrent Networks*, J. F. Kolen and S. C. Kremer, Eds., New York: IEEE Press, 2001, pp. 311–350.

[66] H. G. Zimmermann, R. Grothmann, A. M. Schafer, and C. Tietz, "Modeling large dynamical systems with dynamical consistent neural networks," *New Directions in Statistical Signal Processing: From Systems to Brain*, S. Haykin *et al.*, Eds., Cambridge, MA: MIT Press, 2005.

Chapter 7

Modular Wavelet-Fuzzy Networks

Yuetong Lin and Fei-Yue Wang

Abstract: This chapter presents a novel approach to combining wavelet networks and multilayer feedforward neural network for fuzzy logic control systems. While most of the existing neuro-fuzzy systems focus on implementing the Takagi-Sugeno-Kang (TSK) fuzzy inference model, they fail to keep the knowledge structure that is critical in interpreting the learning process and providing insights to the working mechanism of the underlying systems. It is our intention to utilize individual subnets to implement decision-making process of the fuzzy logic control systems based on the Mamdani model. Center average defuzzification has seen its implementation by a neural network so that a succinct network structure is obtained. More importantly, wavelet networks have been adopted to provide better locality capturing capability and therefore better performance in terms of learning speed and training time. Offline orthogonal least squares method is used for training the wavelet subnets and the overall system is updated using the steepest descent algorithm.

7.1 Introduction

The last few decades have witnessed increasing efforts in the applications of the fuzzy logic (FL) and neural networks (NN) for control of dynamic systems. Yet as indicated by numerous literatures, the most promising perspective is to combine FL and NN in a synergistic way so that both can reap the benefits from the strengths of the other in system modeling and analysis while overcome individual limitations. Research in this field has resulted in many innovative designs that include: network representation of membership functions (MF), fuzzy associative memories, hybrid neurofuzzy controllers, fuzzy set and neural-network-based in-

215

terpolative reasoning, and neural-network-based architecture for fuzzy logic control and decision systems, etc. [1–11, 18, 27, 28].

However, it has been observed that most of these work have led to the loss of the original knowledge structures embedded in a fuzzy logic control systems (FLCS) by its network implementation. The knowledge structure of a system usually provides us important information about the pattern and organization of its decision-making process. Unfortunately, the significance of this idea has been ignored by different hybridization schemes mostly because under mild assumption, a multilayer neural network with two hidden layers is capable of matching any input-output relationship with any desired accuracy, so that the knowledge structure of an FLCS might not strike as important as long as one just wants to capture the input/output behavior of the system by a neural network.

Keeping the knowledge structure in the neural networks implementation is important due to the following reasons:

First of all, it is always desirable to know not only the external input/output relationship of a system, but also the structural information of its internal decision-making process, e.g., how many distinctive stages are involved during the process, and the furthermore, what is the stage(s) responsible for the system performance improvement or degeneration, etc. The structural knowledge must be an indispensable part of the understanding of a system;

Secondly, for supervised learning, one needs a large number of numerical input/output pairs to train a network in order to achieve the desired system performance. For control systems however, such desired input/output are generally not available at least during the early stage of the system development, but common sense, heuristics, or even expert skills can often be acquired easily. This kind of knowledge usually is associated with specific operations or procedures and is organized into some well-conceived structure or pattern. Thus it would be very useful to reserve this structure of knowledge and transplant it into a neural network so that the heuristics can be transferred and refined through the learning capability of the neural network as the system evolves. This also enables a neural network to conduct its learning in a distributed fashion, i.e., its subnets can learn their own functions independently according to the knowledge structure, thus speeding up the network learning. The whole process is very similar to that of training a human operator: at the beginning, he/she is taught with general knowledge and basic procedures in an off-line fashion and then is expected to gain sophisticated task skills later based on his/her on-line, hands-on experience.

For an FLCS, its decision-making process can be broken down into three steps: sensor pattern recognition (i.e., association of sensor reading with membership functions of linguistic sensory patterns), approximate reasoning (i.e., fuzzy implications and compositional rules), and control synthesis (i.e., calculation of fuzzy control actions and the corresponding procedure of defuzzification). The problem to be addressed in this chapter is to show a systematic approach to implementing the FLCS with hybrid networks while preserving the knowledge structure of the FLCS in its network implementation. In other words, our goal is to establish a one-to-one mapping between FLCS and hybrid networks, i.e., 1) a hybrid networks can be set up directly from an FLCS description, and 2) an FLCS can be obtained directly from the hybrid networks [25, 26].

The rest of the chapter is organized as follows. Section 7.2 summarizes briefly the theoretical background for wavelet networks. Section 7.3 introduces the modular design paradigm, followed by an in depth discussion on subnet structure for fuzzy logic systems based on the knowledge structure and fuzzy inference mechanism. Section 7.4 derives the learning algorithms for both offline and online training of the networks. Section 7.5 provides simulation results illustrating the efficacy of the aforementioned approach. Section 7.6 concludes the chapter.

7.2   Wavelet Networks

Wavelet networks [17, 19, 22] were first proposed by Zhang *et al.* [20, 21] and are built on the recent development of the wavelet theory for function approximation. The core of this new type of network is to replace the family of base functions in the widely studied radial basis functions (RBF) network with an orthonormal basis. Unlike the center and the variance of the RBF network, the hierarchy and the thresholds of the wavelet network as well as the number of the hidden nodes are closely related to the wavelet theory. Moreover, contrary to the redundant expression of the RBF network, the optimal expression of the orthogonal wavelet network is unique and thus less computationally intensive.

There are two schemes of constructing wavelet network, both founded on the discrete wavelet transform. For $f \in L^2(R)$, we have

$$f(t) = \sum_{m,n=-\infty}^{\infty} \left(f(t), \psi_{m,n}(t) \right) \psi_{m,n}(t) \tag{7.2.1}$$

where $\psi_{m,n}(x)$ is mother wavelet expressed as

$$\psi_{m,n}(x) = a_0^{-m/2} \psi \left(a_0^{-m} x - n b_0 \right) \tag{7.2.2}$$

with both m and $n \in Z$. For computational efficiency, $a_0 = 2$ and $b_0 = 1$ are commonly chosen so that results lead to a binary dilation of 2^{-m} and a dyadic translation of $n2^m$.

Another popular scheme of decomposing $f(t) \in L^2(R)$ is through father wavelet or the scaling function,

$$f(t) = \sum_n (f(t), \varphi_{M,n}(t)) \, \varphi_{M,n}(t) + \sum_{m \geq M, n} (f(t), \varphi_{m,n}(t)) \, \varphi_{m,n}(t) \quad (7.2.3)$$

where $\varphi(t)$ is the father wavelet that has the following relation with mother wavelet

$$\varphi(t) = \sqrt{2} \sum_k h_k \psi(2t - k),$$

$$\psi(t) = \sqrt{2} \sum_k g_k \varphi(2t - k).$$

The latter scheme demonstrates $f(t)$ can be approximated arbitrarily close by selecting a sufficiently large M such that for any $\varepsilon > 0$

$$\left\| f(t) - \sum_k (f(t), \varphi_{M,k}) \varphi_{M,k} \right\| < \varepsilon. \quad (7.2.4)$$

The approximation by the truncated wavelet decomposition can be expressed by

$$f(t) \approx \sum_k (f, \varphi_{M,k}) \, \varphi_{M,k} = \sum_k C_n \varphi_{M,k}. \quad (7.2.5)$$

It is readily ascertained that this expression has the same structure for a 3-layer neural networks. A prototype of wavelet network can be seen in [19].

7.3   Modular Structure of Wavelet-Fuzzy Networks

Modularity may be viewed as a manifestation of the principle of divide and conquer, which permits us to solve a complex computational task by dividing it into simpler subtasks and then combining their individual solutions [16].

In a typical control application especially in industry environment, control system is required to implement separate sensing, planning and execution phases, and to resolve complexity at various levels of understanding. Specifically, the fuzzy logic control system serves those design requirements by embodying fuzzifier, inference engine and defuzzifier. The process of implementing wavelet-fuzzy networks (WFN) can be divided into three subtasks, i.e., construction of pattern recognition (PR), fuzzy reasoning (FR), and control synthesis (CS) networks. These networks must be integrated in such a way that the structure and decision-making process of the original fuzzy system can be fully retrieved from its network implementation [1].

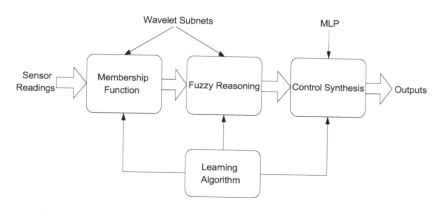

Figure 7.3.1: A hybrid wavelet-fuzzy network.

Figure 7.3.1 shows the hybrid wavelet-fuzzy networks. It consists of three subnets of distinctive functions. The first network identifies patterns of input variables in terms of membership functions of linguistic terms, and the second one conducts fuzzy reasoning (conjunction) by calculating the firing strength of the decision rules. The third carries out the task of control synthesis by generating fuzzy control action and then defuzzifying it. Although three networks are connected sequentially, it is important to point out that the construction and training of these networks can be performed independently and simultaneously, and the decision-making procedure in an FLCS is fully preserved in its network implementation.

7.3.1 Pattern Recognition Wavelet Networks

A wavelet network is constructed to capture the association of each of the sensor readings with signal patterns through membership functions of linguistic variables. An expert's knowledge is needed about the number of linguistic terms and the membership functions of each term for s_i. The expert only needs to provide the value of the membership grades on a subset X that is the collection of sample points in the universe of discourse of s_i. Experience of the designer with these types of systems will be useful in the selection of the points.

For each signal reading s_i, a wavelet network SN_i is constructed to match its values with the linguistic terms in the set of signal patterns A_i. In other words, the function of SN_i is to calculate membership functions $\mu_{S_i^k}(x)$ for $k = 1, \cdots, p_i$, $i = 1, \cdots, m$

The network is an approximation of the membership function μ_{s_i}, for $i =$

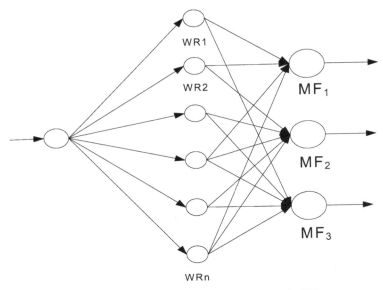

Figure 7.3.2: A two-layer wavelet network SN_i.

$1, \cdots, p_i$. The output of network is $S_i = (s_i^1, \ s_i^2, \cdots, s_i^{p_i})$ for SN_i. Nodes in the output layers are term nodes that act as membership functions to represent the terms of the respective linguistic variables. Figure 7.3.2 shows a two-layer wavelet network SN_i for this purpose.

In our scheme, one subnet is devoted to one sensor reading. At this stage, network SN_i is not required to learn the memberships in high accuracy since these specified membership functions are usually very subjective. Note that if two sensor readings have the identical set of linguistic terms, they can use the same network at the beginning. Through network learning, the membership functions of linguistic terms can be changed adaptively later for better performance. Nodes of input layer in the subnet are the value that represents input linguistic variables.

7.3.2 Fuzzy Reasoning Wavelet Networks

A wavelet network is constructed to calculate the rule strengths for each of the control rules based on the output of the pattern recognition wavelet network. For each decision rule r in the knowledge base of the FLCS, the wavelet network $RN_r, r = 1, \cdots, R$ is used to calculate the firing strength of the rule. Thus, RN_r is actually a network implementation of conjunction operator.

Figure 7.3.3 presents a 2-layer network RN_r. The input of the network is

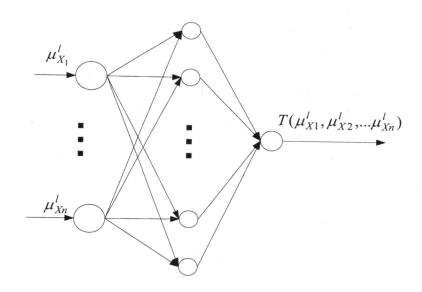

Figure 7.3.3: A two-layer network RN_r.

$X = (\mu_{s_i^1},\ \mu_{s_i^2}, \cdots, \mu_{s_i^{p_i}})$ where $\mu_{s_i^1}$ is the membership function of state variable i. The output of network is $Y = \alpha_r$ that is the rule strength of rule r. The output α_r is used in the construction of control synthesis network to be described. By changing its weights, this network could implement Minimum, Algebraic product, Bounded product, or their triangular norms as a conjunction operator. Therefore the initial training of RN_r can be carried out by using any of these norms, or even their combinations, and the network can be easily modified for new fuzzy reasoning by the use of learning algorithms.

Clearly, as long as every rule has the same number of linguistic terms in its precondition, we can choose the same fuzzy reasoning network for all the control rules at the initial stage. Note no input and output scaling is need for network RN_r since both its input and output are ranging from 0 to 1.

7.3.3 Control Synthesis Wavelet Network

Control synthesis is the process of determining the final crisp control according to the firing strength of rules and membership functions of linguistic terms defined for control actions. It involves steps of deducing consequences for individual rules, generating resultant fuzzy control, and then converting it into a crisp value. The implementation of the defuzzification process especially the widely-used center of average (COA) method however, has always been a daunting task for neural networks. This remains as the dominant cause why the majority of neuro-fuzzy networks favor the Takagi-Sugeno-Kang (TSK) inference model. In our work a variant of the COA, the center average defuzzifier (CA) is considered instead. The CA approximates the COA defuzzification by taking the weighted average of the centers of the M fuzzy sets, with the weights equal the heights of the corresponding fuzzy sets. Specifically, let σ_Y^l be the center of the lth output membership function and w_l be its height, the CA defuzzifier determines the defuzzification output y^* as:

$$y^* = \frac{\sum_{l=1}^{M} w_l \sigma_Y^l}{\sum_{l=1}^{M} w_l}. \tag{7.3.1}$$

Simple as it looks, CA has been proven to be a good approximation of COA [24], but computation is much simpler. Continuity in defuzzification output is another plausible resemblance to human operated control while smooth output is always desirable. The "weighting factor" that is composed of the product of firing strength and height, can more accurately reflect the contribution of a rule to the fuzzy output. Since only two parameters in each MF are involved in defuzzification process, network learning can now be deployed to update MF to accommodate the demand. It is also worth pointing out that the universal approximation property of fuzzy logic system is proved by Wang with CA defuzzification [24].

Here is how the CA defuzzification is implemented in the control synthesis subnet. The key point is to make both height and center of the MF's adjustable. Since (7.3.1) involves the inner product of two vectors, it is more convenient and efficient to use the plain multilayer perceptrons (MLP) in this subnet. Also note that (7.3.1) can be expressed in the form of norm product, i.e.,

$$y^* = \frac{\sum\limits_{l=1}^{M} \sigma_Y^l w_l}{\sum\limits_{l=1}^{M} w_l} = \sum\limits_{l=1}^{M} \bar{w}_l \sigma_Y^l \tag{7.3.2}$$

where

$$\bar{w}_l = \frac{w_l}{\sum\limits_{l=1}^{M} w_l} \tag{7.3.3}$$

represents the normalized weight.

This subnet has two layers, and the normalized product further simplifies the structure: first layer is responsible for calculating the weighting factor w_l, since we use algebraic product as the T-norm for inference, w_l should be the product of the firing strength of the lth rule and the height of the corresponding fuzzy set, namely:

$$w_l = \gamma_l \sup(\mu_Y^l), \tag{7.3.4}$$

where γ_l is the lth output from the fuzzy reasoning subnet or the firing strength of the lth rule, so link weights are set to be $\sup(\mu_Y^l)$, and the neurons in this layers simply perform "purelin" function that directly maps the input to output; second layer only needs one neuron that takes in the weighted input from the preceding layer and generates the norm product, which is also the final defuzzification output. The link weights between these two layers ought to be defined as the centers of the MF's, namely σ_Y^l. Figure 7.3.4 demonstrates this subnet with five rules and one defuzzification output.

7.3.4 Integration of the Sub-Wavelet Networks

Integration of the wavelet networks and MLP's is constructed by connecting the pattern recognition, fuzzy reasoning, and control synthesis networks according to the given set of fuzzy control rules. Once networks SN_i, RN_r and CN_j have been created, the final step toward a structured hybrid networks is to connect those networks appropriately according to the original FLCS.

Figure 7.3.5 presents the modular wavelet-fuzzy networks for a fuzzy logic controller with 9 rules, 2 outputs, and 2 inputs with 3 linguistic terms for each.

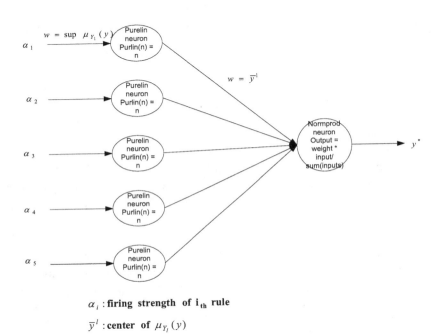

α_i : **firing strength of i_{th} rule**

$\overline{y}^l$: **center of** $\mu_{Y_l}(y)$

Figure 7.3.4: A subnet with five rules and one defuzzification output.

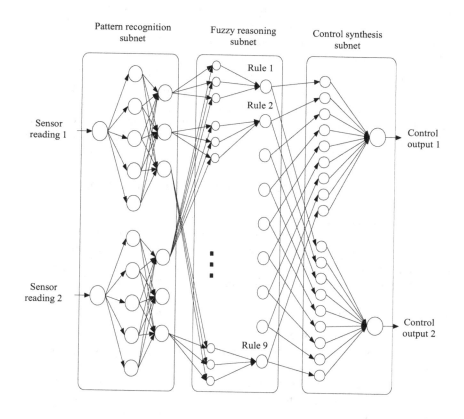

Figure 7.3.5: A modular wavelet-fuzzy networks for a fuzzy logic controller with 9 rules, 2 outputs, and 2 inputs.

It is very clear that the whole computation process of the integrated neural network can be divided into three stages: pattern recognition, fuzzy reasoning, and control synthesis. Therefore, this design scheme leads to a structure resulting in a functional interpretation for each subnet and a structure of flow of knowledge thereof.

Although the structured hybrid networks can be viewed as ordinary multilayer feedforward networks with sensor readings as its input and control actions as its output, the distinctive knowledge structure embedded within this system makes it different from other neural network implementations of fuzzy logic control systems. Note that as multilayer networks, the Wavelet-Fuzzy Networks is not a fully connected one. For example, its subnet SN_i is linked only to input node s_i and has no connection with other input nodes (i.e., connection weights are zero), and subnet RN_r is linked only to one output node of subnet SN_i.

Since the whole decision-making procedure of the FLCS is preserved in the networks, by breaking it up into subnets of pattern recognition, fuzzy reading, and the control synthesis, the actual modifications in membership functions of sensor readings, operators of fuzzy reasoning, membership functions of control actions, and methods of defuzzification, can be recovered separately from this multilayer networks implementation. Therefore, the network-based implementation provides a mechanism for the refinement of fuzzy logic based control systems, a problem that has not been addressed effectively with the original context of fuzzy logic controls. As long as the original FLCS works reasonably, no additional training is required in order to put the structured hybrid system to work, since its performance should be at least as good as the original FLCS from which the networks is constructed. Additional training and online learning by the networks though, can improve its performance and thus lead to the refinement of the existing control rules, and even generation of the new rules for the original FLCS.

7.3.5 Simplified Versions of Wavelet-Fuzzy Networks

The fuzzy reasoning subset mainly implements the T-norm operators. While dedicating one wavelet network for one conjunction brings the flexibility of achieving multiple T-norms with a single structure, it comes at price of increasing network dimensions and learning complexity. In view of the fact that in many applications, the T-norm operator for calculating the firing strength is fixed and needs no adjustment, we replace the fuzzy reasoning subnet with a direct "product" neuron. The "product" neuron uses the multiplication as the node transfer function so that the gradient descent algorithm is feasible for updating associated parameters. Such replacement brings some variations to the whole topology, and thus results in more succinct structures.

We start with Figure 7.3.5 which shows the closest descendant of the prototype topology. Figure 7.3.6 shows the hybrid networks for a two-input, nine-rule, and one-output fuzzy logic controller. The pattern recognition subnet has the identical structure; the last two layers perform the same CA defuzzification as the previous model. However, it can be seen from the diagram that the most significant change takes place in the first layer of the control synthesis subnet in the previous model, where each neuron therein acts as a "purelin" type that takes in the product of the firing strength and the height consequent MF has been spared and merged with the fuzzy reasoning layer (not subnet) in this topology. To keep the functionality unchanged, all synapses that connect to the same neuron in this layer are assigned

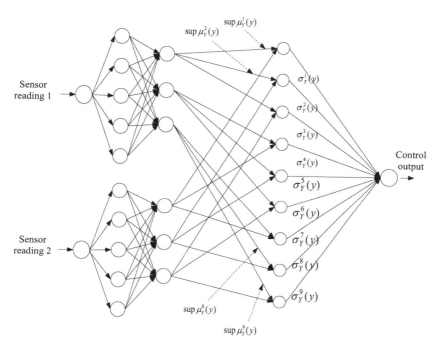

Figure 7.3.6: A hybrid network for a two-input, nine-rule, and one-output fuzzy logic controller.

with the same link weight, $\sup \mu_Y^l(y)$, which is the height for the lth MF of output linguistic variable Y. Each neuron in this layer multiplies the weighted inputs, output can be expressed as:

$$x_{out,l}^3 = (\sup \mu_Y^l(y)) \times \mu_{X_1}^l(x_1) \times \mu_{X_2}^l(x_2) \qquad (7.3.5)$$

where l is the index for the rule base, X_1, X_2, Y are the linguistic variables for two inputs and output respectively.

By constructing fuzzy system in this manner, we preserve the logical structure and empirical knowledge in the hybrid networks due to the physical significance of the link weights. More importantly, a 7-layer network has been simplified to only 5-layer and a faster adaptation of the network parameters is readily expected.

One variation of Figure 7.3.6 is to insert the height of output MF's, $\sup(\mu_Y^l)$ as the bias inputs to the fuzzy reasoning layer, as demonstrated in Figure 7.3.7. Instead of updating both weights connecting the outputs of the two PR subnets and

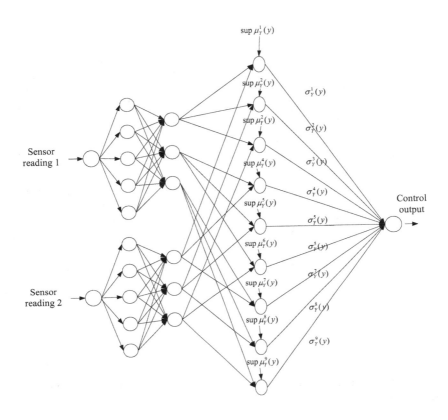

Figure 7.3.7: A variation of Figure 7.3.6.

the neurons in FR layer, this scheme allows only the bias vector to be adjusted. The training advantage can be exemplified by the following numerical example: for the 5-layer structure above that has 9 rules and thus requires 9 neurons in the FR layer, all 18 weights have to be updated in each iteration because each neuron is hooked up with 2 output neurons in FR subnets. However, in the following diagram, even though the number of links is unchanged, but only 9 biases need to be adjusted, and the convenience becomes even significant when the number of rules gets larger.

The next scheme comes from different network implementation of the CA defuzzification. So far we have been using the 'normprod' as the weight function connecting to the output neuron. As will be demonstrated in next chapter, the deepest descent updating rule for 'normprod' function is rather complicated. To

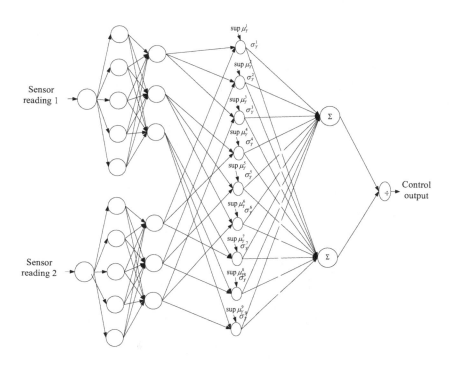

Figure 7.3.8: A structure with additional layer consisting of two neurons.

ease this inconvenience, we thus propose an additional layer consisting of two neurons to take care of this calculation; detail structure is given in Figure 7.3.8.

As shown in the diagram, the two neurons deal with the numerator $\sum_l w^l \sigma_Y^l$ and denominator $\sum_l w^l$ of (7.3.2) for CA defuzzification individually. The link weights are thus set to be σ_Y^l's for the upper neuron, and all 1's for the lower neuron. The output neuron performs a 'division' operation. Following this design, the weighting functions in this subnet are the standard 'dotprod', so their gradients are straightforward. Finally, the derivative of output neuron activation function, 'division', is also readily obtainable when the two inputs and output signals are available.

This final alternation of the original implementation is the most concise one following the 'division' approach for CA defuzzification. Here the output of the lower neuron, $\sum_l w^l$, is not treated as an individual input, but as the bias factor

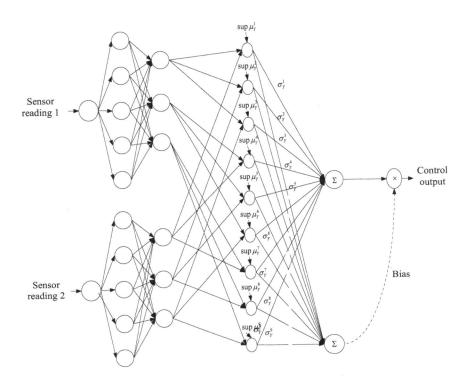

Figure 7.3.9: A hybrid network structure.

to the division neuron. Therefore, the topology is literally simplified to only one summing joint in the first layer of the CS subnet. This scheme combines the brevity of network structure and keeps the deepest descent algorithm feasible, so it will be our model for future discussion including the derivation of learning algorithms and simulation results. Figure 7.3.9 shows this hybrid structure.

7.4 Learning Algorithm for Wavelet-Fuzzy Networks

The hybrid structure with both local receptive field type wavelet networks and sigmoid type MLP involved demands the learning algorithm should have the feature of both types of networks considered, and take advantage of the modular structure that allows the functionally independent subnets to be trained separately.

In general, a modular network fuses supervised and unsupervised learning paradigms in a seamless fashion. In our case, the unsupervised learning is embedded within the decision process and knowledge structure of the fuzzy control: the different subnets work collaboratively instead of competitively, therefore there is no need for additional mediating mechanism.

There are two modes of supervised learning depending upon the availability and source of the training data, namely, off-line and on-line training. In off-line learning (also referred to as batch learning), the update formula for parameter is based on the mean squared error and the update takes place only after the whole training data set has been presented. On-line learning, on the contrary, updates the parameter according to the instant squared error and the update occurs immediately after each input-output pair has been provided. When the two paradigms being applied to our wavelet-fuzzy networks, the subnets can be trained in the off-line mode because the training data should be obtained in a sufficient amount from the a priori knowledge solicited from the domain experts. After the entire system is assembled, the adjusting of the system parameters will be conducted in the on-line fashion.

7.4.1 Orthogonal Least Square Method for Wavelet Subnets Training

The orthogonal least squared (OLS) algorithm [23] is an efficient implementation of the forward stepwise model selection technique for linear regression models. Roughly speaking, in order to select a subset model, all the regressors of the initial model are evaluated to determine how each regressor will contribute to the modeling of the desired system. The regressor selected to be included the subset is the one that will contribute most significantly together with the regressors already selected to model the desired system. The selection process is repeated until the required number of regressors is found. In [20, 21], we can find in-depth introduction on the procedures for applying OLS algorithm to train RBF and wavelet network.

One concern that has to be addressed is the choice between wavelet frame and wavelet bases, or what type of wavelet lattice would be considered as the regressors. Wavelet bases are resulted from the discretization of the continuous mother wavelet. Theoretically, it has been shown a $f \in L^2(R)$ can be expressed as the superposition of the orthonormal wavelet bases with the weighting factors to be the inner product between function itself with wavelet bases. The beauty of orthonormal wavelet bases is the inverse wavelet transform comes with very fast algorithms for estimating the reconstruction coefficients. However, the original mother wavelet before discritization has to satisfy strong conditions in order to guarantee the generation of orthonormal bases. It is more straightforward and convenient to adopt wavelet frame. Normally referred to as "redundant basis," wavelet frame abandons the orthogonality of the wavelet bases to achieve more flexibility on the selection of the wavelet regressors. The tradeoff for such flexibility though, is to lose the fast algorithms associated with orthonormal wavelet bases.

There are two approaches for generating wavelet frame from discretizing continuous mother wavelets: single-scaling and multi-scaling. For single scaling wavelet frame Ψ of $L^2(R^d)$ is of the form

$$\Psi(m,n) = \left\{ \psi_{m,n}(\mathbf{x}) = \alpha^{-\frac{1}{2}dn}\psi(\alpha^{-n}\mathbf{x} - m\beta) : n \in \mathbf{Z}, m \in \mathbf{Z}^d \right\} \quad (7.4.1)$$

where α and β are the scaling and translation step sizes respectively, and ψ represents the continuous mother wavelet. It has been shown it is possible to generate single-scaling wavelet frames of $L^2(R^d)$ with a single mother wavelet, in contrast with $2^d - 1$ mother wavelets required for the orthonormal wavelet bases using single-scaling. It is obvious that single-scaling frame is structurally simpler to implement, not just comparing with wavelet bases, but also with the multi-scaling method. The fact that the most commonly used mother wavelet radial functions are also by nature single-scaling makes it appealing too. However, sticking with one scale factor for all dimensions definitely weakens the multi-resolution property of the wavelet decomposition, and may hamper the approximation accuracy and learning speed when it is applied to wavelet networks. Conditions have been given in [20, 21] on the type of mother wavelet that can be used to construct wavelet frame. The 1-D Mexican Hat function that is defined as

$$\psi(x) = (1 - x^2)e^{-x^2/2}$$

is one frequently referred prototype.

Wavelet networks are adopted in two of the subnets in the proposed wavelet-fuzzy networks, i.e., pattern recognition and fuzzy inference. Both nonlinear mappings of these two subnets are not sophisticated and of predominantly basic shapes, it is unnecessary to complicating the issue with multiple-scaling for different dimensions, therefore only the single-scaling regression is considered for training of these subnets.

Finally, the control synthesis subnet is implemented by a plain two-layer MLP based on the CA defuzzification algorithm. The first layer has the number of neurons equal to that of the fuzzy rules, its main task is to compute the height of the output membership functions weighted by the firing strength coming from the fuzzy inference subnet. The second layer only contains one neuron with norm product capability. No training is necessary because the adjustable parameters in this subnet are just the internal weights that are set to be the centers of the output membership functions, and external weights linking control synthesis and fuzzy inference subnets, which are set to be the heights of the output membership functions.

7.4.2 Steepest Descent Method for Weights Updating

When the individual subnets are constructed, it is up to the steepest descent method to take care of on-line training by adjusting the link weights connecting wavelets or neurons in adjacent layers, and biases related with neurons.

For the sake of simplicity, we restrict our discussion on single output networks. Similar results can be readily generalized to multiple outputs case. We also only focus on the simplified version of the hybrid system as shown in Figure 7.3.9. A summary of notations is first presented below.

- The algorithm is applied to minimize the energy function representing the instantaneous error defined as

$$E = \frac{1}{2}e^2 = \frac{1}{2}(d - y)^2,$$

  where symbol d represents desired output, and y is the output of the network.

- The indices i, j and k refer to different neurons in the networks; with signals propagating through the network from left to right, neuron j lies in a layer to the right of neuron i, and so on.

- The symbol $w_{ji}^{(l)}$ denotes the synaptic weight of link connecting the ith neuron in the $(l-1)$th layer to the jth neuron in the lth layer; in particular for layer 3, we use the following symbols to distinguish the neurons in different PR subnets: $w_{ji_x}^{(3)}$ denotes the weight connecting jth neuron in layer 3, and i_xth neuron in the upper subnet in layer 2 where $i_x = 1, 2$, or 3; similarly $w_{ji_y}^{(3)}$ denotes the weight connecting jth neuron in layer 3, and i_yth neuron in the lower subnet in layer 2 where $i_y = 1, 2$, or 3. For output layer 4, the symbol is simplified to $w_{1i}^{(4)}$.

- x_i, $i = 1, 2$: WFN inputs.

- φ^l: node transfer function (activation function) of lth layer.

- The induced local field (i.e., weighted sum of all synaptic inputs plus bias) of neuron j is denoted by v_i^l; it constitutes the signal applied to the activation function associated with neuron j.

- u_j^l: output signal from the jth neuron in the lth layer.

- Symbol m_l denotes the size in layer l.

A. Using 'Normprod' for CA Defuzzification

We revisit the system diagram shown in Figure 7.3.9. For convenience of derivation, all the parameters have been standardized to the notations adopted above.

1. For output layer weights $\omega_{1i}^{(4)}$

$$\Delta\omega_{1i}^{(4)} = -\eta\frac{\partial E}{\partial\omega_{1i}^{(4)}} = -\eta\frac{\partial E}{\partial y}\frac{\partial y}{\partial v^4}\frac{\partial v_4}{\partial\omega_{1i}^{(4)}}.$$

Individually,

$$\frac{\partial E}{\partial y} = \partial\left(\frac{1}{2}e^2\right)\Big/\partial y = e\frac{\partial e}{\partial y} = -e$$

$$\frac{\partial y}{\partial v^4} = \frac{\partial\varphi^4(v^4)}{\partial v^4} = \varphi'(v^4) = 1, \quad\text{since } \varphi^4(x) = x$$

$$\frac{\partial v^4}{\partial\omega_{1i}^{(4)}} = \frac{\partial\left(\sum_k\omega_{1k}^{(4)}u_k^3\Big/\sum_k u_k^3\right)}{\partial\omega_{1i}^{(4)}} = \frac{u_i^3}{\sum_k u_k^3}.$$

Therefore,

$$\Delta\omega_{1i}^{(4)} = \eta e\frac{u_i^3}{\sum_k u_k^3}.$$

2. For hidden layer 3 weights $\omega_{ji_x}^{(3)}$ and $\omega_{ji_y}^{(3)}$

$$\Delta\omega_{ji_x}^{(3)} = -\eta\frac{\partial E}{\partial\omega_{ji_x}^{(3)}} = -\eta\frac{\partial E}{\partial y}\frac{\partial y}{\partial v^4}\frac{\partial v^4}{\partial u_j^3}\frac{\partial u_j^3}{\partial v_{ji_x}^3}\frac{\partial v_{ji_x}^3}{\partial\omega_{ji_x}^{(3)}}.$$

Individually,

$$\frac{\partial E}{\partial y} = -e, \quad\text{and } \frac{\partial y}{\partial v^4} = 1$$

$$\frac{\partial u_j^3}{\partial v_{ji_x}^3} = \frac{\partial\varphi_j^3(v_{ji_x}^3, v_{ji_y}^3)}{\partial v_{ji_x}^3} = \frac{\partial(v_{ji_x}^3 v_{ji_y}^3)}{\partial v_{ji_x}^3} = v_{ji_y}^3, \quad\text{because } \varphi^3 = v_{i_x}^3 v_{i_y}^3$$

$$\frac{\partial v^4}{\partial u_j^3} = \frac{1}{\partial u_j^3} \frac{\sum_k \omega_{1k}^4 u_k^3}{\sum_k u_k^3} = \frac{\omega_{1j}^4 \sum_k u_k^3 - \sum_k \omega_{1k}^4 u_k^3}{\left(\sum_k u_k^3\right)^2}$$

$$= \frac{\omega_{1j}^4}{\sum_k u_k^3} - \frac{\sum_k \omega_{1k}^4 u_k^3}{\left(\sum_k u_k^3\right)^2} = \frac{\omega_{1j}^4}{\sum_k u_k^3} - \frac{v^4}{\sum_k u_k^3}$$

$$= \frac{\omega_{1j}^4 - v^4}{\sum_k u_k^3}.$$

Note: The weight function connecting layer 3 and 4 is set to be the so-called "normprod" defined as $\sum_k \omega_k p_k \big/ \sum_k p_k$, where ω_k's are the weights, and p_k's represent the source inputs. We have

$$\frac{\partial v_{ji_x}^3}{\partial w_{ji_x}^3} = \frac{\partial (w_{ji_x}^3 u_{i_x}^2)}{\partial w_{ji_x}^3} = u_{i_x}^2.$$

Therefore,

$$\Delta \omega_{ji_x}^{(3)} = -\eta \frac{\partial E}{\partial \omega_{ji_x}^{(3)}} = \eta e v_{ji_y}^3 u_{i_x}^2 \frac{\omega_{1j}^4 - v^4}{\sum_k u_k^3} = \eta e \frac{(\omega_{1j}^4 - v^4) u_j^3}{\omega_{ji_x}^{(3)} \sum_k u_k^3}.$$

3. Local error signals at output of layer 2 neurons

To update the pattern recognition subnets, the local error information is needed. Since PR subnets are all of feedforward structure, the local error signal at each neuron, $\delta_j^{(i)}$ is defined as:

$$\delta_j^{(i)} = \frac{\partial E}{\partial v_j^{(i)}}$$

where $v_j^{(i)}$ denotes the induced input signal to the jth neuron in the ith layer.

In order to derive $\delta_j^{(i)}$ for layer 2, we need to obtain $\delta_j^{(i)}$ for layer 3 in advance. From

$$\frac{\partial E}{\partial v_{ji_x}^3} = \frac{\partial E}{\partial y} \frac{\partial y}{\partial v^4} \frac{\partial v^4}{\partial u_j^3} \frac{\partial u_j^3}{\partial v_{ji_x}^3}$$

and based on the previous results, we have

$$\frac{\partial E}{\partial v_{ji_x}^3} = -e \frac{\omega_{1j}^4 - v^4}{\sum_k u_k^3} v_{ji_y}^3.$$

So, the gradient with respect to the output from i_xth in layer 2 to jth in layer 3 is:

$$\frac{\partial E}{\partial u_{ji_x}^2} = \frac{\partial E}{\partial v_{ji_x}^3} \frac{\partial v_{ji_x}^3}{\partial u_{ji_x}^2} = \frac{\partial E}{\partial v_{ji_x}^3} w_{ji_x}^3 = -e \frac{(\omega_{1j}^4 - v^4)u_j^3}{\left(\sum_k u_k^3\right) u_{ji_x}^2}.$$

However, from Figure 7.3.9 we know every output neuron in layer 2 has connections with two neurons in layer 3. Therefore, the overall output gradient at these neurons should be:

$$\frac{\partial E}{\partial u_{i_x}^2} = \sum_m \frac{\partial E}{\partial u_{mi_x}^2}$$

where $m \in$ neurons in layer 3 that receive input from i_xth neuron in layer 2.

Now we can derive the local error signal $\delta_{i_x}^2$:

$$\delta_{i_x}^2 = -\frac{\partial E}{\partial v_{i_x}^2} = -\frac{\partial E}{\partial u_{i_x}^2} \frac{\partial u_{i_x}^2}{\partial v_{i_x}^2} = -\left(\sum_m \frac{\partial E}{\partial u_{mi_x}^2}\right)(\varphi^2)' = -\sum_m \frac{\partial E}{\partial u_{mi_x}^2}$$

since the net transfer function $\varphi^2(x) = x$.

4. *For hidden layer 2 weights $\omega_{j_x i}^2$ and input weights ω_{ji}^1*

We can follow the standard backpropagation (BP) algorithm for the weights updating rules for these two layers, i.e., for $\omega_{j_x i}^2$

$$\Delta\omega_{j_x i}^2 = \eta\delta_{j_x}^2 u_{j_x i}^2,$$

$$\omega_{j_x i}^2(k+1) = \omega_{j_x i}^2(k) + \Delta\omega_{j_x i}^2 = \omega_{j_x i}^2(k) + \eta\delta_{j_x}^2 u_{j_x i}^2,$$

and for ω_j^1,

$$\Delta\omega_j^1 = \eta\delta_j^1 u_j^1,$$

$$\delta_j^1 = -\frac{\partial E}{\partial v_j^1} = -\sum_{k=1}^{n(i_x)} \delta_k^2 \omega_{kj}^2 (\varphi^1)',$$

$$\omega_j^1(k+1) = \omega_j^1(k) + \Delta\omega_j^1 = \omega_j^1(k) + \eta\delta_j^1 u_j^1.$$

Note: the for layer 1 weights, subscript i that indicating the source node is all omitted because for every PR subnet, the input layer only has one neuron.

5. *For bias b_j^3*

$$\Delta b_j^3 = -\eta\frac{\partial E}{\partial b_j^3} = -\eta\frac{\partial E}{\partial y}\frac{\partial y}{\partial v^4}\frac{\partial v^4}{\partial u_j^3}\frac{\partial u_j^3}{\partial b_j^3}.$$

We have already know most of the terms in the chain. The only new term is $\partial u_j^3/\partial b_j^3$,

$$\frac{\partial u_j^3}{\partial b_j^3} = \frac{\partial\varphi(v_{ji_x}, v_{ji_y}, b_j^3)}{\partial b_j^3} = \frac{\partial(v_{ji_x} v_{ji_y} b_j^3)}{\partial b_j^3} = v_{ji_x} v_{ji_y} = v_j^3.$$

Therefore,

$$\Delta b_j^3 = -\eta \frac{\partial E}{\partial b_j^3} = \eta e \frac{\omega_{1j}^4 - v^4}{\sum_k u_k^3} v_j^3.$$

Accordingly, the following changes are made regarding the updating rules for hidden layer 2 weights $\omega_{j_x i}^2$ and input weights ω_{ji}^1. Still, for layer 2:

$$\delta_{i_x}^2 = -\frac{\partial E}{\partial v_{i_x}^2} = -\frac{\partial E}{\partial u_{i_x}^2}\frac{\partial u_{i_x}^2}{\partial v_{i_x}^2} = -\sum_m \frac{\partial E}{\partial u_{mi_x}^2}(\varphi^2)' = -\sum_m \frac{\partial E}{\partial u_{mi_x}^2}$$

and

$$\frac{\partial E}{\partial u_{ji_x}^2} = \frac{\partial E}{\partial v_{ji_x}^3}\frac{\partial v_{ji_x}^3}{\partial u_{ji_x}^2}.$$

But, $\partial v_{ji_x}^3 / \partial u_{ji_x}^3 = 1$ because $v_{ji_x}^3 = u_{ji_x}^3$ under this scheme, and

$$\frac{\partial E}{\partial v_{ji_x}^3} = \frac{\partial E}{\partial y}\frac{\partial y}{\partial v^4}\frac{\partial v^4}{\partial u_j^3}\frac{\partial u_j^3}{\partial v_{ji_x}^3} = -e\frac{\omega_{1j}^4 - v^4}{\sum_k u_k^3}\frac{\partial u_j^3}{\partial v_{ji_x}^3}$$

with $u_j^3 = \varphi(v_{ji_x}^3, v_{ji_y}^3, b_j^3) = v_{ji_x}^3 v_{ji_y}^3 b_j^3$,

$$\frac{\partial u_j^3}{\partial v_{ji_x}^3} = v_{ji_y}^3 b_j^3.$$

Therefore,

$$\delta_{i_x}^2 = -\sum_m \frac{\partial E}{\partial u_{mi_x}^2} = -\sum_m -e\frac{\omega_{1m}^4 - v^4}{\sum_k u_k^3}(v_{mi_y}^3 b_m^3).$$

B. Using Division for CA Defuzzification

We only seek to derive the BP algorithm for the scenario where the division operation of the CA is done by inserting the denominator, $\sum_l \gamma_l \sup(\mu_Y^l)$ as the bias input to the output neuron. However, this bias needs not to be updated using the gradient method. In fact, there is no error signal backpropagated through the bias path to the lower summing joint in layer 4, and further through the all 1's synaptic weights to the neurons in layer 3 because they are either fixed (all 1's weights) or computed in the forward training process (the bias input to layer 4) in every epoch.

Most results in the above are still valid for this approach, and we only need to apply the steepest descent method for the weights updating in layer 3, namely $\partial E / \partial w_{1j}^3$ and the local error at layer 3, δ_j^3

1. For $\partial E/\partial w_{1i}^3$

$$\frac{\partial E}{\partial w_{1i}^4} = \frac{\partial E}{\partial y} \frac{\partial y}{\partial v^5} \frac{\partial v^5}{\partial u^5} \frac{\partial u^5}{\partial v^4} \frac{\partial v^4}{\partial w_{1i}^4}.$$

We know from the above that $\partial E/\partial y = -1$, and because $\varphi^4(x) = x$, we have $v^4 = u^5$. From $y = \varphi(v^5, b^5) = v^5/b^5$, we obtain

$$\frac{\partial y}{\partial v^5} = \frac{1}{b^5}.$$

Also, it is straightforward to have

$$\frac{\partial v^5}{\partial u^5} = 1,$$

because $w^5 = 1$ leads to $v^5 = w^5 u^5$.

Finally, from $v^4 = \sum_l w_{1l}^4 u_l^4$, we have

$$\frac{\partial v^4}{\partial w_{1i}^4} = u_i^4.$$

Therefore, to summarize, we obtain

$$\frac{\partial E}{\partial w_{1i}^4} = -\frac{1}{b^5} u_i^4.$$

2. For δ_j^3

From definition,

$$\delta_j^3 = -\frac{\partial E}{\partial u_j^3},$$

and

$$\frac{\partial E}{\partial u_j^3} = \frac{\partial E}{\partial v^4} \frac{\partial v^4}{\partial u_j^3},$$

it is readily obtainable

$$\frac{\partial v^4}{\partial u_j^3} = w_{1j}^3.$$

So,

$$\frac{\partial E}{\partial v^4} = -\frac{1}{b^5}.$$

Therefore,

$$\delta_j^3 = -\frac{\partial E}{\partial u_j^3} = \frac{w_{1j}^3}{b^5}.$$

7.5 Simulation Results

We now demonstrate simulation results for applying wavelet-fuzzy networks.

As an illustrative example, consider the problem of modeling a nonlinear function known as sinc or the Mexican Hat function shown in Figure 7.5.1. Using grid points in the range $[-5, 5] \times [-5, 5]$ within the input space, 121 training data are obtained first.

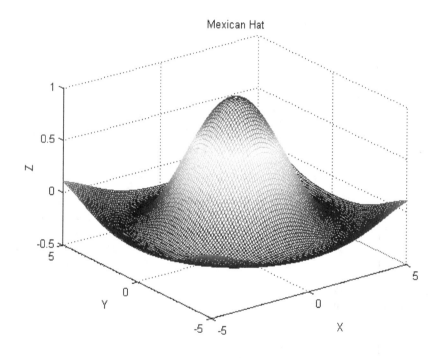

Figure 7.5.1: The Mexican Hat function.

With the heuristic knowledge from this figure, a fuzzy logic system is constructed as followings. First we use three fuzzy sets {SM = Small, ME = Medium, LA = Large} for x and y, and five fuzzy sets {NE = Negative, NS = Negative Small, ZE = Zero, PM = Positive Medium, PL = Positive Large} for z. The domain intervals for x and y are assumed to be $[-5, 5]$, and $[-0.2, 1.0]$ for z. The membership functions for these are defined in Figure 7.5.2. Then, the relationship between (x, y) and z are described with a rough rule set:

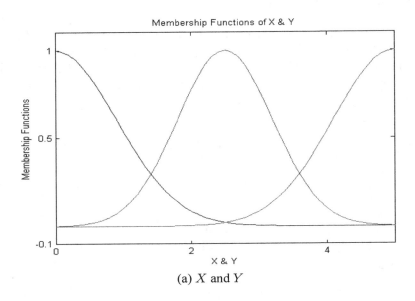

(a) X and Y

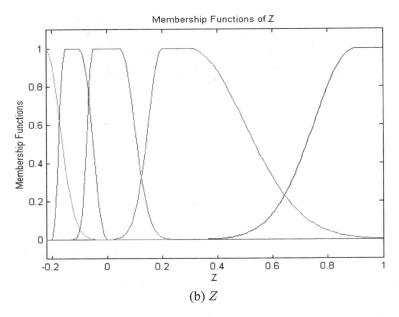

(b) Z

Figure 7.5.2: Membership functions for input and output linguistic variables.

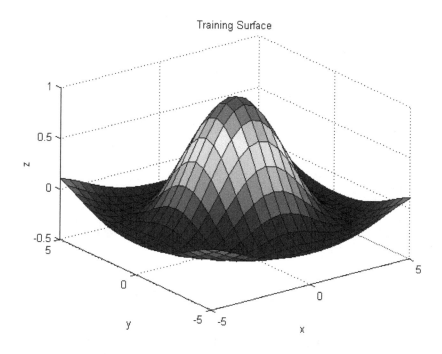

Figure 7.5.3: Reconstructed Mexican Hat.

Rule 1: If x is SM and y is SM then z is PL;
Rule 2: If x is SM and y is ME then z is PM;
Rule 3: If x is SM and y is LA then z is NE;
Rule 4: If x is ME and y is SM then z is PM;
Rule 5: If x is ME and y is ME then z is NE;
Rule 6: If x is ME and y is LA then z is NS;
Rule 7: If x is LA and y is SM then z is NE;
Rule 8: If x is LA and y is ME then z is NS;
Rule 9: If x is LA and y is LA then z is ZE.

Using the wavelet-fuzzy networks, we can reconstruct the surface of Mexican Hat as illustrated in Figure 7.5.3. After 100 epochs of training, the reconstruction fairly resembles the original surface with SSE is brought down to 0.18. The slope is matched pretty well. Nonetheless, there is visible inaccuracy on the top of the hat. Also the recovery of low value range of z is a little rough too. These can be

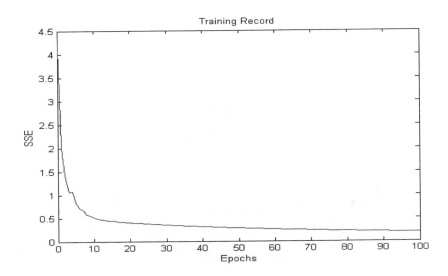

Figure 7.5.4: Convergence of the training error.

attributed to the inadequacy of the original knowledge solicited about the function itself. For the top of the hat area where z varies drastically, and the low boundary where values of z are very subtle, more linguistic variables and rules are needed to catch the fine details and nature of the underlying model. This is illustrated in Figure 7.5.4 by the training record where the convergence starts off pretty fast but makes little progress as it approaches the set goal.

We now show the updated membership functions of the input and output linguistic variables after training in Figure 7.5.5. There are some seemingly violations of the properties for the updated membership functions, but the trends of change are adequately visible. We can see the three MF's are not significantly different as we previously defined because the overlap regions become larger, but MF for 'small' shows a faster decreasing orientation as x approaches larger values. This proves again we need finer partition of the boundary areas.

Since we use 'centroid' of membership functions of the output linguistic variable in the CA defuzzification, the simulation results in Figure 7.5.6 only show the updated 'centroid'. The Table 7.5.1 reflects the changes.

What we can interpret from Table 7.5.1 and Figure 7.5.6 is: there is no significant difference in 'NE' and 'NS'; a new linguistic variable, which could be named as 'PS' or positive small, should be inserted to describe the consequent of rule 5

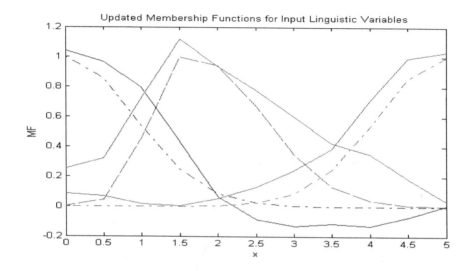

Figure 7.5.5: Updated membership functions of the input and output linguistic variables after training.

Table 7.5.1: Numerical values for 'centroid' before and after training

	PL	PM	NE	PM	NE
Before	0.8420	0.3633	−0.1112	0.3633	−0.1112
After	1.3050	0.5795	−0.2232	0.5795	0.0435
	NS	NE	NS	ZE	
Before	−0.1832	−0.1112	−0.1832	0.0212	
After	−0.2087	−0.2232	−0.2087	0.0713	

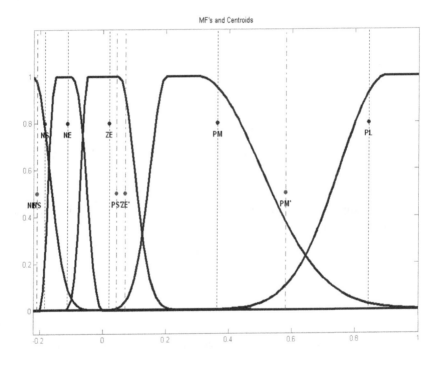

Figure 7.5.6: The updated 'centroid'.

that is originally labeled as 'NE'; and more variables should be added in the high value range of z as both 'PM' and 'PL' have been moved toward right.

7.6   Conclusions

In this chapter, a hybrid network implementation for fuzzy logic controller (FLCS) referred to as wavelet-fuzzy networks is proposed based on the modular networks doctrine and the internal knowledge structure of FLCS. Wavelet network has been chosen to construct the pattern recognition and fuzzy inference subnets where both exhibit highly localized property, and plain MLP is used for the control synthesis subnet where output empirical knowledge is well represented by the synaptic weights. Orthogonal least square method is adopted for the offline training of wavelet subnets. The entire network is updated by the steepest descent algorithm to optimize the performance.

Bibliography

[1] L. A. Zadeh, "Soft computing and fuzzy logic," *IEEE Transaction on Software*, vol. 11, no. 6, pp. 48–56, 1994.

[2] N. Sinha and M. M. Gupta, *Soft Computing and Intelligent Systems: Theory and Applications*, New York: Academic Press, 2000.

[3] D. Rustkowska, *Neuro-Fuzzy Architectures and Hybrid Learning*, Physica-Verlag, 2003.

[4] C.-T. Lin and C. S. G. Lee, *Neural Fuzzy Systems: A Neuro-Fuzzy Synergism to Intelligent Systems*, Englewood Cliffs, NJ: Prentice Hall, 1995.

[5] C.-T. Lin and C. S. G. Lee, "Neural-network-based fuzzy logic control and decision system," *IEEE Trans. Computers*, vol. 2, no. 1, pp. 46–63, 1994.

[6] J.-S. Jang, C.-T. Sun, and E. Mizutani, *Neuro-Fuzzy and Soft Computing: A Computational Approach to Learning and Machine Intelligence*, Englewood Cliffs, NJ: Prentice Hall, 1993.

[7] B. Kosko, *Neural Networks and Fuzzy Systems: A Dynamical Systems Approach to Machine Intelligence*, Englewood Cliffs, NJ: Prentice-Hall, 1990.

[8] E. Czogala and J. Leski, *Fuzzy and Neuro-Fuzzy Intelligent Systems*, Physica-Verlag, 1998.

[9] S. Mitra and Y. Hayashi, "Neuro-fuzzy rule generation: survey in soft computing framework," *IEEE Trans. Neural Networks*, vol. 11, no. 3, pp. 748–768, 2000.

[10] J. J. Shann and H. C. Fu, "A fuzzy neural network for rule acquiring on fuzzy control systems," *Fuzzy Sets and Systems*, vol. 71, no. 3, pp. 345–357, 1995.

[11] J. J. Buckley and Y. Hayashi, "Fuzzy neural networks: A survey," *Fuzzy Sets and Systems*, vol. 66, no. 1, pp. 1–13, 1994.

[12] H. Ishibuchi, R. Fujioka, and H. Tanaka, "Neural networks that learn from fuzzy if-then rules," *IEEE Trans. Fuzzy Systems*, vol. 1, no. 1, pp. 85–97, 1993.

[13] C. C. Lee, "Fuzzy logic in control systems: Fuzzy logic controller-Part I," *IEEE Trans. Systems, Man and Cybernetics*, vol. 20, no. 2, pp. 404–418, 1990.

[14] C. C. Lee, "Fuzzy logic in control systems: Fuzzy logic controller-Part II," *IEEE Trans. Systems, Man and Cybernetics*, vol. 20, no. 2, pp. 419–435, 1990.

[15] E. H. Mamdani and S. Assilian, "An experiment in linguistic synthesis with a fuzzy logic controller," *International Journal of Man-Machine Studies*, vol. 7, no. 1, pp. 1–13.

[16] S. Haykin, *Neural Networks A Comprehensive Foundation*, 2nd Ed., Englewood Cliffs, NJ: Prentice Hall, 2000.

[17] D. Ho, P.-A. Zhang, and J. Xu, "Fuzzy wavelet networks for function learning," *IEEE Trans. Fuzzy Systems*, vol. 9, no. 1, pp. 200–211, 2001.

[18] W. Pedrycz, *Computational Intelligence: An Introduction*, CRC Press, 1998.

[19] J. Zhang, G. G. Walter, Y. Miao, and W. N. W. Lee, "Wavelet neural networks for function learning," *IEEE Trans. Signal Processing*, vol. 43, no. 6, pp. 1485–1499, 1995.

[20] Q. Zhang, "Using wavelet networks in nonparametric estimation," *IEEE Transactions On Neural Networks*, vol. 8, no. 2, pp. 227–236, 1997.

[21] Q. Zhang and A. Benvenise, "Wavelet networks," *IEEE Trans. Neural Networks*, vol. 3, no. 6, pp. 889–898, 1992.

[22] L. Jiao, J. Pan, and Y. Fang, "Multiwavelet neural network and its approximation properties," *IEEE Trans. Neural Networks*, vol. 12, no. 5, pp. 1060–1066. 2001.

[23] S. Chen, C. F. N. Cowan, and P. M. Grant, "Orthogonal least squares learning algorithm for radial basis function networks," *IEEE Trans. Neural Networks*, vol. 2, no. 2, pp. 302–309, 1991.

[24] L.-X. Wang, *A Course in Fuzzy Systems and Control*, Englewood Cliffs, NJ: Prentice Hall, 1998.

[25] Y. Lin, F.-Y. Wang, and Y. Gao, "A modular hybrid networks for fuzzy logic control systems," *Proc. 2005 IEEE International Conference on Networking, Sensing, and Control*, Tucson, AZ, Mar. 2005, pp. 671–676.

[26] Y. Lin and F.-Y. Wang, "Predicting chaotic time series using adaptive wavelet-fuzzy inference systems," *Proc. 2005 IEEE Intelligent Vehicles Symposium*, Las Vegas, NV, June 2005, pp. 888–893.

[27] F.-Y. Wang, "Building knowledge structure in neural nets using fuzzy logic," in *Robotics and Manufacturing: Recent Trends in Research, Education and Applications*, M. Jamshidi, Ed., New York: ASME Press, 1992.

[28] F.-Y. Wang and H.-M. Kim, "Implementing adaptive fuzzy logic controllers with neural networks: A design paradigm," *Journal of Intelligent and Fuzzy Systems*, vol. 3, no. 2, pp. 165–180, 1995.

Chapter 8

Ant Colony Algorithms: The State-of-the-Art

Jihui Zhang, Junqin Xu, and Siying Zhang

Abstract: Ant colony optimization (ACO) is a new meta-heuristic method based on the observation of real ant colonies' behavior. It provides a new way to solve many complicated optimization problems such as traveling salesman problem (TSP), quadratic assignment problem (QAP) and many others. A detailed overview on the principle of ACO and its applications are presented in this chapter. This survey should be helpful to the readers who are interested in ACO.

8.1 Introduction

There are many problems with combinatorial nature in our complex world. Well known examples include: traveling salesman problem (TSP), quadratic assignment problem (QAP), vehicle routing problem (VRP), communication network design, scheduling, and many others. All these are abstract models of some real world problems. Many research results have shown that most of these problems are NP-hard and there are no polynomial algorithms for them. In addition, the size of instances met in the real world is usually very large, so that it is impossible to solve them to optimality in an acceptable time. Requirement for computer resources such as memory increases rapidly with the increase of the problem size; this is the so-called "combinatorial explosion." Because of these real difficulties, we have to accept suboptimal solutions, provided that they are of acceptable quality and can be obtained in an acceptable time with acceptable computing effort. The concept of "satisfactory solution" is in accordance with human practice. Several approaches have been proposed for designing "acceptable quality" solutions within "acceptable time." Generally speaking, these methods can be divided into three classes: *approximation algorithms*, which guarantee to find a solution within a known gap from optimality; *probabilistic algorithms*, which guarantee that, for instances big enough, the probability of getting a bad solution is very small; and *heuristic algorithms*, which offers no guarantees. However, heuristic algorithms historically, on average, have the best track record on the quality/time trade-off for problems of interest. The third class includes three main heuristic algorithms, i.e., those focus on solution structure in order to quickly find a good feasible solution, which include *constructive heuristics* (e.g., farthest insertion for TSP) and *local search heuristics* (e.g., 2-opt and 3-opt for TSP); those focus on heuristic guidance and try to make a good trade-off between search "diversification" and "intensification" in order to escape from local optima, called "meta-heuristics" (e.g., simulated annealing, tabu search, genetic algorithms, ant system, macroevolutionary algorithm, particle swarm optimization (PSO)); and those focus on mathematical programming contribution (e.g., bionomic algorithms, scatter search techniques). For more details, please refer to [1].

Ant colony optimization (ACO) is a novel meta-heuristic proposed in recent years. It has been used successfully to solve some difficult discrete optimization problems. Many experimental results show that it has some good properties such that it is worthy of further research. Therefore, a detailed survey is given in this chapter. This survey will be helpful to the readers who are interested in this optimization algorithm.

8.2 Principles of ACO

The ant colony algorithm was first proposed by Dorigo and his colleagues [3, 4, 5] as a multi-agent approach to some difficult combinatorial optimization problems. Up to now, there has been much research to extend and apply ant-based algorithms to many different discrete optimization problems, and also there are some successful applications.

Ant colony optimization is inspired by the behavior of real ant colonies. Ants are social insects living in colony whose behavior is directed more to the survival of the colony as a whole than to that of a single component of the colony. An important and interesting behavior of ant colonies is their forging behavior, and in particular, how ants can find the shortest paths between food sources and their nest. Many observations show that while walking from food source to the nest and vice versa, ants deposit on the ground substances called *pheromones*, that form an pheromone trail. Ants use this trail to direct their movement. Ants can smell pheromone and choose, in probability, paths with strong pheromone concentration. Experiments [6] also show that this pheromone trail following behavior can direct ants to find the shortest path from food source to the nest. There are two factors playing an important role in the collective behavior of ant colonies. Ants perform this specific behavior using a simple form of indirect communication mediated by pheromone laying known as *stigmergy*. Stigmergy is a class of mechanisms that mediate animal-animal interactions. It consists of indirect communication that is taking place between individuals of an insect society by local modifications induced by these insects on their environment. Another important aspect of real ants' foraging behavior is the combination of *autocatalytic* (positive feedback) with *implicit evaluation* of "solutions" [4].

8.3 Ant Colony Optimization

In ant colony optimization meta-heuristics, a colony of artificial ants cooperate to find a good solution to some difficult discrete optimization problems. Cooperation is a key in the design of ACO algorithms: the choice is to allocate the computational resources to a set of relatively simple agent that communicate indirectly by stigmergy. Good solutions are an emergent property of the agents' cooperative interaction. Artificial ants are an abstraction of the behavioral traits of real ants, and on the other hand, they are enriched with some capabilities which cannot be found in real ants.

The main underlying idea is that of parallel searches over several constructive computational threads, all based on a dynamic memory structure incorporating information on the effectiveness of previously obtained results. Thus the behavior of each single agent is inspired by the behavior of real ants. An ant is defined as a simple computational agent, which iteratively constructs a solution to the problem considered. Partial solutions are seen as *states*: Each ant moves from state ι to another one ψ, forming a more complete partial solution. At each step σ, each ant k, computes a set $A_k^\sigma(\iota)$ of feasible expansions to its current state, and moves to one of these, according to a probability distribution specified as follows:

1. The *attractiveness*, denoted by η, of the move, is computed by some heuristic indicating an *a priori* desirability of that move;

2. The *trail level*, denoted by τ, of the move, indicating how proficient it has been in the past making that particular move, and thus it represents an *a posteriori* indication of the desirability of that move.

Trials are updated at each iteration, increasing the level of those that facilitate moves which are part of "good" solutions, while decreasing all others. The specific formula defining the probability distribution at each move uses a $tabu_k$ set, which indicates a problem-dependent set of infeasible moves for ant k. In general, $p_{\iota\psi}^k$ has the form of the following equation:

$$p_{\iota\psi}^k = \begin{cases} 0, & \text{if } \psi \in tabu_k; \\ [\alpha\tau_{\iota\psi} + (1-\alpha)\eta_{\iota\psi}] \Big/ \sum_{\iota\upsilon\notin tabu_k} [\alpha\tau_{\iota\upsilon} + (1-\alpha)\eta_{\iota\upsilon}], & \text{otherwise;} \end{cases}$$

$$(8.3.1)$$

where $\alpha \in [0,1]$ defines the relative importance of trail with respect to attractiveness. After each iteration t of the algorithm, i.e., when all ants have completed a solution, pheromone trails are updated using formula in the next expression:

$$\tau_{\iota\psi}(t) = \rho\tau_{\iota\psi}(t-1) + \Delta\tau_{\iota\psi} \tag{8.3.2}$$

where ρ is a user-defined coefficient and $\Delta\tau_{\iota\psi}$ represents the sum of the contributions of all ants that used the move $(\iota \rightarrow \psi)$ to construct their solution. The ants' contributions are proportional to the quality of the achieved solutions, i.e., the better an ant's solution is, the higher the trail contribution added to the move will be. The general structure of an ACO can be expressed as in Figure 8.3.1.

Step 1. (initialization)
    Initialize $\tau_{\iota\psi}$ for $\forall(\iota, \chi)$

Step 2. (construction)
    For each ant k **do**
        repeat
            compute $\eta_{\iota\psi}$, for $\forall(\iota, \psi)$;
            choose the state to move into, with probability given by (8.3.1);
            append the chosen move to the tabu list of ant k;
        until ant k completes its solution.
            carry the solution to its local optimum.
    End do

Step 3. (trail update)
    For each ant move $(\iota \rightarrow \psi)$ **do**
        compute $\Delta\tau_{\iota\psi}$;
        update the trail matrix by means of (8.3.2).
    End do

Step 4. (termination condition)
    If not (end condition) **go to** Step 2.

Step 5. (end)

Figure 8.3.1: Fundamental framework of an ACO.

During recent years, many variants of ACO algorithms have been proposed by different authors and used to solve different combinatorial optimization problems, but their main ideas are similar [77, 78, 79], especially in the use of (1) a colony of cooperating "ants;" (2) an artificial "pheromone" trail for local stigmergetic communication; (3) a sequence of local moves and (4) stochastic decision policy using local information. On the other hand, in order to get a satisfactory searching performance, some additional features that cannot be posed by real ants are added. For details, see [7].

8.4   Applications of ACO Algorithms

There are many successful applications of ACO meta-heuristic algorithms to some hard combinatorial optimization problems, as well as some continuous-space optimization problems. This section gives a brief overview about this. According to the classification in [7], these applications can be classified as static and dynamic combinatorial optimizations.

Static problems are those in which the characteristics of the problem are given when the problem is defined, and no change occurs while the problem is being solved. On the contrary, dynamic problems are defined as a function of some quantities whose values are set by the dynamics of an underlying system. Thus problem changes at run-time. Therefore optimization algorithms must be able to adapt online to changing environment. In the following subsections, a brief survey about these applications is given.

8.4.1 Traveling Salesman Problem

A.   Ant System (AS)

The first successful application of ant colony optimization is the traveling salesman problem. It has been used as a good example to introduce ant colony algorithm by Dorigo [5]. Ant System (AS) is the first ACO algorithm proposed by Dorigo. In AS, some artificial ants build solutions of TSP by moving on the problem graph from one city to the next in an iterative process. During each cycle, after each tour, each ant deposits "pheromone" (some kind of information) on the trail it visited to make it more desirable. The amount of "pheromone trail" $\tau_{ij}(t)$ associated with the arc (i, j) is intended to represent the learned desirability of choosing city j when ant is in city i. This information will be changed during problem solution to accept the experience acquired by ants during the searching process. It represents a "positive feedback mechanism": *the shorter the tour generated by an ant, the more "pheromone" it has and thus more desirable for other ants to follow.* In order to avoid invalid solution (i.e., solution in which one or more cities are visited more than once), each ant has a *tabu list* which contains the cities each ant has visited. The decision table, $A_i = [a_{ij}(t)]_{|N_i|}$, of node i is obtained by the composition of the local "pheromone trail" values with local heuristic values as follows:

$$a_{ij}(t) = \frac{\tau_{ij}^{\alpha}(t)\eta_{ij}^{\beta}}{\sum_{\iota \in N_i} \tau_{i\iota}^{\alpha}(t)\eta_{i\iota}^{\beta}}$$

where $\tau_{ij}(t)$ is the amount of pheromone trail on arc (i, j) at time t, $\eta_{ij} = 1/d_{ij}$ is the heuristic value of moving from i to j, N_i is the set of neighbors of node i, α and β are parameters controlling the relative weight of "pheromone" and heuristic value. Suppose at cycle t, ant k is in city i. Then it will choose city j with probability $p_{ij}^k(t)$ as its next stop, where

$$p_{ij}^k(t) = \frac{a_{ij}(t)}{\sum_{\iota \in N_i^k} a_{i\iota}(t)}$$

and $N_i^k \subseteq N_i$ is the set of nodes in the neighborhood of node i that ant k has not visited yet. After all ants have completed their tours, "pheromone evaporation" is triggered, and each ant deposits "pheromone" $\Delta\tau_{ij}^k(t)$ on arc (i, j), and

$$\Delta\tau_{ij}^k(t) = \begin{cases} 1/L^k(t), & \text{if } (i, j) \in T^k(t) \\ 0, & \text{if } (i, j) \notin T^k(t) \end{cases}$$

where $T^k(t)$ is the tour done by ant k at cycle t and $L^k(t)$ is its length. The addition of new "pheromone" and "pheromone evaporation" are implemented as follows:

$$\tau_{ij}(t) \leftarrow (1 - \rho)\tau_{ij}(t) + \Delta\tau_{ij}(t)$$

and

$$\Delta\tau_{ij}(t) = \sum_{k=1}^{m} \Delta\tau_{ij}^k(t)$$

where, m is the number of ants and $\rho \in [0, 1]$ is the "pheromone" decay rate. The above process will be continued until the termination condition is met. Many experimental results [5] have shown that AS has a good performance.

B. *Other AS-Like Approaches*

There are many other AS-like approaches for solving the TSP. A brief survey will be given in this subsection.

Max-Min AS (MMAS) was proposed by Stüzle and Hoos in 1997 [12, 13, 20]. The main idea of MMAS is similar to that of AS, but it has its own special characteristics:

- "pheromone" is updated off-line, and only the "pheromone" information on the "best trail" is updated after each iteration;
- "pheromone" values are restricted to an interval $[\tau_{\min}, \tau_{\max}]$;
- "pheromone trails" are initialized to $\tau_{\max}$;
- in order to avoid *stagnation* behavior, a "trail smoothing mechanism" is used, i.e., $\Delta\tau_{ij} \propto (\tau_{\max} - \tau_{ij}(t))$.

AS_{rank} is another approach with the following characteristics:

- the m ants are ranked by tour length $(L_1(t), \cdots, L_m(t))$ and the arcs which are visited by one of the first $\sigma - 1$ ants in the ranking receive an amount of "pheromone" proportional to the visiting ant rank;
- the arc used by the ant that generated the best tour from beginning of the trail receive additional "pheromone";
- the dynamics of the amount of "pheromone" $\tau_{ij}(t)$ is given by

$$\tau_{ij}(t) \leftarrow (1 - \rho)\tau_{ij}(t) + \sigma\Delta\tau_{ij}^+(t) + \Delta\tau_{ij}^r(t)$$

where $\Delta\tau_{ij}^+(t) = 1/L^+(t)$,

$$\Delta\tau_{ij}^r(t) = \sum_{\mu=1}^{\sigma-1} \Delta\tau_{ij}^\mu(t) = \begin{cases} (\sigma-\mu)/L^\mu(t), & \text{if ant with rank } \mu \text{ uses arc } (i,j), \\ 0, & \text{otherwise,} \end{cases}$$

L^+ is the length of the best solution from the beginning of the search, and $L^\mu(t)$ is the length of the tour performed by ant with rank μ at iteration t.

C. Ant Colony System (ACS)

Ant colony system (ACS) was proposed by Dorigo and Gambradella [8–10] as an improvement to AS.

- Pheromone trails are updated off-line: at the end of an iteration, pheromone trail is added to the arcs used by the ant that found the best tour from the beginning of the trail. The off-line pheromone trail is updated as: $\tau_{ij}(t) \leftarrow (1 - \rho)\tau_{ij}(t) + \rho\Delta\tau_{ij}(t)$, where $\rho \in [0, 1]$, $\Delta\tau_{ij}(t) = 1/L^+$, and L^+ is the length of the best tour since the beginning of the trail.
- Ants use a new decision rule called *pseudo-random-proportional*, i.e., ant k in city i chooses city $j \in N_i^k$ to move as follows: Let $A_i = [a_{ij}(t)]_{|N_i|}$ be the decision table with

$$a_{ij}(t) = \frac{\tau_{ij}(t)\eta_{ij}^\beta}{\sum_{\iota \in N_i} \tau_{i\iota}(t)\eta_{i\iota}^\beta} \quad \forall j \in N_i.$$

Let $q \in U(0,1)$ be a random variable uniformly distributed over $[0,1]$, and $q_0 \in [0,1]$ be a tunable parameter. The *pseudo-random-proportional* rule can be described as follows: ant k in city i choose $j \in N_i$ to move to with probability $p_{ij}^k(t)$, where

$$p_{ij}^k(t) = \begin{cases} \begin{cases} 1 & \text{if } j = \arg\max\{a_{ij}\} \\ 0 & \text{otherwise} \end{cases} & \text{if } q \le q_0, \\ a_{ij}(t)/\sum_{\iota \in N_i^k} a_{i\iota}(t) & \text{if } q > q_0. \end{cases}$$

- Ants perform online step-by-step pheromone update according to the criterion: $\tau_{ij}(t) \leftarrow (1 - \varphi)\tau_{ij}(t) + \varphi\tau_0$, where $\varphi \in (0,1]$.
- A data structure called a *candidate list* is used to provide additional local heuristic information. A candidate list is a list of preferred cities to be visited from a given city.

8.4.2 Quadratic Assignment Problem (QAP)

QAP is the problem of assigning n facilities to n locations such that the cost of the assignment, which is a function of the way facilities have been assigned to locations, is minimized. QAP can be viewed as a generalization of TSP, so ACO algorithms can be used to deal with it. Maniezzo, Colorni and Dorigo [15] applied exactly the same algorithm as AS using the QAP-specific min-max heuristic to compute η values. Experimental results show that AS-QAP has a good performance. Recently, Maniezzo and Colorni proposed two variants of AS-QAP [18, 19] and added local optimization to them, the resulting algorithms are compared with some of the best heuristics for QAP, and their AS-QAP has a better performance. Stützle and Hoos use their MMAS to solve QAP where some good performances are obtained. A detailed overview on applications of ACO to QAP can be found in [50].

8.4.3 Vehicle Routing Problem (VRP)

Hartl and Strauss proposed an AS-like algorithm to one instance of VRP. In their algorithm, heuristics for VRP and local optimizer are used, and the tabu list is built by considering the constraints on the maximum total tour length L of a vehicle and maximum capacity D. Marc *et al.* [86] proposed D-ants for VRP using divide and conquer technique. Experimental results on a set of standard problems showed that the performance of AS-VRP is interesting.

Gambradella *et al.* also studied VRP using ACO algorithm. In their method, the problem is reformulated by adding to the city set of $M - 1$ depots (M is the number of vehicles) as a TSP with additional constraints. They defined an algorithm called HAS-VRP in which each ant builds a complete tour without violating vehicle capacity constraints. A complete tour is composed of many subtours connecting depots, and each subtour corresponds to the tour associated to one of the vehicles. "Pheromone trail" is updated off-line. In addition, a local optimizer based on edge exchange is applied. They also studied the vehicle routing problem with time windows (VRPTW). The approach has been proved to be competitive compared to the best known methods in literature.

8.4.4 Shortest Common Supersequence Problem

Shortest common supersequence (SCS) problem can be stated as follows: given a set L of strings over an alphabet σ, find a string of minimum length that is a supersequence of all strings in L. A string S is a supersequence of a string A if S can can be obtained from A by inserting in A zero or more characters. AS-SCS was proposed by Michel and Middendorf [27, 28] to deal with this problem. In AS-SCS, a *look-ahead function* which takes into consideration the influence of the choice of the next symbol to append at the next iteration. The heuristic value η is replaced by the value returned from the look-ahead function. A simple heuristic called LM is factorized in the pheromone trail. Experimental results show that AS-SCS-LM has the best performance for majority of the test problems.

8.4.5 Graph Coloring Problem

Costa and Hertz proposed AS-ATP [26] for assignment type problems. In order to solve this kind of problem, ants must make two choices: (1) choose an item; and (2) choose a resource assigning to the chosen item. These two choices are made by means of two probabilistic rules which function at two distinct pheromone trails τ_1 and τ_2, and two heuristic values η_1 and η_2. The use of two pheromone trails is the main novelty of the AS-ATP.

8.4.6 Sequential Ordering Problem

Sequential ordering problem (SOP) is to find a minimum weight Hamiltonial path on a directed graph with weights on arcs and on nodes subjects to precedence constraints among nodes. SOP is similar to an asymmetric TSP and is NP-hard. Gambardella and Dorigo [25] attacked SOP by HAS-SOP. HAS-SOP is similar to AS except for the set of feasible nodes, which is built by considering additional precedence constraints. The local optimizer is based on a well known heuristic 3-opt. Experimental results obtained by HAS-SOP are very good in terms of solution quality and of computing time.

8.4.7 Scheduling Problem

AS was applied to the job shop scheduling problem by Colorni, Dorigo and Maniezzo [21]. In their method, η is obtained using the longest remaining processing time (LRPT) heuristic, and the tabu list is also different from that for TSP. Ying *et al.* [81] proposed an ant colony algorithm for the job shop scheduling problem. In his method, the scheduling problem is represented by a disjunctive graph model, so it can be solved by ACS as similar to a TSP-like problem. Li *et al.* [83] proposed a nested hybrid ant colony algorithm for a kind of hybrid production scheduling problem. Ying *et al.* [82] proposed an ant colony system for permutation flow shop sequencing problem. Simulation results show that ACS is a more effective meta-heuristic for that. Caroline *et al.* [85] suggested an ant colony optimization algorithm with a look-ahead mechanism for an industrial scheduling problem. In constructing a candidate solution, ants use a transition rule that incorporate complete information on past decisions (*the trail*) and local information (*visibility*) on the immediate decision that is to be made. The look-ahead mechanism allows the incorporation of information on the anticipated decisions that are beyond the immediate choice horizon. Christian *et al.* [84] proposed an ant colony optimization method for FOP shop scheduling using a different pheromone representation. Other scheduling problems such as bus driver scheduling [45], batch process scheduling [42] and thermal unit scheduling [46] are also studied. Although some results are not very good, it provides a new way for these kinds of problems and further careful research is meaningful and necessary.

8.4.8 Network Optimization

Network optimization problems have some special characteristics such as inherent information, computation distribution, non-stationary stochastic dynamics and asynchronous evolution. Thus, it belongs to dynamic optimization. ACO approach has been applied to network routing problems. Routing is one of the most important components of network control, and concerns the network-wide distributed activity of building and using a *routing table* to direct data traffic. The routing table of node i is a data structure that directs data packets entering node i, which should be the next node to move to among the set N_i of neighbors of i. The generic routing problem can be stated as the problem *to build routing tables so that some measure of network performance is optimized*. Communication networks can be classified into connection-oriented and connectionless networks. In the former, all packets of the same session follow a common path selected by a preliminary setup phase. On the contrary, in the latter case, data packets of the same session can follow through different paths.

A. Connection-Oriented Network Routing

Ant-based control (ABC) [29, 30] was proposed by Schoonderwoerd *et al.* and applied to a telephone network model which is modeled by a graph $G = (N, A)$. In this model, each node i has a set N_i of neighbors and is characterized by a total capacity C_i and a spare capacity S_i. C_i represents the maximum number of connections node i can establish, and S_i represents the percentage of capacity which is still available for new connections. Each link (i,j) has an associated vector of pheromone trail value τ_{ijd} $d = 1, \cdots, i-1, i+1, \cdots, n)$ and represents a measure of the desirability of choosing link (i,j) when the destination node is d. In the algorithm of ABC, no local heuristics are used. The ant decision table is obtained using only pheromone values. Thus, $a_{\mathrm{ind}}(t) = t_{\mathrm{ind}}(t)$. An exploration mechanism is used for ants' decision: with some low probability ants can choose one neighbor to move to following a uniformly random scheme over all the current neighbors. Routing tables for calls are obtained using ant-decision tables in a deterministic way: at setup time, a route from node s to node d is built by choosing sequentially and deterministically, starting from node s, the neighboring node with the highest probability value until node d is reached. Once the call is setup this way, the spare capacity S_i of each node i on the selected route is decreased by a fixed amount. If, at call setup time, any of the nodes along the route under construction has no spare capacity left, then the call is rejected. When a call terminates, the corresponding reserved capacity of nodes on its route is available again. Ants are launched at regular temporal intervals from all nodes towards destination nodes selected in a uniform manner. Ants deposit pheromone trail online on the links they visited. The pheromone trail is updated according to the following formula:

$$\tau_{ijs}(t) \leftarrow \tau_{ijs} + \Delta \tau^k(t).$$

After the visited entry has been updated, the pheromone value of all entries relative to the destination s decays according the following formula:

$$\tau_{ins} \leftarrow \frac{\tau_{ins}}{1 + \Delta \tau^k(t)}, \forall n \in N.$$

White, Pagurek and Oppacher [31] use ACO algorithm for routing in connection-oriented point-to-point and point-to-multipoint networks. Their algorithm is similar to AS. The heuristic information η is locally estimated by link costs and pheromone trail is updated online. Bonabeau *et al.* [33] improves the performance of ABC by introducing dynamic programming mechanism.

B. *Connectionless Networking*

Di Caro and Dorigo [34] used AntNet for distributed adaptive routing in best-effort connectionless data networks. Its main characteristics are listed as follows.

- Real trip times experienced by ants and local statistical models are used to evaluate the goodness of the paths.

- Pheromone is deposited once a complete path is built.

– Local heuristic information η about the current traffic status is used for decision making.

They also proposed an enhanced version of AntNet called AntNet-FA in which

– forward ants are substituted by "flying ants": while building a path from source to destination node, "flying ants" make use of high priority queues and do not store the trip time;

– each node maintains a simple local model of the local link queue depletion process.

In addition, other methods such as *regular ants algorithms, co-operative asymmetric forward* and *ABC-backward*, etc. are also discussed. Please refer [7] for detailed information about each of them.

8.4.9 Frequency Assignment Problem

The frequency assignment problem (FAP) is a problem that arises when a region is covered, for wireless communications, by cells centered on base stations and transmitters scattered around the region that want to connect with the antenna of the base stations. Each connection (link) between a transmitter and a base station can be made on a frequency supported by the antenna. However, the frequency concurrently operated by overlapping cells must be separated in order to minimize the interference on the communications taking place in the cells. ACO was used by Maniezzo *et al.* [44] to solve the FAP. Computational results show that ants heuristic is competitive with the best approaches so far presented.

8.4.10 Continuous Space Optimization Problem

Generally speaking, ACO provides a good method for many hard combinatorial optimization problems. In some recent papers [66, 80] it was used to solve some complicated continuous space optimization problems. The key is to produce a suitable form for ACO [77–79, 91, 92].

8.4.11 Machine Learning

This is a new application area. In [88], the recently developed ant colony optimization meta-heuristic procedure has been recast as a rule based machine learning method, called the ant colony classifier system, and applied to three process engineering examples. The learning algorithm addresses the problem of knowledge acquisition in terms of rules from example cases by developing and maintaining the knowledge base through the use of simple mechanism, pheromone trail information matrix and the use of available heuristic information. The performance of an ant colony classifier is compared with the well-known decision tree based C4.5 algorithm in terms of the predictive accuracy on test cases and the simplicity of rules discovered. The results indicate that the ant classifier is able to discover rules in the data sets with better predictive accuracy than the C4.5 algorithm. Richard *et al.* [87] proposed a new feature selection mechanism based on ant colony optimization for fuzzy rough data reduction.

8.5 Theoretical Research

Last but not the least is the theoretical aspects such as modelling and convergence of the ant colony algorithm. Compared with application results, there is very little on these aspects up to now. Gutjahr [93] gives a proof to the convergence of ACO. Daniel [90] studied the dynamics of ACO using a deterministic model. In order to get a deep understanding about ACO, more theoretical research is definitely necessary.

8.6 Conclusions

A detailed overview about ACO meta-heuristics, including its main principles, various versions of ACO and their applications, is given in this chapter. ACO is a new interesting meta-heuristic for hard optimization problems, especially, for combinatorial optimization problems. Compared with other methods, ACO is not yet fruitful, and further careful research is meaningful.

Acknowledgements

This work was supported by the Shandong Natural Science Fund (2002G01) and Qingdao Natural Science Fund (03-2-jz-19).

Bibliography

[1] V. Maniezzo and A. Carbonaro, "Ant colony optimization: An overview," in *Essays and Surveys in Metaheuristics*, C. Ribeiro and P. Hansen, Eds., Boston: Kluwer, 2001, pp. 21–44.

[2] E. Bonabeau, M. Dorigo, and G. Theraulaz, "Inspiration for optimization from social insect behavior," *Nature*, vol. 406, no. 6, pp. 39–42, 2000.

[3] M. Dorigo, *Optimization, Learning and Natural Algorithms*, Ph. D. Thesis, Politecnico di Milano, Italy, 1992.

[4] M. Dorigo, V. Maniezzo, and A. Colorni, "The ant system: An autocatalytic optimizing process," Technical Report No. 91-016 (Revised), Politecnico di Milano, Italy, 1991.

[5] M. Dorigo, V. Maniezzo, and A. Colorni, "The ant system: Optimization by a colony of cooperating agents," *IEEE Transactions on Systems, Man, and Cybernetics-Part B*, vol. 26, no. 1, pp. 29–41, 1996.

[6] J. L. Deneubourg, S. Aron, S. Goss, and J. M. Pasteels, "The self-organizing exploratory pattern of the Argentine ant," *Journal of Insect Behavior*, vol. 3, no. 2, pp. 159–168, 1990.

[7] M. Dorigo and G. Di Caro, "Ant algorithms for discrete optimization," *Artificial Intelligence*, vol. 5, no. 3, pp. 137–172, 1999.

[8] L. M. Gambardella and M. Dorigo, "Ant-Q: A reinforcement learning approach to the traveling salesman problem," *Proc. 12th International Conference on Machine Learning*, Palo Alto, CA, July, 1995, pp. 252–260.

[9] M. Dorigo and L. M. Gambardella, "Ant colonies for the traveling salesman problem," *BioSystems*, vol. 43, no. 2, pp. 73–81, 1997.

[10] M. Dorigo and L. M. Gambardella, "Ant colony system: A cooperative learning approach to the traveling salesman problem," *IEEE Transaction on Evolutionary Computation*, vol. 1, no. 1, pp. 53–66, 1997.

[11] L. M. Gambardella and M. Dorigo, "Solving symmetric and asymmetric TSPs by ant colonies," *Proc. IEEE Conference on Evolutionary Computation*, Nagoya, Japan, May 1996, pp. 622–627.

[12] T. Stützle and H. Hoos, "The $MAX-MIN$ ant system and local search for the traveling salesman problem," *Proc. IEEE International Conference on Evolutionary Computation and Evolutionary Programming*, Indianapolis, IN, Apr. 1997, pp. 309–314.

[13] T. Stützle and H. Hoos, "Improvements on the ant system: Introducing MAX - MIN ant system," *Proc. International Conference on Artificial Neural Networks and Genetic Algorithms*, Norwich, UK, Apr. 1997, pp. 245–249.

[14] B. Bullnheimer, R. F. Hartl, and C. Strauss, "A new rank-based version of ant system: a computational study," Tech. Report POM-03, Institute of Management Science, Univ. of Vienna, 1997.

[15] V. Maniezzo, A. Colorni, and D. Dorigo, "The ant system applied to the quadratic assignment problem," Tech. Report IRIDIA/94-28, Univ. Libre de Bruxelles, Belgium, 1994.

[16] L. M. Gambradella, E. D. Taillard, and M. Dorigo, "Ant colonies for QAP," Tech. Report 4-97, IDSIA, Lugano, Switerland, 1997.

[17] L. M. Gambradella, E. D. Taillard, and M. Dorigo, "Ant colonies for QAP," *Journal of the Operations Research Society*, vol. 50, no. 2, pp. 167–176, 1997.

[18] V. Maniezzo, "Exact and approximate nondeterministic tree-search procedures for the quadratic assignment problem," Tech. Report CSR 98-1, Univ. di Bologna, Italy, 1998.

[19] V. Maniezzo and A. Colorni, "The ant system applied to the quadratic assignment problem," *IEEE Transactions on Knowledge and Data Engineering*, vol. 11, no. 5, pp. 769–778, 1999.

[20] T. Stützle and H. Hoos, "MAX - MIN ant system and local search for combinatorial optimization problems," in *Meta-Heuristics: Advances and Trends in Local Search Paradigms for Optimization*, Edited by V. Stefan, S. Martello, and I. H. Osman, Boston: Kluwer, 1998, pp. 137–154.

[21] A. Colorni, M. Dorigo, V. Maniezzo, and M. Trubian, "Ant system for job-shop scheduling," *Belgian Journal of Operational Research, Statistics and Computer Science*, vol. 34, no. 1, pp. 39–53, 1994.

[22] B. Bullnheimer, R. F. Hartl, and C. Strauss, "An improved ant system for the vehicle routing problem," Tech. Report POM 10-97, Institute of Management Science, Univ. of Vienna, 1997.

[23] B. Bullnheimer, R. F. Hartl, and C. Strauss, "Applying the ant system to the vehicle routing problem," in *Meta-Heuristics: Advances and Trends in Local Search Paradigms for Optimization*, Edited by V. Stefan, S. Martello, and I. H. Osman, Boston: Kluwer, 1998, pp. 109–120.

[24] L. M. Gambardella, E. Taillard, and G. Agazzi, "Macs-VRPTW: A multiple ant colony system for vehicle routing problem with time windows," in *New methods in Optimization*, Edited by D. Corne, M. Dorigo, and F. Glover, London: McGraw-Hill, 1999, pp. 63–76.

[25] L. M. Gambardella and M. Dorigo, "HAS-SOP: A hybrid ant system for the sequential ordering problem," Tech. Report 11-97, IDSIA, Lugano, CH, 1997.

[26] D. Costa and A. Hertz, "Ants can color graphs," *Journal of the Operational Research Society*, vol. 48, no. 3, pp. 295–305, 1997.

[27] R. Michel and M. Middendorf, "An island model-based ant system with look-ahead for the shortest supersequence problem," in *Lecture Notes in Computer Science*, vol. 1498 (*Proc. 5th Int. Conf. on Parallel Problem Solving from Nature*), Edited by A. E. Eiben, T. Bäck, M. Schoenauer, and H.-P. Schwefel, Berlin: Springer, 1998, pp. 692–701.

[28] R. Michel and M. Middendorf, "An ACO algorithm for the shortest common supersequence problem," in *New Methods in Optimization*, Edited by D. Corne, M. Dorigo, and F. Glover, New York: McGraw-Hill, 1999, pp. 51–61.

[29] R. Schoonderwoerd, O. Holland, J. Bruten, and L. Rothkrantz, "Ant-based load balancing in telecommunications networks," *Adaptive Behavior*, vol. 5, no. 2, pp. 169–207, 1997.

[30] R. Schoonderwoerd, O. Holland, and J. Bruten, "Ant-like agents for load balancing in telecommunications networks," *Proc. first Int. Conf. on Autonomous Agents*, Marina del Rey, California, USA, May, 1997, pp. 209–216.

[31] T. White, B. Pagurek, and F. Oppacher, "Connection management using adaptive mobile agents," *Proc. Int. Conf. on parallel and distributed processing techniques and applications (PDPTA98)*, Las Vegas, Vevada, USA, July, 1998, pp. 802–809.

[32] G. Di Caro and M. Dorigo, "Extending AntNet for best-effort quality-of-service routing," *Proc. ANTS'98–From Ant Colonies to Artificial Ants: First Int. Workshop on Ant Colony Optimization*, Brussels, Belgium, October, 1998, pp. 26–30.

[33] E. Bonabeau, F. Henaux, S. Guerin, D. Snyers, P. Kuntz, and G. Theraulaz, "Routing in telecommunication networks with 'smart' ant-like agents," *Proc. 2nd International Workshop on Intelligent Agents for Telecommunications Applications*, Paris, France, June 1999, pp. 60–71.

[34] G. Di Caro and M. Dorigo, "AntNet: A mobile agents approach to adaptive routing," Tech. Report 97-12, IRIDIA, Univ. Libre de Bruxelles, 1997.

[35] G. Di Caro and M. Dorigo, "AntNet: Distributed stigmergetic control for communications networks," *Journal of Artificial Intelligence Research*, vol. 9, no. 3, pp. 317–365, 1998.

[36] G. Di Caro and M. Dorigo, "Two ant colony algorithms for best-effort routing in datagram networks," *Proc. 10th IASTED International Conference on Parallel and Distributed Computing and Systems (PDCS'98)*, Las Vegas, NV, 1998, pp. 541–546.

[37] D. Subramanian, P. Druschel, and J. Chen, "Ants and reinforcement learning: A case study in routing in dynamic networks," *Proc. IJCAI-97, International Joint Conference on Artificial Intelligence*, Nagoya, Japan, Aug. 1997, pp. 832–838.

[38] M. Heusse, S. Guerlin, D. Snyers, and P. Kuntz, "Adaptive agent-driven routing and local balancing in communication networks," Tech. Rep. RR-98001-IASC. Department of Artificial Intelligence and Science Congnitives, ENST Bretagne, 1998.

[39] R. Van der Put, "Routing in the fax factory using mobile agents," Tech. Rep. R &D-SV-98-276, KPN Research, 1998.

[40] R. van der Put and L. Rothkrantz, "Routing in packet switched networks using agents," *Simulation Practice and Theory*, vol. 7, no. 3, pp. 634–645, 1999.

[41] M. H. Botee and E. Bonabeau, "Evolving ANT colony optimization," *Advances in Complex Systems*, vol. 1, no.2/3, pp. 149–159, 1998.

[42] V. K. Jayaraman, B. D. Kulkarni, S. Karale, and P. Shelokar, "Ant colony framework for optimal design and scheduling of batch plants," *Computers and Chemical Engineering*, vol. 24, no. 8, pp. 1901–1912, 2000.

[43] G. Leguizamon and Z. Michalewicz, "A new version of ant system for subset problems," *Proc. Congress on Evolutionary Computation (CEC'99)*, Piscataway, NJ, 1999, pp. 1459–1464.

[44] V. Maniezzo and A. Carbonaro, "An ANTS heuristic for the frequency assignment problem," *Journal of Future Generation of Computing System*, vol. 16, no. 8, pp. 927-935, 2000.

[45] P. Forsyth and A. Wren, "An ant system for bus driver scheduling," *Proc. 7th International Workshop on Computer-Aided Scheduling of Public Transport*, Boston, MA, Aug. 1997, pp. 85–90.

[46] A. Bauer, B. Bullnheimer, R. F. Hartl, and C. Strauss, "An ant colony optimization approach for the single machine total tardiness problem," *Proc. Congress on Evolutionary Computation (CEC'99)*, Piscataway, NJ, 1999, pp. 1445–1450.

[47] P. Kuntz and D. Snyers, "Emergent colonization and graph partitioning," *Proc. 3rd International Conference on Simulation of Adaptive Behavior: From Animals to Animats*, Cambridge, MA, July 1994, pp. 125–130.

[48] P. Kuntz, P. Layzell, and D. Snyers, "A colony of ant-like agents for partitioning in VLSI technology," *Proc. 4th European Conference on Artificial Life*, Cambridge, MA, 1997, pp. 417–424.

[49] G. Bilchev and I. C. Parmee, "Searching heavily constrained design spaces," *Proc. 22nd Int. Conf. on Computer Aided Design CAD-95*, Yelta, Ukraine, 1995, pp. 230–235.

[50] T. Stutzle and M. Dorigo, "ACO algorithms for the quadratic assignment problem," in *New Methods in Optimization*, Edited by D. Corne, M. Dorigo, and F. Glover, Boston: McGraw-Hill, 1999.

[51] I. -K. Yu, C. S. Chou, and Y. H. Song, "Application of the ant colony algorithm to short-term generation scheduling problem of thermal units," *Proc. IEEE 1998 International Conference on Power System Technology*, New York, NY, 1998, pp. 552–556.

[52] S. Li, Z. Liu, Z. Zhou, "An ant algorithm based VC routing method in ATM Networks", *Journal of China Institute of Communications*, vol. 21, no. 1, pp. 22–28, 2000.

[53] L. Ma, "Hybrid heuristic algorithm for the TSP and its extended problems," *Journal of University of Shanghai for Science and Technology*, vol. 21, no. 1, pp. 25–28, 1999.

[54] L. Ma and L. Wang, "Ant optimization algorithm for knapsack problem," *Computer Applications*, vol. 21, no. 8, pp. 4–5, 2001.

[55] Y. Chen, "An ant colony algorithm with crossover operator," *Computer Engineering*, vol. 27, no. 12, pp. 74–77, 2001.

[56] L. Ma and F. Jiang, "Ant algorithm for the degree-contrained minimum spanning tree," *Journal of Systems Engineering*, vol. 14, no.3, pp. 211–214, 1999.

[57] L. Ma and F. Jiang, "Solving multi-criteria traveling salesman problem by ant algorithm," *System Engineering, Theory, Methodology Applications*, vol. 8, no. 4, pp. 23–27, 1999.

[58] Y. Ding, P. Liu and Q. Su, "Ant colony algorithm in chemistry and its application in spectra analyzing", *Computers and Applied Chemistry*, vol. 19, no. 3, pp. 326–328, 2002.

[59] S. Zhang, G. Lu and Z. Liu, "QoS routing based on ant algorithm," *Journal of Circuits and Systems*, vol. 5, no. 1, pp. 1–5.

[60] Z. Zhang, R. He and Y. Zhang, "Ant-based dynamic logical topologies reconfiguration for WDM networks", *Journal of China Institute of Communications*, vol. 22, no. 11, pp. 42–49, 2001.

[61] G. Lü, Z. Liu and Z. Zhou, "A distributed QoS routing algorithm based on ant-algorithm", *Journal of China Institute of Communications*, vol. 22, no. 9, pp. 34–42, 2001.

[62] S. Zhang and Z. Liu, "A distributed delay-bounded constraint multicast routing algorithm based on ant-algorithm," *Journal of China Institute of Communications*, vol. 22, no. 3, pp. 71–74, 2001.

[63] S. Li, L. Pan, and H. Zhu, "Ant colony based multicast routing," *Computer Engineering*, vol. 27, no. 4, pp. 63–65, 2001.

[64] G. Chen, L. Wang, and G. Tang, "Distributed network reconfiguration for loss reduction using ant colony algorithm," *Proc. EPSA*, vol. 21, no. 1, pp. 48–53, 2001.

[65] G. Chen, L. Wang, and G. Tang, "An ant colony optimization method for transmission network expansion planning," *Power System Technology*, vol. 25, no. 6, pp. 21–24, 2001.

[66] Y. Li and T. Wu, "An adaptive ant colony system algorithm for continuous-space optimization problem," *Pattern Recognition and Artificial Intelligence*, vol. 14, no. 4, pp. 423–427, 2001.

[67] C. Chen and X. Xie, "Ant colony algorithm used to search for critical slip surface of open-pit slope," *J. Xiangtan Min. Inst.*, vol. 17, no. 3, pp. 62–64, 2002.

[68] J. He, R. Xiao, and H. Shi, "Implementation of ant algorithm for isomorphism identification of mechanisms," *Pattern Recognition and Artificial Intelligence*, vol. 14, no. 4, pp. 406–412, 2001.

[69] L. Ma and P. Xiang, "Applications of ant colony algorithms to combinatorial optimization," *Journal of Management Sciences in China*, vol. 4, no. 2, pp. 32–37, 2001.

[70] Y. Wang and J. Xie, "An ant colony algorithm for multicast routing," *System Engineering & Electronics*, vol. 23, no. 8, pp. 98–101, 2001.

[71] B. Wu and Z. Shi, "An ant colony algorithm based partition algorithm for TSP," *Chinese J. Computers*, vol. 24, no. 12, pp. 1328–1333, 2001.

[72] G. Qin and J. Yang, "An improved ant colony algorithm based on adaptively adjusting pheromone," *Information and Control*, vol. 31, no. 3, pp. 198–210, 2002.

[73] Q. Wu, J. Zhang, and X. Xu, "An ant colony algorithm with mutation features," *Journal of Computer Research & Development*, vol. 36, no. 10, pp. 1240–1245, 1999.

[74] J. Zhang, Q. Gao, and X. Xu, "Ant Colony algorithm: A survey," *Control Theory & Application*, vol. 8, no. 1, pp. 1–5, 2000.

[75] J. Zhang and X. Xu, "An ant colony algorithm with forgetting factors," *Journal of System Simulation*, vol. 11, no. 6 pp. 409–412, 1999.

[76] J. Zhang and X. Xu, "A new evolutionary algorithm-ant colony algorithm," *System Engineering Theory and Practice*, vol. 19, no. 3, pp. 84–87, 1999.

[77] R. Dai, "Research On System Science and System Complexity," *Journal of System Simulation*, vol. 14, no. 11, pp. 1411–1416, 2002.

[78] L. Cao and R. Dai, "On Metasynthesis and Decision Making," *Journal of Computer Research and Development*, vol. 40, no. 4, pp. 531–537, 2003.

[79] X. Li and R. Dai, "System science and complexity(II)," *ACTA Automatica Sinica*, vol. 24, no. 4, pp. 476–483, 1998.

[80] L. Wang and Q. Wu, "Ant system algorithm based system identification," *ACTA Automatica Sinica*, vol. 29, no. 1, pp. 102–109, 2003.

[81] K. Ying and C. Liao, "An ant colony system approach for scheduling problems," *Production Planning & Control*, vol. 14, no. 1, pp. 68–75, 2003.

[82] K.-C. Ying and C.-J. Liao, "An ant colony system for permutation flow-shop sequencing," *Computers & Operations Research*, vol. 31, no. 6, pp. 791–801, 2004.

[83] Y. Li and T. Wu, "A nested ant colony algorithm for hybrid production scheduling problems," *ACTA Automatica Sinica*, vol. 29, no. 1, pp. 95–101, 2003.

[84] B. Christian and S. Michael, "Ant colony optimization for FOP shop scheduling: a case study on different pheromone representations," *Proc. 2002 Congress on Evolutionary Computation*, Honolulu, HI, 2002, pp. 1558–1563.

[85] G. Caroline, G. Marc, and L. Wilson, "A look-ahead addition to the ant colony optimization meta-heuristic and its application to an industrial scheduling problem," *Proc. 4th Int. Conf. on Metaheuristics MIC'2001*, Porto, Portugal, 2001, pp. 1–6.

[86] R. Marc, D. Karl, and F. Richard, "D-ants: Savings based ants divide and conquer the vehicle routing problem," *Computers & Operations Research*, vol. 31, no. 4, pp. 563–591, 2004.

[87] J. Richard and S. Qiang, "Fuzzy-rough data reduction with ant colony optimization," *Fuzzy Sets and Systems*, vol. 149, no. 1, pp. 5–20, 2005.

[88] P. S. Shelokar, V. K. Jayaraman, and B. D. Kulkarni, "An ant colony classifier system: Application to some process engineering problems," *Computers and Chemical Engineering*, vol. 28, no. 12, pp. 1577–1584, 2004.

[89] M. Birattari, G. Di Caro, and M. Dorigo, "Toward the formal foundation of ant programming," in *Ant Algorithms, The Third International Workshop, LNCS 2463*, Edited by M. Dorigo, G. Di Caro, and M. Samples, Berlin: Springer, 2002, pp. 181–201.

[90] M. Daniel and M. Martin, "Modelling the dynamics of ant colony optimization," *Evolutionary Computation*, vol. 10, no. 3, pp. 235–265, 2002.

[91] J. Dreo and P. Siarry, "Continuous interacting ant colony algorithm based on dense hierarchy," *Future Generation Computer Systems*, vol. 20, no. 5, pp. 841–854, 2004.

[92] H. Valckenaers, M. Kollingbaum, and O. Bochmann, "Multi-agent coordination and control using stigmergy," *Computers in Industry*, vol. 53, no. 1, pp. 75–96, 2004

[93] W. J. Gutjahr, "A graph-based ant system and its convergence," *Future Generation Computer Systems*, vol. 16, no. 8, pp. 873–888, 2000.

Chapter 9

Motif Discoveries in DNA and Protein Sequences Using Self-Organizing Neural Networks

Derong Liu and Xiaoxu Xiong

Abstract: In this chapter, we study the problem of motif discoveries in unaligned DNA and protein sequences. The problem of motif identification in DNA and protein sequences has been studied for many years. Current popular algorithms for motif identification in protein sequences face two difficulties: high computational cost and the possibility of insertions and deletions. We propose a self-organizing neural network structure for solving the problem of motif identification in DNA and protein sequences. Our network contains several subnetworks with each performing classifications at different levels. The top level divides the input space into a small number of regions and the bottom level classifies all input patterns into motifs and non-motif patterns. Depending on the number of input patterns to be classified, several levels between the top level and the bottom level are needed to perform intermediate classifications. We maintain a low computational complexity through the use of the layered structure so that each pattern's classification is performed with respect to a small subspace of the whole input space. Our self-organizing neural network will grow as needed (e.g., when more motif patterns are classified). It will give the same amount of attention to each input pattern and it will not omit any potential motif patterns. The definition of pairwise distance between patterns provided in this chapter can deal with up to two insertions/deletions allowed in a motif, while other existing algorithm can only deal with one insertion or deletion. Finally, simulation results show that our algorithm outperforms existing algorithms in certain aspects. In particular, simulation results show that our algorithm can identify motifs with more mutations than existing algorithms and our algorithm works well for long DNA sequences as well.

9.1 Introduction

DNA, RNA and proteins are important molecules that support life on Earth. There are 4 different kinds of nucleotides (A, G, C, and T) that make up the DNA of all organisms. These are the four base letters that constitute the alphabets of DNA. The four base letters of RNA are A, G, C, and U, where the T in DNA is replaced by U in RNA. On the other hand, proteins of all organisms are made up of 20 different kinds of amino acids (letters).

A. Motifs in DNA and Protein Sequences

DNA, RNA and protein sequences can be thought of as being composed of motifs interspersed in relatively unconstrained sequence. A motif is a short stretch of a molecule that forms a highly constrained sequence [2]. The expression of a motif can be in one of the following forms.

1. Use an actual sequence as the description of a motif. Such a sequence is also called a consensus sequence [12, 28, 58]. Each position of the consensus sequence is the letter that appears most frequently in all known examples of that motif, e.g., $ACTTATAA$ and $AGTTATAA$ are two examples of consensus sequence of a motif.

2. Use a so-called "degenerate" expression to show all possible letters for each position of a motif [18, 26]. For example, the expression

$$A - [CG] - T - T - [AC] - [TCG] - A - A \qquad (9.1.1)$$

   indicates that $AGTTCTAA$ and $ACTTAGAA$ are two of the possible occurrences; see, for example, [39] for similar concepts used in the design of degenerate primers [32].

3. Use a more biologically plausible representation to describe a motif. In this case, a probability matrix can be used to assign a different probability to each possible letter at each position in the motif [4, 5, 20]. For example, Table 9.1.1 shows a probability matrix representation of the motif given by (9.1.1). This matrix representation not only gives the possibility of which letter can appear in each position of the motif, but also shows the probability of their appearances. For example, the sixth position of this motif will have letters G, C, and T appearing with probabilities of 20%, 30%, and 50%, respectively.

4. Hidden Markov model (HMM) can also be used to describe motifs [15, 21]. An HMM is obtained by a slight modification of the Markov model. Based on HMM algorithm, an output matrix Π can be formed by the state transition matrix and the probability vector of A, G, C, and T associated

Table 9.1.1: Frequency of each letter appearing in every position of a motif

	1	2	3	4	5	6	7	8
A	1.0	0.0	0.0	0.0	0.67	0.0	1.0	1.0
G	0.0	0.5	0.0	0.0	0.0	0.3	0.0	0.0
C	0.0	0.5	0.0	0.0	0.33	0.2	0.0	0.0
T	0.0	0.0	1.0	1.0	0.0	0.5	0.0	0.0

with each state [35, 50]. It is a probabilistic model for motifs when we have prealigned sequences [57] that are known to share some common blocks.

Understanding what motifs mean is a major part of research in bioinformatics. In order to understand motifs, one needs first to identify and locate them in DNA and protein sequences. By one way or another, biologists have identified some motifs. They can explain their structures, common locations and certain functions. They are usually the beginning of translation of DNA to protein [45]. A protein binds optimally to places with some specific patterns (e.g., motifs) and it can still bind effectively even if one or more positions in the binding site sequence deviate from its ideal binding site sequence [1, 22, 33]. This means that a motif may have slightly different appearances at different locations [40]. The goal of this chapter is to develop algorithms that can identify and locate motifs, if any, given a set of DNA or protein sequences.

Generally speaking, the motif finding problem in DNA sequences can be described as follows: Given a set of unaligned DNA or protein sequences, project the length of motifs and locate all motifs with the projected length that these sequences hold. It is not necessary for all the sequences to have the same motif. Some sequences may have more than one repetition of a motif and some motifs may not show up in every sequence. The appearances of the same motif in the sequences are not necessarily the same. A subsequence* is determined to be a motif if it matches a possible appearance indicated by (9.1.1) or by the matrix representation in Table 9.1.1. Obviously, information provided in Table 9.1.1 is more than that in (9.1.1). Here the frequency or probability of letters in each position of a motif is in $[0, 1]$. Usually the frequency of the letter that appeared most frequently should be larger than 40% [31, 49].

References [16, 34, 46] presented an unsupervised learning method for finding contiguous motifs. This kind of motifs has some biological properties of interest such as being DNA binding sites for a regulatory protein. The work

---

*By subsequence we mean a *contiguous* part of the sequence; this is more commonly called "substring" in the string matching research community (e.g., see [24, 48]).

in [16, 34, 46] showed that unsupervised learning method is a good approach for dealing with the problem of finding motifs. An algorithm called MEME is proposed in [2, 3] for identifying contiguous motifs. This algorithm is an extension to the expectation maximization algorithm for motif finding. The Gibbs sampling algorithm [37, 47, 58] uses a Monte Carlo procedure and it assumes motifs are ungapped sequence blocks. The algorithm tries to converge to a conserved block if it exists. Experimental results showed that the Gibbs sampling method misses motifs when the number of mutations is relatively large [57]. In this chapter, we will develop an algorithm based on a new structure of self-organizing neural networks [19] and we will compare the performance of our algorithm with that of [2, 37]. For motif identification, we will project the length of motifs as well as the maximum number of letters that can be mismatched in a pattern [49]. In this case, the target patterns to be found are described by a given length and by how many letters that can be mismatched.

B. *Multiple Sequence Alignment Based on Motif Discovery*

Multiple sequence alignment (MSA) is another basic problem in computational biology. One of the potential applications of motif discovery is multiple sequence alignment in which identified motifs are used as marks for sequence alignments. The alignment of a set of sequences is basically a matrix where the rows correspond to the sequences in the set, possibly with some spaces inserted, or some gaps in between [53]. Figure 9.1.1 shows an example of alignment of 6 protein sequences which are obtained from the Swiss-Prot protein library [60]. By aligning protein sequences, we can discover similarities and changes in the group of sequences, which may help make further decision including gene classification as well as finding cause of disease.

Multiple sequence alignment method such as CLUSTAL W [53], ITERALIGN [6] and PROBE [43] can also serve as motif identification tools. CLUSTAL W aligns multiple sequences by calculating the global similarity among sequences. ITERALIGN and PROBE produce aligned blocks that are separated by variable-length unaligned segments. Sequence blocks in the alignment results of these methods can be treated as motif sets [43]. Usually these methods work on pre-aligned sequences and the conserved blocks they find may have some limits, such as that the blocks must be in alignable position and at most one pattern from each sequence can be included in a motif set.

C. *Protein Sequence with Insertions and Deletions*

Popular MSA algorithms are progressive alignment [13, 53], iterative alignment [51], Markov model [21] and other stochastic algorithms [2, 23, 26, 57]. Based on these algorithms, several MSA tools have been developed, such as ClustalW [13, 53] and Gibbs sampling [37]. Among existing algorithms and tools,

```
SRC_RSVP    FPIKWTAPEAALYGRFTI...GRVPYPGMVNREVLDQVERG
YES_AVISY   FPIKWTAPEAALYGRFTI...GRVPYPGMVNREVLEQVERG
ABL_MLVAB   FPIKWTAPESLAYNKFSI...GMSPYPGIDLSQVYELLEKD
FES_FSVGA   QVPVKWTAPEALNYGRYSS...GASPYPNLSNQQTREFVEKG
FPS_FUJSV   QIPVKWTAPEALNYGWYSSE...GAVPYANLSNQQTREAIEQG
KRAF_MSV36  TGSVLWMAPEVIRMQDNPFSF...GELPYAHINNRDQIFMVGRG
```

After alignment, we get

```
SRC_RSVP    -FPIKWTAPEAALY--GRFTI...GRVPYPGMVNR-EVLDQVERG
YES_AVISY   -FPIKWTAPEAALY--GRFTI...GRVPYPGMVNR-EVLEQVERG
ABL_MLVAB   -FPIKWTAPESLAY--NKFSI...GMSPYPGIDLS-QVYELLEKD
FES_FSVGA   QVPVKWTAPEALNY--GRYSS...GASPYPNLSNQ-QTREFVEKG
FPS_FUJSV   QIPVKWTAPEALNY--GWYSS...GAVPYANLSNQ-QTREAIEQG
KRAF_MSV36  TGSVLWMAPEVIRMQDNPFSF...GELPYAHINNRDQI.FMVGRG
```

Motif patterns with insertions and deletions

Figure 9.1.1: The alignment of protein sequences.

algorithms based on progressive alignment are the most widely used. However, the computational cost of these algorithms is up to $O(W^2)$, where W is the sum of the length of all sequences. The reason that these algorithms have high computational cost is that the pairwise alignment algorithm is applied on every column of each sequence. For example, in progressive alignment, one has to find a dynamic programming table that includes every column of each sequence. Such a table is huge and not necessary. As shown in Figure 9.1.1, after identifying motifs in the given sequences, we can align the same motif in each sequence, then align the rest in between the motifs.

In the motif discovery problem, we have to deal with motifs with mutations, insertions and deletions. Current motif finding algorithm such as MEME [2], Gibbs sampling [37] and WINNOWER [49], perform well in finding motifs with only mutations. When dealing with insertions and deletions, especially when there are more than one consecutive insertion or deletion in the motif patterns, these algorithms fail at identifying motifs. The Bayesian algorithm [57] can deal with cases with insertion and deletion, but not with more than one consecutive insertions or deletions. Insertions and deletions bring great difficulties to the motif discovery problem because they make the result less predictable and more variable. The motif discovery problem in protein sequences can be described as finding similar fields with certain length, with certain maximum number of columns mutated and with certain number of tolerable insertions or deletions. In this chapter, we consider the case with at most two insertions or deletions or their combinations in a single motif pattern. Both of the two insertions or deletions can be consecutive.

9.2   Subsequences and Encoding

9.2.1 Encoding of DNA and Protein Sequences

We consider the case where all motifs to be identified from a given set of DNA or protein sequences have the same length [42]. In general, the consensus sequence of a motif and the motif itself are not known *a priori* and we have to obtain them by using identification algorithms. What one obtains after the use of identification algorithms are specific appearances of a motif, usually with a few mismatched letter positions comparing to the motif consensus sequence. For a given set of DNA or protein sequences, in order to identify motifs in these sequences, we have to specify the maximum number of letter mismatches that can be tolerated (comparing to the consensus form) in addition to projecting the length of motifs to be found.

Test subsequences, which we call input sequences or input patterns, can be obtained from the given set of DNA or protein sequences once the projected length of motifs is given. Figure 9.2.1 shows a sketch of how input patterns are obtained from a DNA sequence. In the figure, the projected length of motifs is $M = 7$. All subsequences of seven connected letters obtained using a sliding window (see Figure 9.2.1) from the given DNA or protein sequences will form the set of input patterns. For a DNA sequence of length W, we can obtain $W - M + 1$ input patterns if the projected length of motifs is M.

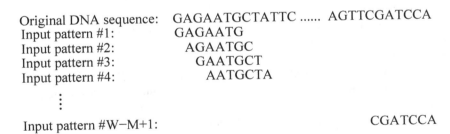

Original DNA sequence: GAGAATGCTATTC ...... AGTTCGATCCA
Input pattern #1:   GAGAATG
Input pattern #2:   AGAATGC
Input pattern #3:   GAATGCT
Input pattern #4:   AATGCTA
                             ⋮
Input pattern #W−M+1:   CGATCCA

Figure 9.2.1: An illustration on how to obtain input patterns ($M = 7$) from a given DNA sequence.

Letters used in DNA or protein sequences will be encoded using binary numbers. All letters will be encoded using binary code with the same length, for example, four for DNA and RNA sequences and 20 for protein sequences. Table 9.2.1

Table 9.2.1: Encoder table for DNA letters

Standard	1	1	0	0
A	1	1	0	1
C	1	1	1	0
G	1	0	0	0
T	0	1	0	0

shows an example of binary codes designed for DNA sequences. There are four letters in this case and each letter is encoded by flipping one bit of the standard code '1 1 0 0.' Letters coded this way will have exactly the same Hamming distance between any pair of letters [17, 54]. Also, the scheme shown in Table 9.2.1 can also guarantee that 1's and 0's will appear on average the same number of times. The coding scheme we used in the present chapter is similar to [29] even though in reality certain pairs of letters may appear closer than others, e.g., in protein sequences, L and I are more similar than L and R [52].

9.2.2 Pairwise Distance Calculation with Insertions and Deletions

In the preceding section, we assume that there are at most two consecutive letter insertions or deletions in a motif pattern. Under this assumption, we analyze a group of motif patterns from protein sequences, e.g., patterns in Figure 9.2.2. We observe that in these patterns, when they are aligned according to identified motifs, column i of a pattern can be aligned to one of the columns in the range from $i-4$ to $i+4$ of other patterns. The two aligned columns can have a maximum index difference of 4. The extreme case happens when one of the pattern has two insertions and the other pattern has two deletions, e.g., pattern #1 and pattern #2 in Figure 9.2.2. Before we align the patterns, pattern #1 has insertions at column 4 (letter "A") and column 14 (letter "C"), while pattern #2 has deletions at column 2 (between two letter "D"s) and column 4 (between letters "M" and "S"). We notice that columns 15 to 17 of pattern #1 should be aligned to columns 11 to 13 of pattern #2. In the illustration of Figure 9.2.2, we assume the length of motif to be 15. Since we allow each appearance of motif to have a maximum of 2 letter insertions, we can choose test subsequences of length 17.

To identify motifs with up to two insertions and/or deletions, test subsequences are also obtained from the given protein sequences using a sliding window as shown in Figure 9.2.1. Let W be the length of a protein sequence, and let M be the length of the motif to be identified. The length of the sliding window will be $M' = M + 2$. Placing the sliding window at the beginning of the sequence, we get

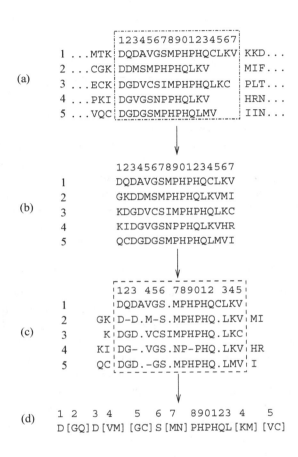

```
                    12345678901234567
    1 ...MTK DQDAVGSMPHPHQCLKV KKD...
    2 ...CGK DDMSMPHPHQLKV       MIF...
(a) 3 ...ECK DGDVCSIMPHPHQLKC   PLT...
    4 ...PKI DGVGSNPPHQLKV       HRN...
    5 ...VQC DGDGSMPHPHQLMV     IIN...
```

```
                12345678901234567
    1           DQDAVGSMPHPHQCLKV
    2           GKDDMSMPHPHQLKVMI
(b) 3           KDGDVCSIMPHPHQLKC
    4           KIDGVGSNPPHQLKVHR
    5           QCDGDGSMPHPHQLMVI
```

```
                123 456 789012 345
    1           DQDAVGS.MPHPHQCLKV
    2      GK D-D.M-S.MPHPHQ.LKV MI
(c) 3       K DGD.VCSIMPHPHQ.LKC
    4      KI DG-.VGS.NP-PHQ.LKV HR
    5      QC DGD.-GS.MPHPHQ.LMV I
```

```
        1 2   3 4    5   6 7   890123 4    5
(d)     D[GQ]D[VM][GC]S[MN]PHPHQL[KM][VC]
```

Figure 9.2.2: The aligned motif patterns with at most 2 insertions/deletions ('−' indicates a deletion in this sequence and '.' indicates an insertion in another sequence). (a) Protein sequence segments. (b) Test patterns of length = 17. (c) Alignment. (d) Motif expression, length = 15.

the first input pattern with length of $M + 2$. Shift the window one column at a time to get all the input patterns. The total number of input patterns we get from the sequence will be $W - M' + 1 = W - M - 1$. Due to the effects of sliding window, when input patterns are finally aligned, the maximum column index difference becomes 2 (a positive difference of 2 or a negative difference of 2) between any pairs of input patterns. In order to define the pairwise similarity value (or distance) between a pair of input patterns that may have column index difference up to 2 due to insertions and deletions, we put every 3 consecutive letters into a group. We will now consider the distance between groups of letters. Comparing between two appearances of a motif, every group of three consecutive letters in one appearance will have at least one letter in common with the corresponding group in the other appearance, assuming a maximum of 2 letter insertions or deletions when there is no mutations. These groups can be obtained by applying sliding window of length 3 to each input pattern. For each input pattern with length of M', we will get $M = M' - 2$ groups of letters, each group with length of 3. For example, for input pattern #2 of Figure 9.2.2 (b), we get the following groups of 3 letters:

$$\{GKD, KDD, DDM, DMS, MSM, SMP, MPH, PHP, HPH, PHQ,$$
$$HQL, QLK, LKV, KVM, VMI\}.$$

Next, we will encode all the input patterns using binary digits. Each protein letter will be encoded using a 20-digit binary vector, with one digit flipped from

$$\{1, 1, \cdots, 1, 0, 0, \cdots, 0\}$$

(ten 1s followed by ten 0s). This coding strategy guarantees that the coded vectors of any two different protein letters have exactly two digits that are different. Encoding a protein pattern means encoding all the letter groups of that pattern. For example, each of the above groups of 3 letters will be encoded by a $60 = 20 \times 3$ digit binary code.

First we study the 3-letter groups from corresponding positions of two input patterns. We want to determine the distance or the similarity value between two 3-letter groups. We consider two possibilities. The first one is that there is one insertion or one deletion in either of the groups. If this is the case, a sub-group of 2 letters from one of the groups may match a 2-letter sub-group from the other group. All these 2-letter sub-groups should follow the same order as they appear in the 3-letter groups. For example, 2-letter sub-groups are AC, CG or AG from ACG. Such 2-letter sub-groups from a 3-letter group will always have 3 possibilities. Second, there are two insertions or two deletions in either one of the 3-letter groups. In this case, as long as there is a common letter appearing in both groups, we would grant a relatively smaller pairwise distance value between these 3-letter groups. The above strategy can be expressed by the following mathematical description. Each input pattern (length $M' = M + 2$) is converted into a test pattern with M 3-letter groups. After encoding, each input pattern will be in the form of binary vectors $P = \{p_1, p_2, \cdots, p_M\}$, where the length of each binary vector p_i is 60. Each binary vector p_i has 3 letters. Each binary vector can be expressed as $p_{i,1-20}$, $p_{i,21-40}$ and $p_{i,41-60}$. Now we form sub-groups using two of the three vectors following the same order as they have in the 60-digit 3-letter group vector. We will get 3 sub-groups and we denote them as

$$p_i' = \left(p_{i[1]}', p_{i[2]}', p_{i[3]}'\right)$$
$$= \left(\{p_{i,1-20}p_{i,21-40}\}, \{p_{i,1-20}p_{i,41-60}\}, \{p_{i,21-40}p_{i,41-60}\}\right).$$

The distance between any two given vectors p_i and q_i will be defined as

$$d_i = \min_{j,k,l \in \{1,2,3\}, j \neq k \neq l} \left(\sum_{r=1}^{40} |p_{i[1],r}' - q_{i[j],r}'| \right.$$
$$+ \sum_{r=1}^{40} |p_{i[2],r}' - q_{i[k],r}'| + \sum_{r=1}^{40} |p_{i[3],r}' - q_{i[l],r}'| \right)$$
$$+ \sum_{k=1}^{20} \left| (p_{i,k} + p_{i,k+20} + p_{i,k+40}) \right.$$
$$\left. - (q_{i,k} + q_{i,k+20} + q_{i,k+40}) \right|. \tag{9.2.1}$$

Let $\mathcal{D}(P, Q)$ denotes the pairwise distance of any two input patterns P and Q, each with $m + 2$ letters. We get

$$\mathcal{D}(P, Q) = \sum_{i=1}^{m} d_i. \tag{9.2.2}$$

Equation (9.2.1) shows the distance between two 3-letter groups. In the equation, the first part reflects the minimum sum of distances between the sub-groups obtained from both 3-letter groups. For each 3-letter group, we get 3 possible sub-groups. Then there are a total of $6 = 3!$ possible cases of putting together any two sub-groups from each group. The second part of (9.2.1) is an absolute value that reflects the sum of distances between single letters from each group. For example, if we want to calculate the pairwise distance between two 3-letter groups ACG and tag, the first part of (9.2.1) reflects the sum of the distances between AG and ag, CG and tg, and AC and ta. The second part reflects the sum of the distances between A and a, G and g, and C and t. We use this strategy because the given two 3-letter groups are no longer alignable due to insertions and deletions. The present definition of distance in (9.2.1) gives a good representation of the similarity between a pair of 3-letter groups.

9.3 Self-Organizing Neural Networks for Motif Identification

9.3.1 A New Structure of Self-Organizing Neural Networks

This subsection describes the structure of our self-organizing neural network for motif discovery. The basic structure forms the subnetworks used in our self-organizing neural networks and contains two layers, i.e., an input layer and an output layer [7–11]. The number of output neurons of a subnetwork is the same as the number of categories classified by this subnetwork. The number of input neurons is determined by the projected length of motifs after encoding [56], e.g., $m' \times 20$, where $m' = m + 2$ and m is the length of projected motifs. The input patterns are obtained from the given protein sequences by taking all subsequences with the same length of m'. Each output neuron represents a category that has been classified by a subnetwork and each output category is represented by the connection weights from all input neurons to the corresponding output neuron. Subnetworks perform the function of classification in a hierarchical manner. The first subnetwork is placed at the top level and it performs a very rough classification, e.g., dividing the input space into 4–8 categories. The second subnetwork is placed at the next level and it usually divides the input space into 16–32 categories, which indicates a slightly more detailed classification of the input space. The last subnetwork in our self-organizing neural network will be placed at the lowest level and it classifies all the input patterns into either a motif or a non-motif category with one or a few patterns [36, 41]. Typically, the number of output neurons will be very large for the last subnetwork and gradually reduced to a small number for the first subnetwork. Figure 9.3.1 shows the structure of our self-organizing neural network with three subnetworks. In the structure shown in the figure, there are four input neurons and three subnetworks. The first subnetwork has 2 output neurons, the second subnetwork has 3 output neurons, and the third subnetwork has 4 output neurons. Each of the output neurons represents a category that has been created and it is represented by the connection weights to the output neuron. The output category α of the first subnetwork contains two patterns (a and b) and the other contains one pattern (c). The output category a of the second subnetwork contains two patterns (1 and 2) and the other two categories each contains just one pattern. The output categories 1 and 2 of the third subnetwork represent two motifs while categories 3 and 4 are not motifs (if we desire to have at least three patterns for each motif identified).

We can also illustrate the structure in Figure 9.3.1 using a tree of sorting bins as shown in Figure 9.3.2. In the figure, there are sorting bins at each level of the tree. From one level down to the next, the number of bins increases. At the lowest level, bins will be divided into motifs and non-motif categories. Figure 9.3.2 also shows an example of how a new input pattern is sorted into a category. The new input

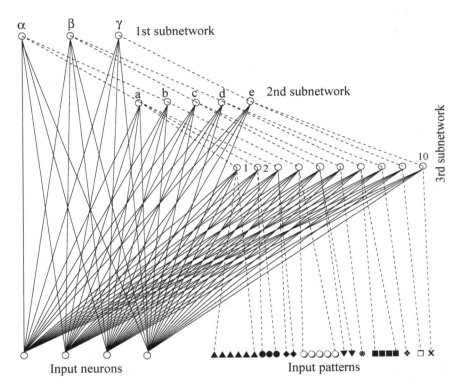

Figure 9.3.1: Structure of the self-organizing neural networks.

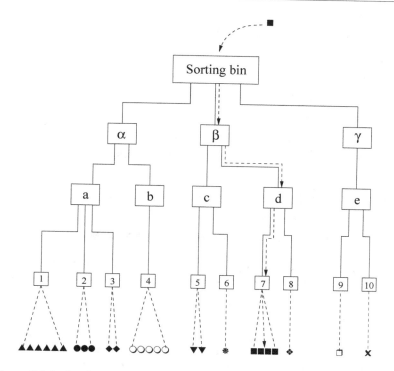

Figure 9.3.2: Sorting strategy of the self-organizing neural networks method.

pattern is first sorted by the bin at the top level. Then it is distributed to a suitable bin at the next level, and this process continues until the pattern reaches the lowest level where it is classified into a motif category or a non-motif category. By using the present neural network structure, the identification of motifs can be completed in one cycle of sorting (presenting all input patterns to the network). Multiple categories (at the lowest level) as shown in Figure 9.3.2 will be generated in one cycle. On the other hand, existing methods for motif discoveries, such as MEME and Gibbs sampling methods, only sort the input patterns in each cycle into two groups: a motif category and a group containing all other patterns, as shown in Figure 9.3.3. Using these algorithm, multiple trials will have to be employed so that multiple motifs can be discovered.

9.3.2 Rules for Weight Update and Output Node Creation

When an input pattern is applied to our self-organizing neural network, it will be classified to an output category by every subnetwork. An output category of a

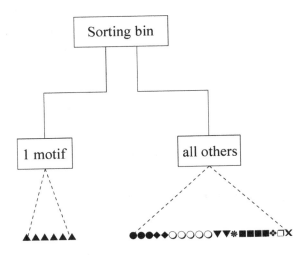

Figure 9.3.3: Sorting strategy of the MEME and Gibbs methods.

lower level subnetwork is said to belong to an output category of a higher level subnetwork if one or more input patterns are classified to belong to these two output categories. The connection weights for each category of the last subnetwork (at the lowest level) are calculated as the center of the category, i.e., the geometric center of all input patterns that are currently classified into the category associated with the corresponding output neuron. The connection weights for an output category of all other subnetworks (except the last subnetwork) are calculated as the geometric center of all categories from the lower level of subnetwork that belong to this category.

When a new input pattern is applied to a subnetwork, its classification to an output category of every subnetwork involves the following two steps.

1) The distance between the input pattern and each output category is calculated by comparing the input pattern with the connection weights from the input neurons to that category. The minimum of these distances is determined and thus a winning category is also determined. This step works similarly to the winner-take-all networks [25]. These winning neurons form the tree of classification as in Figure 9.3.1. For the example network shown in Figure 9.3.1, an input pattern will be first compared to the three categories $\{\alpha\}$, $\{\beta\}$ and $\{\gamma\}$ at the first level. At the next level, it will be either compared to $\{a, b\}$, $\{c, d\}$ or $\{e\}$ depending on which of the three output categories at the first level becomes the winning category.

2) Within the winning category, the similarity of all patterns in this category including the new pattern will be calculated and compared to a threshold value. If the similarity value is less than the threshold, the new pattern will be classified into the winning category. Otherwise, the new pattern cannot be classified into the winning category.

Based on the above two steps, we can describe the discovery algorithm for the general case that there is no insertion and deletion column in the motifs. Assume there are a total of L subnetworks for $l = 1, 2, \cdots, L$. Assume that there are M input neurons and the lth subnetwork has N_l output neurons. The input patterns obtained form the given DNA or protein sequences are used as motif candidates and are provided to each subnetwork of our self-organizing neural network. The outputs of the last subnetwork correspond to classifications of all input patterns into motifs and non-motif categories. The projected length of motifs possibly existing in the input sequences is the same as M.

We first describe the case for identification of motifs without insertions and deletions. We denote the input patterns as x^i, $i = 1, 2, \cdots$. Suppose that t input patterns have been presented to the network and have been classified. When a new input pattern, i.e., the $(t + 1)$st pattern x^{t+1}, is introduced to the lth subnetwork, the distances from the new input pattern to those categories of the lth subnetwork that belong to the $(l - 1)$st winning category W_q^{l-1} is calculated as

$$y_n^l = \sum_{m=1}^{M} |x_m^{t+1} - w_{mn}^l|, \text{ for } n \in W_q^{l-1},$$

where x_m^{t+1} is the mth component of the input pattern x^{t+1} and w_{mn}^l is the connection weight of the lth subnetwork from the mth input neuron to the nth output neuron after the presentation of the tth input sequence. Denote

$$y_q^l = \min_{n \in W_q^{l-1}} \{y_n^l\},$$

i.e., the qth output category of the lth subnetwork is the winning category that has the smallest distance to the new input pattern. Assume that the qth output category of the lth subnetwork contains p_q^l patterns from the $(l+1)$st subnetwork. Within this winning category q, we will calculate the similarity value of all the $p_q^l + 1$ patterns including the new input pattern. The similarity value of a group of patterns is calculated as the maximum of the pairwise distance [27] between all pairs of patterns in the group.

For the winning category q determined above, we calculate the distances from the new input pattern to all other patterns in the category as

$$d_j^l = \sum_{m=1}^{M} |x_m^{t+1} - e_{mj}^{l+1}|, \ j = 1, 2, \cdots, p_q, \tag{9.3.2}$$

where

$$e_{mj}^{l+1} = \begin{cases} x_m^j, & \text{if } l = L \text{ and } x_m^j \text{ belongs to the category } q \text{ of the } (l-1)\text{st level,} \\ w_{mj}^{l+1}, & \text{if } 1 \le l < L \text{ and } w_{mj}^{l+1} \text{ belongs to the category } q \text{ of the } l\text{th level.} \end{cases} \tag{9.3.3}$$

We then perform the following threshold tests. If

$$\max_{1 \le j \le p_q} \{d_j^l\} < \rho_l, \tag{9.3.4}$$

this new input pattern will be considered to match the category q of the lth subnetwork. The threshold value ρ_l in (9.3.4) takes different values for different level of subnetwork. We note that all pairwise distances in this category will be less than the threshold ρ_l if (9.3.4) is satisfied for the new input pattern since all other patterns are previously classified into this category using the same threshold value.

If we consider insertions and deletions in the motif set, we will use the pairwise distance defined in Subsection 9.2.2. Equation (9.3.2) will be replaced by:

$$d_j^l = \mathcal{D}(x_m^{t+1}, e_{mj}^{l+1}), \ j = 1, 2, \cdots, p_q, \tag{9.3.5}$$

where e_{mj}^{l+1} is defined in (9.3.3), $\mathcal{D}$ is defined in (9.2.2) and p_q is the number of patterns in category q. We then perform the threshold tests as in (9.3.4).

If there is only one category wins the match, this new input pattern will be classified into the category q of the lth subnetwork. Otherwise we need to perform further similarity test described in Step 2. If the set of winning category is Q, we calculate the distance from the new input pattern to the center of each category in Q. For all if $n \in Q$, we calculate

$$y_n^l = \mathcal{D}(x^{t+1}, w_n^l),$$

where w_n^l is the connection weight vector from the input neurons to the nth output neuron after the presentation of the tth input pattern. Denote

$$y_q^l = \min_{n \in Q}\{y_n^l\}, \tag{9.3.7}$$

if the qth output category is the winning category that has the smallest distance to the new input pattern. If there are more than one category having the same smallest value, we will pick the category that has the most number of patterns in it as the winning category.

If there is a category q wins in both (9.3.4) and (9.3.7), this new input pattern will be classified into this winning category of the lth subnetwork. Otherwise, the new input pattern cannot be classified into any existing category at this level.

We describe in the following some more details about our calculation procedure.

a) We start from the top level, i.e., the first subnetwork, and work down the level one by one, when classifying a new input pattern. After a winning category has been determined at the lth level, the input pattern will only be compared to those patterns at the $(l+1)$st level that are classified to belong to the winning category at the lth level and the winning category is denoted by W_q^l.

b) If the threshold tests in (9.3.4) are successful for $l = 1, 2, \cdots, L$, we perform the following updates for the Lth subnetwork:

$$w_{mq}^L := \frac{1}{p_q^L + 1} \sum_{j=1}^{p_q^L + 1} x_m^j = \frac{1}{p_q^L + 1} \left[p_q^L \times w_{mq}^L + x_m^{t+1} \right],$$

$$m = 1, 2, \cdots, M, \ p_q^L := p_q^L + 1,$$

where $x^{p_q^L + 1}$ indicates the new input pattern x^{t+1} for convenience. We perform the following updates for the rest of subnetworks:

$$w_{mq}^l := \frac{1}{p_q^L} \sum_{j=1}^{p_q^L} w_{mj}^{l+1}, \ m = 1, 2, \cdots, M, \ l = L - 1, L - 2, \cdots, 2, 1.$$

c) If the threshold tests in (9.3.4) are successful for $l = 1, 2, \cdots, L_1$, where $L_1 < L$, we will add an output neuron to subnetworks $L_1+1, L_1+2, \cdots, L$. Each of these newly added categories will contain only one pattern and the weights of the new categories are chosen as

$$w_{mn}^l = x_m^{t+1}, \ m = 1, 2, \cdots, M, \ n = N_l + 1, \ l = L_1 + 1, L_1 + 2, \cdots, L.$$

We also update the number of output neurons for these subnetworks as

$$N_l := N_l + 1, \ p_{N_l}^l = 1, \ l = L_1 + 1, L_1 + 2, \cdots, L.$$

In this case, it is not necessary to perform threshold tests for subnetworks $L_1 + 1, L_1 + 2, \cdots, L$ anymore. For subnetworks $1, 2, \cdots, L_1$, we will perform the following updates:

$$p_q^{L_1} := p_q^{L_1} + 1,$$

$$w_{mq}^l := \frac{1}{p_q^l} \sum_{j=1}^{p_q^l} w_{mj}^{l+1}, \ m = 1, 2, \cdots, M, \ l = L_1, L_1 - 1, \cdots, 2, 1.$$

9.3.3 Optimal Choice of Consecutive Patterns

Consecutive patterns are input patterns from the same protein sequence and with their location differences being smaller than the desired length of motif. For example, $x^t, x^{t+1}, x^{t+2}, \cdots$ are consecutive patterns. According to the pairwise distance defined earlier in Subsection 9.2.2, consecutive patterns can win the similarity test in the same category. Apparently, they should not be included in the same category. We need to determine which of the consecutive patterns should really be in the category. During the classification procedure of a new input pattern, if there is a consecutive pattern to this new input pattern in a category, we need to determine whether this new input pattern matches better with the category than its consecutive pattern that is already in the category. To achieve this, we use the average pairwise distance within a category. When we find a consecutive pattern in the category to be checked, we make pairwise similarity tests between the new input pattern and all patterns in the category except its consecutive one. If (9.3.4) is satisfied for this category, we need to examine whether the new input pattern is a better choice than its consecutive pattern in the category. We compare two average pairwise distance values. The first one is the average of all the pairwise distance values from the new input to all the patterns in the category except the consecutive pattern. The second one is the average of all the pairwise distance values from the consecutive pattern to the rest of patterns in the category. For example if the qth category in the lth subnetwork wins, meanwhile the σth pattern in this category, x^σ is a consecutive pattern to the new input pattern x^{t+1} ($\sigma \neq t + 1$), we use the following equation to calculate the average pairwise distance values for x^k, $k = \sigma$ or $t + 1$,

$$d_q^{\text{avg}}(x^k) = \frac{1}{p_q - 1} \sum_{j=1, j \neq \sigma}^{p_q} \mathcal{D}(x^k, x^j). \tag{9.3.8}$$

If

$$d_q^{\text{avg}}(x^{t+1}) < d_q^{\text{avg}}(x^\sigma),$$

we choose the new pattern as a replacement of the consecutive pattern classified earlier. We delete the old pattern from the category and add the new one. Change the weights of the winning category accordingly. If

$$d_q^{\text{avg}}(x^{t+1}) > d_q^{\text{avg}}(x^\sigma),$$

we skip this category and do the similarity test for the rest of categories.

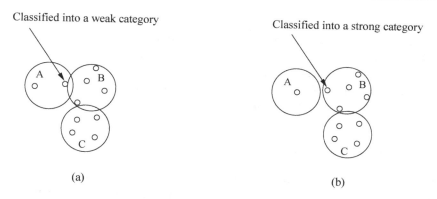

Figure 9.3.4: (a) A new input pattern fails to be classified into the category indicated by C. (b) The classification succeeded in a different trial.

9.3.4 Order Randomization and Recycling of Input Patterns

After one cycle of the motif identification procedure, our neural network is able to identify most of the patterns belonging to some motifs. However, there might still be some missing ones. That is because that the classification of an input pattern to a category using the present self-organizing neural network will be affected by the order in which input patterns are presented to the network. The new input pattern will only be tested in existing categories in the network. If the pairwise test wins in an earlier category, the pattern will not be included in categories built later. Figure 9.3.4 (a) shows a case where an input pattern is placed in a non-motif category A (e.g., a category with less than two members). After that, the same pattern may not be considered to belong to a motif category B that is created later. In Figure 9.3.4 (b), the same input pattern is classified to the motif category B since in this case the category B is created before A.

To avoid the problem shown in Figure 9.3.4 (a), we use the following procedure. After the first trial, we keep all motif categories and recycle all input patterns in non-motif categories to determine whether we have misclassified any patterns during the first trial. (1) Initial trial: We randomly selected the order of presentation of all input patterns, and run our algorithm to identify motif categories. (2) Recycling input patterns: Keep all motif categories including all patterns belonging to these categories, remove all non-motif categories, randomly selected the order of presentation of input patterns from non-motif categories, and run our algorithm. After the second trial with recycled input patterns, the problem shown in Figure 9.3.4 (a) if any, will be resolved. Thus, it is likely some motif categories will get new members to join. It is also likely that some more motif categories will be created.

In our simulation studies, we have used 3 as the threshold to determine whether a category is a motif or not, i.e., if a category contain three members or more, it is classified as a motif category and otherwise, it is not. Our simulation results also indicate that two trials is enough to identify all motif categories since additional trials have not produced anything new.

9.4   Simulation Results

We will compare our algorithm with existing algorithms in the present simulation studies. We will use both randomly generated and real DNA sequences to test our algorithm. In each example, input patterns to our self-organizing neural network will be obtained from DNA or protein sequences as described in Section 9.2. The first three examples consider the case of motifs without insertions and deletions and the last two examples deal with motifs with up to two insertions and/or deletions.

Example 9.4.1. In this example, we will apply our algorithm to motif discoveries in Ornithine Carbamoyltransferase family protein sequences (OTCase family). We choose 9 OTC samples from Swiss-Prot gene library [60]. The lengths of these sequences are between 305 and 340 letters. The average length is 322. We project the length of the target motif to be 17 and the maximum number of mismatched letters to be 4. A total of 2754 input patterns are obtained from the 9 protein sequences. We consider in this example the number of tolerable mismatches to be 4 letters. We choose to use three levels of subnetworks with one output neuron initially at each level. After the presentation of all input patterns in random order to the network, we obtain 8 motif sets each has at least 5 appearances. Results are shown in Figure 9.4.1. The motif sets are marked with different underlines or blocks. The consensus forms of the motifs are used to summarize the results.   □

Example 9.4.2. In this example, we will test our algorithm on a group of DNA sequences that share strong and weak motifs [30, 38]. The target samples are ancient conserved untranslated sequences (ACUTS). The DNA samples are obtained from the ACUTS database [59]. The ACUTS DNA sequences are usually used in identifying new regulatory elements in untranslated regions of protein-coding genes [14]. We pick the ACTAC_3UT entries which included 7 pieces of sequences. The lengths of the sequences are between 98 and 1866 residues and the average length is 525. The projected length of target motifs is 17 and the maximum number of mismatched letters is 6. A total of 3225 input patterns are obtained from the 7 DNA sequences. After applying the input patterns

```
              1
OTC2_ECOLI    SDLYKKHFLKLLDFTPAQFTSLLTLAAQLKADKKNGKEVQKLTGKNIALIFEKDSTRTRCSFEVAAFDQGARVTYL
OTC1_PSESH    NARHFLSMMDYTPDELLGLIRRGVELKDLRIRGELFEPLKNRVLGMIFEKSSTRTRLSFEAGMIQLGGQAIFLSHR
OTC1_ECOLI    SGFYHKHFLKLLDFTPAELNSLLQLAAKLKADKKSGKEEAKLTGKNIALIFEKDSTRTRCSFEVAAYDQGARVTYL
OTC1_LACLA    MFQGRSFLKEIDFSKDELLYLIDFAIHLKKLKKEHIQHKYLLDKNIALIFEKTSTRTRAAFTTAAVDLGAHPEFLG
OTC2_LACLA    MVTTNKRDFITTEDYTKEEILDIVTLGLKIKAAIKNGYYPPLLKNKSLGMIFQQTSTRTRVSFETAMTQLGGHAEY
OTC2_PSESF    KITSLKNRNLLTMNEFNQSELSHLIDRAIECKRLKKDRIFNLGLNHLNICIIFLKPSGRTSTSFVVASYDEGAHFQ
OTC_BACAN     MSTVQVPKLNTKDLLTLEELTQEEIISLIEFAIYLKKNKQEPLLQGKILGLIFDKHSTRTRVSFEAGMVQLGGHGM
OTC_ANASP     MAALLGRDLLSLADLTPTELQELLQLATQLKSQQLKLRCNKVLGLLFSKASTRTRVSFTVAMYQLGGQVIDLNPNV
OTC_AQUAE     MKRDFVDLWDLSPKEAWEIVKKTLKVKKGEEELGKPLSGKTIALLFTKPSTRTRVSFEVGIYQLGGNSLFFQEKEL

              77
OTC2_ECOLI    GPSGSQIGHKESIKDTARVLGRMYDGIQYRGHGQEVVETLAQYAGVPVWNGLTNEFHPTQLLADLMTMQEHLPGKA
OTC1_PSESH    DTQLGRGEPIADSAKVMSRMLDAVMIRTYAHSNLTEFAANSRVPVINGLSDDLHPCQLLADMQTFLEHRGSIKGKT
OTC1_ECOLI    GPSGSQIGHKESIKDTARVLGRMYDGIQYRGIVETLAEYASVPVWNGLTNEFHPTQLLADLLTMQEHLPGKA
OTC1_LACLA    PNDIQLGKKESISDTAKVLGSMFDGIEFRGFKQSDVEILAKDSGRPVWNGLTDVWHPTQMLADFMTIKEHFGHLQD
OTC2_LACLA    LAPGGQIQLGGHETIEDTSTVLSRLLDIIMARVDRHESVNNLAKHTTIPVLNGMSDYNHPTQEVGDLTTMIEHLPAG
OTC2_PSESF    FFPADNIRFGHKESIKDFARVVGRLFDGIAFRGFEHEVAEELAKHSGIPVWNALTDTHHPTQVLADVMTVKEEFGR
OTC_BACAN     FLNGKEMQMGRGETVSDTAKVLSHYIDGIMIRTFSHADVEELAKESSIPVINGLTDDHHPCQALADLMTIYEETNT
OTC_ANASP     TQVSRGEPVQDTARVLERYLDVLAIRTFEQQELATFAEYAKIPVINALTDLEHPCQILADLLTVQECFDSISGLTL
OTC_AQUAE     QVSRGEDVRDTARTLSKYVDGVIVRNHSHTWLKEFANFASVPVINALTNMSHPCQILSDVFTLYEHYGEELKNLKV

              153
OTC2_ECOLI    FNEMTLVYAGDARNNMGNSMLEAAALTGLDLRLLAPKACWPEESLVAECSALAEKHGGKITLTEDVAAGVKGADFI
OTC1_PSESH    VAWIGDGNNMCNSYIEAAIQFDFQLRVACPAGYEPNPEFLALAGERVTIVRDPKAAVAGAHLVSTDVNTSMGQEEE
OTC1_ECOLI    FNEMTLVYAGDARNNMGNSMLEAAALTGLDLRLVAPKACWPEEALVTECRALAQQNGGNITLTEDVAKGVEGADFI
OTC1_LACLA    LTLAYVGDGRNNVANSLLVTGAILGVNITIISPESLQPALEIQKLARKYAMKSRSKISIRTDLNGLENADIVYTDV
OTC2_LACLA    KKLEDCKVVFVGDATQVCFSLGLIATKMGMHFVHFGPKGYQLNEEHQAKLAANCEVSGGTYEVTDDEESIVGADFL
OTC2_PSESF    IEGVTIAYVGDGRNNMVTSLAIGALKFGYNLRIIAPNALHPTDAVLAGIYEQTPERNGSIEIFTEVAAGVHQADVI
OTC_BACAN     FKGIKLAYVGDGNNVCHSLLLASAKVGMHMTVATPVGYKPNEEIVKKALAIAKETGAEIEILHNPELAVNEADFIY
OTC_ANASP     TYVGDGNNVANSLMLGCALAGMNVRIATPSGYEPNPQVVAQAQAIADGKTEILLTNDPDLATKGASVLYTDVWASM
OTC_AQUAE     AYVGDGNNVCNTLMVGAGMFGLKLFVATPEGYEPNSYYYKKALEFSKENGGSVELTNNPVESVKDADVVYTDVWVS

              229
OTC2_ECOLI    YTDVWVSMGEAKEKWAERIALLRGYQVNAQMMALTDNPNVKFLHCLPAFHDDQTTLGKQMAKEFDLHGGMEVTDEV
OTC1_PSESH    TARRMALFAPFQVTRASLDLAEKDVLFMHCLPAHRGEEISVDLLDDSRSVAWDQAENRLHAQKALLEFLVAPSHQR
OTC1_ECOLI    YTDVWVSMGEAKEKWAERIALLREYQVNSKMMQLTGNPEVKFLHCLPAFHDDQTTLGKKMAEEFGLHGGMEVTDEV
OTC1_LACLA    WVSMGEEAQTAKRIKLLKSYQINQKVVEKIINKNFIFMHCLPSFHDLNTEVMKEIKENYNLNELEVTDEVFNSKNS
OTC2_LACLA    YTDVWISMGESVSVEERIALLKPYKVTEKMMALTGKADTIFMHCLPASRGEEVVDAVIDGPNSICFDEAENRLTS
OTC2_PSESF    TDVWMSMGQEGEEEKYTLEQFYQINKELVKHAKQTYHFLHCLPAHREEEVTGEIIDGPQSIVFEQAGNRLHAQKAL
OTC_BACAN     GQEAEADDRFPIFQPYQISEQLLSLAEPNAIVLHCLPAHRGEEITEEVIEGSQSRVWQQAENRLHVQKALLASILG
OTC_AQUAE     MGEENKNIEAFLPYQVNEKLLSFAKSSVKVMHCLPAKKGQEITEEVFEKNADFIFTQAENRLHTQKTLMEFLFREP

              305
OTC2_ECOLI    FESAASIVFDQAENRMHTIKAVMMATLGE
OTC1_PSESH    A
OTC1_ECOLI    FESAASIVFDQAENRMHTIKAVMVATLSK
OTC1_LACLA    VVFEQAENRMHTIKEVMAATLGDLFIPKI
OTC2_LACLA    IRALLVWLMSDYAEKNPYDLKAQAKAKAELEAYLAK
OTC2_PSESF    VFDQGENRMHTIKALMLETVVP
OTC_BACAN     LVSLFKNVEELS
OTC_ANASP     AE
OTC_AQUAE     QA
```

Motif consensus form:

Motif 1:	PDELLHLIDRAIELKRL	Motif 5:	GLTLAYVGDGRNNMNNS
Motif 2:	NKNIGLIFEKPSTRTRV	Motif 6:	KGADVIYTDVWVSMGEE
Motif 3:	QFGHKESIKDTARVLGR	Motif 7:	EEEKRIALFRPYQVNKK
Motif 4:	WNGLTDDHHPTQLLADL	Motif 8:	VKFMHCLPAFHDDETTE

Figure 9.4.1: The motif discovery results in OTCase family proteins.

Table 9.4.1: Comparison of the motif sets from self-organizing neural network method and MEME method

ID	Our Result		MEME Result	
	Consensus	NOS	Consensus	NOS
1	GTACAGTTTGTTTATAC	9	GTACAGTTTGTTTATAC	9
2	ACCTTCCACTCAGGATG	11	ACCTTCCACTCAGGATG	11
3	CAAAAGAACAATAATCA	8	CAACATATTCATAGTCT	7
4	CACAGAGGCACCAATTT	7	CACACCCGAGCATTCT	6
5	CTTCTTGGAATCCTCTG	7	CTTCTTGGAATCCTCTG	7
6	TTCTAATATTTATTGCT	7	TTCTAATATTTATTGCT	4
7	TGGACTTGGAACCTATA	7	GACTCGCAACCTACAAA	4
8	AGCTTTCTGAATAAAAG	7	ACTTTCTGAATAAAAGA	6
9	TACAATAACAATACCTT	7	TACAGGAAAGATACCTT	5
10	GGTGAGTCTGTGCATTT	7	AATGAGTCTGTGCATAT	7
11	TTCCAATACATTAATAT	7	CCCTTTTCTCTACATTT	7
12	CCATCGATCGAACGATT	7	CGATCAATCGAACGATT	4
13	CCCCCACCTCTCATCAG	7	CATCTCCCATCAGTCAT	4
14	ACTTTCTGAATAAAAGA	6	GACAATCAAAGGAAACA	5
15	GGGTTCGGCAGATGTTT	6	GGGTGGGGCAGTTGTTT	4
16	TTTGTGGTTCAAAATAT	13		
17	AGGAATTTAAAACAAAT	11		
18	AATAATAAAGTAAAAAA	11		
19	ATATTTTTTTTCTTCAG	10		
20	ACCTTTCGGATAAAACC	6		

NOS = number of samples in the motif set.

to our neural networks, we obtain a total of 19 motif sets. In order to compare our algorithm with MEME method, we apply the same DNA sequences to the MEME online server (http://meme.sdsc.edu/meme/website/meme.html). The MEME method finds 15 motif sets. Table 9.4.1 shows a comparison between the motif sets we found and MEME results. In the table we list the consensus sequence of each motif set obtained by both self-organizing neural network method and MEME method. We list the number of patterns that are found for each motif set. The first 15 motif sets are those found by both our algorithm and MEME method. Compared with MEME method, our algorithm finds more patterns for most of these motif sets. Motif sets 16 to 20 are found by our method only.   □

Example 9.4.3. In this example, following [55], we generate i.i.d. samples of DNA sequences with certain lengths. Motifs with random mismatch letters at randomly chosen positions are implanted in these sequences. The performance of the algorithm is defined as follows:

$$P_{erf} = \frac{|R \cap T|}{|R \cup T|} \tag{9.4.1}$$

where R is the motif set generated, T is the motif set identified, and $| \cdot |$ indicates the cardinality of a set. The numerator of the performance represents the number of motifs we found that are really motifs. The denominator represents the whole set of any motifs that are generated or found. In the figures shown in this example, the horizontal axis represents the percentage of mismatch of the motifs (i.e., ϵ/M, where ϵ is the number of letters that is tolerable as the representation of a motif), and the vertical axis indicates the performance averaged over 8 such simulations. A result that is closer to 1 implies a better performance.

Figures 9.4.2 and 9.4.3 show the performance of the system on finding motifs of lengths 13 and 15. From the figures we can see that the results are still acceptable even with the mismatch letters up to 30%. After that, the performances drops sharply. The reason of the sharp drop is that for the 4 letter DNA case, the total number of randomly generated sequences is not large enough, which makes the generated patterns to be often similar to noise. Comparing to the results obtained using MEME in [2] and using Gibbs in [37], our simulation results can find motifs with at least one more mismatch letter than the other two algorithms. For example, for motif length of 15, our algorithm achieved 100% performance (i.e., identified all motifs) when there are four letter mismatches allowed, while MEME and Gibbs algorithms both achieved less than 20% performace. We can conclude that in this aspect our algorithm outperforms the MEME and Gibbs algorithms. In this simulation example, we generated 10 DNA sequences with 200 letters in each sequence. The computation time of our algorithm is 3 minutes on a SUN Ultra 60 workstation. Compared with MEME (15 minutes) and Gibbs (12 minutes), our algorithm demands less computation time. □

Example 9.4.4. Similar to Example 9.4.3, we show the performance of the system on finding motif with insertions and deletions in Figure 9.4.4. The length of the projected motif is 17. The horizontal axis represents the percentage of mismatch of the motifs (i.e., ϵ/M, where ϵ is the number of letters that is tolerable in the representation of a motif), and vertical axis indicates the average of the performances in 8 simulation runs defined above. In this simulation, we fix the distribution of the number of insertions and deletions. 30 percent of the motif appearances we generated have two insertions or deletions or their combination.

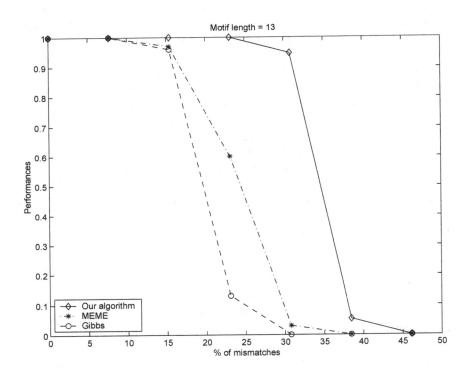

Figure 9.4.2: Comparison results for motif length = 13.

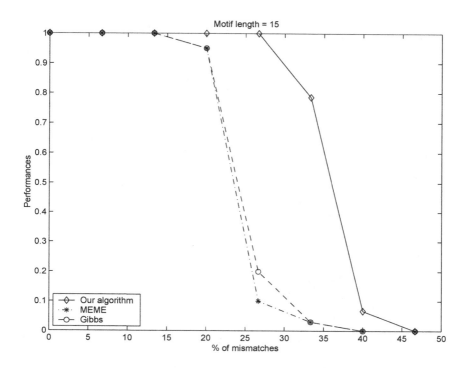

Figure 9.4.3: Comparison results for motif length =15.

40 percent of them have only one insertion or deletion. The rest appearances do not have insertion or deletion. Besides these insertions and deletions, all the appearances have a certain number of columns mutated according to the horizontal axis of the figure. In Figure 9.4.4, we compare our results to that of the Gibbs algorithm.

From the figure we can see that compared with the Gibbs algorithm, our algorithm has much better performance. Let (17, 5) denote the case of setting motif length to 17 and number of mismatch columns to 5. Our neural network algorithm performs well and finds nearly all the patterns generated in the case of (17, 5). In the case of having insertions and deletions, for (17, 7), our algorithm finds nearly 60% of all patterns while the Gibbs algorithm missed all patterns. □

Example 9.4.5. In this example we will apply our algorithm to motif discovery problem in DNA Repair Protein RAD51 homolog protein sequences. The 10 samples are collected from Swiss-Prot Genes Library. Their names and descriptions are listed in Table 9.4.2. We apply these 10 sequences into our network, and get

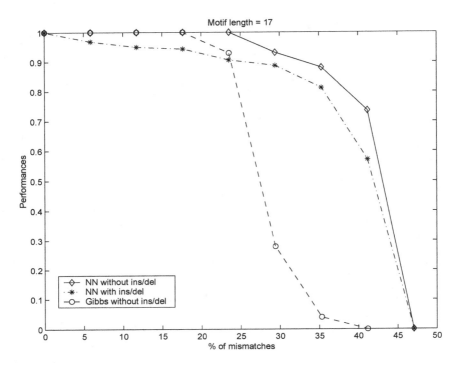

Figure 9.4.4: Comparison result for motif length = 17.

Table 9.4.2: Entry name, accession number and length of DNA repair protein RAD51 sequences

Index	Protein name	Accession number	Length
1	RA51_CHICK	P37383	339
2	RA51_CRIGR	P70099	339
3	RA51_DROME	Q27297	336
4	RA51_HUMAN	Q06609	339
5	RA51_LYCES	Q40134	342
6	RA51_MOUSE	Q08297	339
7	RA51_RABIT	O77507	339
8	RA51_SCHPO	P36601	365
9	RA51_USTMA	Q99133	339
10	RA51_YEAST	P25454	400

Table 9.4.3: Patterns and their locations in the RA_51 protein sequences

Pattern	Content	Sequence	Column	Mismatch	Insertion+Deletion
Consensus	LQGG.IETGSITELF.GEF	-	-	-	-
1	LQGG.IETGSITELF.GEF	1	113	0	0
2	LQGG.IETGSITEMF.GEF	2	113	1	0
3	L-GG.IETGSITEIF.GEF	3	110	1	1
4	LQGG.IETGSITEMF.GEF	4	113	1	0
5	LEGG.IETGSITEIFYGEF	5	116	2	1
6	LQGG.IETGSITEMF.GEF	6	113	1	0
7	LQGG.IETGSITEMF.GEF	7	113	1	0
8	LQGG.VETGSITELF.GEF	8	135	1	0
9	L-GG.METGSITEL-.GEF	9	113	1	2
10	L-GGKVETGSITELF.GEF	10	171	1	2

'—' indicates a deletion in this sequence and '.' indicates an insertion in another sequence.

17 motif sets that have at least 5 appearances in each set. A typical output of the alignment of a motif set is shown in Table 9.4.3.

In the Table 9.4.3, a motif set with 10 appearances is listed, along with a consensus sequence. All the appearances are aligned in the format we introduced in Figure 9.2.2 (c). These motif appearances can be located by the sequence and column indexes listed in the table. The total number of mutations of each appearance to the consensus sequence is also shown in the table. In the last column of the table, we list the total number of insertions and deletions in each appearance. □

9.5 Conclusions

In this chapter, we studied the problem of motif discoveries in unaligned DNA and protein sequences. We developed a self-organizing neural network structure for solving the problem of motif identification in DNA and protein sequences. Our network contains several subnetworks with each subnetwork performing classifications at different level. We maintain a low computational complexity through the use of the layered structure so that each pattern's classification is performed with respect to a small subspace of the whole input space. We also maintain a high reliability using our self-organizing neural network since it will grow as needed to make sure that all input patterns are considered and are given the same amount of attention. Our algorithm can find motifs with up to 2 insertions, deletions or their combinations. Simulation results show that our algorithm outperforms existing algorithms MEME and Gibbs' in certain aspects. Our algorithm works well for long DNA sequences as well.

Bibliography

[1] T. L. Bailey and C. Elkan, "The value of prior knowledge in discovering motifs with MEME," *Proc. 3rd International Conference on Intelligent Systems for Molecular Biology*, Cambridge, UK, July 1995, pp. 21–29.

[2] T. L. Bailey and C. Elkan, "Unsupervised learning of multiple motifs in biopolymers using expectation maximization," *Machine Learning*, vol. 21, no. 1–2, pp. 51–83, Oct./Nov. 1995.

[3] T. L. Bailey and M. Gribskov, "Combining evidence using p-values: Application to sequence homology searches," *Bioinformatics*, vol. 14, no. 1, pp. 48–54, Feb. 1998.

[4] A. Basu, P. Chaudhuri, and P. P. Majumder, "Identification of polymorphic motifs using probabilistic search algorithms," *Genome Research*, vol. 15, no. 1, pp. 67–77, Jan. 2005.

[5] K. Blekas, D. I. Fotiadis, and A. Likas, "A sequential method for discovering probabilistic motifs in proteins," *Methods of Information in Medicine*, vol. 43, no. 1, pp. 9–12, 2004.

[6] L. Brocchieri and S. Karlin, "A symmetric-iterated multiple alignment of protein sequences," *Journal of Molecular Biology*, vol. 276, no. 1, pp. 249–264, Feb. 1998.

[7] G. A. Carpenter and S. Grossberg, "A massively parallel architecture for a self-organizing neural pattern-recognition machine," *Computer Vision, Graphics, and Image Processing*, vol. 37, no. 1, pp. 54–115, Jan. 1987.

[8] G. A. Carpenter and S. Grossberg, "Search mechanisms for adaptive resonance theory (ART) architectures," *Proc. International Joint Conference on Neural Networks*, Washington, DC, June 1989, vol. 1, pp. 210–205.

[9] G. A. Carpenter and S. Grossberg, "ART-3: Hierarchical search using chemical transmitters in self-organizing pattern-recognition architectures," *Neural Networks*, vol. 3, no. 2, pp. 129–152, Mar. 1990.

[10] G. A. Carpenter and S. Grossberg, "A self-organizing neural network for supervised learning, recognition, and prediction," *IEEE Communications Magazine*, vol. 30, no. 9, pp. 38–49, Sept. 1992.

[11] G. A. Carpenter, S. Grossberg, and D. Rosen, "ART 2-A: An adaptive resonance algorithm for rapid category learning and recognition," *Proc. International Joint Conference on Neural Networks*, Seattle, WA, July 1991, vol. 2, pp. 151–156.

[12] B. C. H. Chang, A. Ratnaweera, S. K. Halgamuge, and H. C. Watson, "Particle swarm optimisation for protein motif discovery," *Genetic Programming and Evolvable Machines*, vol. 5, no. 2, pp. 203–214, June 2004.

[13] R. Chenna, H. Sugawara, T. Koike, R. Lopez, T. J. Gibson, D. G. Higgins, and J. D. Thompson, "Multiple sequence alignment with the clustal series of programs," *Nucleic Acids Research*, vol. 31, no. 13, pp. 3497–3500, July 2003.

[14] L. Duret and P. Bucher, "Searching for regulatory elements in human noncoding sequences," *Current Opinion in Structural Biology*, vol. 7, no. 3, pp. 399–406, June 1997.

[15] S. R. Eddy, "Profile hidden Markov models," *Bioinformatics*, vol. 14, no. 9, pp. 755–763, Oct. 1998.

[16] O. Emanuelsson, H. Nielsen, and G. von Heijne, "ChloroP, A neural network-based method for predicting chloroplast transit peptides and their cleavage sites," *Protein Science*, vol. 8, no. 5, pp. 978–984, May 1999.

[17] D. Frishman and P. Argos, "A neural network for recognizing distantly related protein sequences," in *Handbook of Neural Computation*, E. Fiesler and R. Beale, Eds., New York: IOP Publishing and Oxford University Press, pp. G4.4:1–8, 1997.

[18] M. C. Frith, Y. Fu, L. Yu, J. F. Chen, U. Hansen, and Z. Weng, "Detection of functional DNA motifs via statistical over-representation," *Nucleic Acids Research*, vol. 32, no. 4, pp. 1372–1381, Feb. 2004.

[19] Y. Gdalyahu, D. Weinshall, and M. Werman, "Self-organization in vision: Stochastic clustering for image segmentation, perceptual grouping, and image database organization," *IEEE Trans. Pattern Analysis and Machine Intelligence*, vol. 23, no. 10, pp. 1053–1074, Oct. 2001.

[20] P. Gonnet and F. Lisacek, "Probabilistic alignment of motifs with sequences," *Bioinformatics*, vol. 18, no. 8, pp. 1091–1101, Aug. 2002.

[21] W. N. Grundy, T. L. Bailey, C. P. Elkan, and M. E. Baker, "Meta-MEME: Motif-based hidden Markov models of protein families," *Computer Applications in the Biosciences*, vol. 13, no. 4, pp. 397–406, Aug. 1997.

[22] K. Gulukota, J. Sidney, A. Sette, and C. DeLisi, "Two complementary methods for predicting peptides binding major histocompatibility complex molecules," *Journal of Molecular Biology*, vol. 267, no. 5, pp. 1258–1267 Apr. 1997.

[23] M. Gupta and J. S. Liu, "Discovery of conserved sequence patterns using a stochastic dictionary model," *Journal of the American Statistical Association*, vol. 98, no. 461, pp. 55–66, Mar. 2003.

[24] D. Gusfield, *Algorithms on Strings, Trees, and Sequences: Computer Science and Computational Biology*, New York: Cambridge University Press, 1997.

[25] S. Haykin, *Neural Networks: A Comprehensive Foundation*, Upper Saddle River, NJ: Prentice Hall, 1999.

[26] G. Z. Hertz and G. D. Stormo, "Identifying DNA and protein patterns with statistically significant alignments of multiple sequences," *Bioinformatics*, vol. 15, no. 7/8, pp. 563–577, July/Aug. 1999.

[27] T. Hofmann and J. M. Buhmann, "Pairwise data clustering by deterministic annealing," *IEEE Trans. Pattern Analysis and Machine Intelligence*, vol. 19, no. 1, pp. 1–14, Jan. 1997.

[28] S. T. Jensen and J. S. Liu, "BioOptimizer: a Bayesian scoring function approach to motif discovery," *Bioinformatics*, vol. 20, no. 10, pp. 1557–1564, July 2004.

[29] E. Jeong, I. F. Chung, and S. Miyano, "A neural network method for identification of RNA-interacting residues in protein," *Proc. 15th International Conference on Genome Informatics*, Yokohama, Japan, Dec. 2004, pp. 105–116.

[30] U. Keich and P. A. Pevzner, "Finding motifs in the twilight zone," *Bioinformatics*, vol. 18, no. 10, pp. 1374–1381, Oct. 2002.

[31] U. Keich and P. A. Pevzner, "Subtle motifs: Defining the limits of motif finding algorithms," *Bioinformatics*, vol. 18, no. 10, pp. 1382–1390, Oct. 2002.

[32] S. Keles, M. J. van der Laan, and C. Vulpe, "Regulatory motif finding by logic regression," *Bioinformatics*, vol. 20, no. 16, pp. 2799–2811, Nov. 2004.

[33] J. T. Kim, J. E. Gewehr, and T. Martinetz, "Binding matrix: A novel approach for binding site recognition," *Journal of Bioinformatics and Computational Biology*, vol. 2, no. 2, pp. 289–307, June 2004.

[34] S. Knudsen, "Promoter2.0: For the recognition of PolII promoter sequences," *Bioinformatics*, vol. 15, no. 5, pp. 356–361, May 1999.

[35] A. Krogh, M. Brown, I. S. Mian, K. Sjolander, and D. Haussler, "Hidden Markov models in computational biology: Applications to protein modeling," *Journal of Molecular Biology*, vol. 235, no. 5, pp. 1501–1531, Feb. 1994.

[36] P. Lavoie, J.-F. Crespo, and Y. Savaria, "Generalization, discrimination, and multiple categorization using adaptive resonance theory," *IEEE Transactions on Neural Networks*, vol. 10, no. 4, pp. 757–767, July 1999.

[37] C. E. Lawrence, S. F. Altschul, M. S. Boguski, J. S. Liu, A. F. Neuwald, and J. C. Wootton, "Detecting subtle sequence signals: A Gibbs sampling strategy for multiple alignment," *Science*, vol. 262, no. 5131 pp. 208–214, Oct. 1993.

[38] S. Liang, M. P. Samanta, and B. A. Biegel, "cWINNOWER algorithm for finding fuzzy DNA motifs," *Journal of Bioinformatics and Computational Biology*, vol. 2, no. 1, pp. 47–60, Mar. 2004.

[39] C. Linhart and R. Shamir, "The degenerate primer design problem," *Bioinformatics*, vol. 18, Suppl. 1, pp. S172–S180, 2002.

[40] Y. Liu, X. S. Liu, L. Wei, R. B. Altman, and S. Batzoglou, "Eukaryotic regulatory element conservation analysis and identification using comparative genomics," *Genome Research*, vol. 14, no. 3, pp. 451–458, Mar. 2004.

[41] D. Liu, X. Xiong, Z. G. Hou, and B. DasGupta, "Identification of motifs with insertions and deletions in protein sequences using self-organizing neural networks," *Neural Networks*, vol. 18, no. 5-6, pp. 835–842, June-July 2005.

[42] A. M. Moses, D. Y. Chiang, and M. B. Eisen, "Phylogenetic motif detection by expectation-maximization on evolutionary mixtures," *Proc. Pacific Symposium on Biocomputing*, Fairmont Orchid, HI, Jan. 2004, pp. 324–335.

[43] A. F. Neuwald, J. S. Liu, D. J. Lipman, and C. E. Lawrence, "Extracting protein alignment models from the sequence database," *Nucleic Acids Research*, vol. 25, no. 9, pp. 1665–1677, May 1997.

[44] A. F. Neuwald and A. Poleksic, "PSI-BLAST searches using hidden Markov models of structural repeats: prediction of an unusual sliding DNA clamp and of β-propellers in UV-damaged DNA-binding protein," *Nucleic Acids Research*, vol. 28, no. 18, pp. 3570–3580, Sept. 2000.

[45] H. Nielsen, J. Engelbrecht, S. Brunak, and G. von Heijne, "Identification of prokaryotic and eukaryotic signal peptides and prediction of their cleavage sites," *Protein Engineering*, vol. 10, no. 1, pp. 1–6, Jan. 1997.

[46] H. Nielsen, J. Engelbrecht, S. Brunak, and G. von Heijne, "A neural network method for identification of prokaryotic and eukaryotic signal peptides and prediction of their cleavage sites," *International Journal of Neural Systems*, vol. 8, no. 5-6, pp. 581–599, Oct./Dec. 1997.

[47] A. R. Ortiz, A. Kolinski, and J. Skolnick, "Nativelike topology assembly of small proteins using predicted restraints in Monte Carlo folding simulations," *Proc. National Academy of Sciences of the USA*, vol. 95, no. 3, pp. 1020–1025, Feb. 1998.

[48] B. Padmanabhan and A. Tuzhilin, "Pattern discovery in temporal databases: A temporal logic approach," *Proc. Second International Conf. Knowledge Discovery and Data Mining,* Portland, Oregon, Aug. 1996, pp. 351–354.

[49] P. A. Pevzner and S.-H. Sze, "Combinatorial approaches to finding subtle signals in DNA sequences," *Proc. 8th International Conference on Intelligent Systems for Molecular Biology,* San Diego, CA, Aug. 2000, pp. 269–278.

[50] L. R. Rabiner, "A tutorial on hidden Markov models and selected applications in speech recognition," *Proc. IEEE,* vol. 77, no. 2, pp. 257–286, Feb. 1989.

[51] K. Reinert, J. Stoye, and T. Will, "An iterative method for faster sum-of-pairs multiple sequence alignment," *Bioinformatics,* vol. 16, no. 9, pp. 808–814, Sept. 2000.

[52] W. R. Taylor and D. T. Jones, "Deriving an amino acid distance matrix," *Journal of Theoretical Biology,* vol. 164, no. 1, pp. 65–83, Sept. 1993.

[53] J. D. Thompson, D. G. Higgins, and T. J. Gibson, "CLUSTAL W: Improving the sensitivity of progressive multiple sequence alignment through sequence weighting, position-specific gap penalties and weight matrix choice," *Nucleic Acids Research,* vol. 22, no. 22, pp. 4673–4680, Nov. 1994.

[54] G. White and W. Seffens, "Using a neural network to backtranslate amino acid sequences," *Electronic Journal of Biotechnology,* vol. 1, no. 3, Dec. 1998. Available from: http://www.ejbiotechnology.info/content/vol1/issue3/full/5/index.html.

[55] C. T. Workman and G. D. Stormo, "ANN-Spec: A method for discovering transcription factor binding sites with improved specificity," *Proc. Pacific Symposium on Biocomputing,* Honolulu, HI, Jan. 2000, pp. 467–478.

[56] C. Wu, S. Shivakumar, H. P. Lin, S. Veldurti, and Y. Bhatikar, "Neural networks for molecular sequence classification," *Mathematics and Computers in Simulation,* vol. 40, no. 1-2, pp. 23–33, Dec. 1995.

[57] J. Xie, K. C. Li, and M. Bina, "A Bayesian insertion/deletion algorithm for distant protein motif searching via entropy filtering," *Journal of the American Statistical Association,* vol. 99, no. 466, pp. 409–420, June 2004.

[58] E. P. Xing, W. Wu, M. I. Jordan, and R. M. Karp, "Logos: A modular Bayesian model for de novo motif detection," *Journal of Bioinformatics and Computational Biology,* vol. 2, no. 1, pp. 127–154, Mar. 2004.

[59] http://pbil.univ-lyon1.fr/acuts/ACUTS_home.html.

[60] Swiss-Prot, http://us.expasy.org/sprot/.

Chapter 10

Computational Complexities of Combinatorial Problems With Applications to Reverse Engineering of Biological Networks

Piotr Berman, Bhaskar DasGupta and Eduardo Sontag

Abstract: In this chapter, we discuss some computational problems that arise in the reverse engineering of protein and gene networks. We discuss the biological motivations, provide precise formulations of the combinatorial questions that follow from these motivations and finally describe the computational complexity issue, namely efficient approximation algorithms, for these problems.

10.1   Introduction

The problems discussed here are motivated by a central concern of contemporary cell biology, that of unraveling (or "reverse engineering") the web of interactions among the components of complex protein and genetic regulatory networks. Notwithstanding the remarkable progress in genetics and molecular biology in the sequencing of the genomes of a number of species, the inference and quantification of interconnections in signaling and genetic networks that are critical to cell function is still a challenging practical and theoretical problem. High-throughput technologies allow the monitoring the expression levels of sets of genes, and the activity states of signaling proteins, providing snapshots of the transcriptional and signaling behavior of living cells. Statistical and machine learning techniques, such as clustering, are often used in order to group genes into co-expression patterns, but they are less able to explain functional interactions.

An intrinsic difficulty in capturing such interactions in intact cells by traditional genetic experiments or pharmacological interventions is that any perturbation to a particular gene or signaling component may rapidly propagate throughout the network, causing global changes. The question thus arises of how to use the observed global changes to derive interactions between individual nodes. In this chapter we discuss some computational problems that arises in the context of experimental design for reverse engineering of protein and gene networks. Biological networks may have a very large number of species and parameters. For example, the *E. coli* transcription network identified in [15] has 577 interactions involving 116 transcription factors and 419 operons. For such large-scale networks, exhaustive calculations are not practically possible due to combinatorial explosion and this necessitates the design of *provably efficient* approximation algorithms.

10.2 Motivations

We will first pose our problems in linear algebra terms, and then recast it as a combinatorial question. After that, we will discuss its motivations from systems biology.

10.2.1 Linear Algebraic Formulations and the Combinatorial Questions

Our problem is described in terms of two matrices $A \in \mathbb{R}^{n \times n}$ and $B \in \mathbb{R}^{n \times m}$ such that:

- A is *unknown*;
- B is *initially unknown*, but each of its columns $B_1, B_2, \cdots, B_m$ can be retrieved with a *unit-cost query*;
- the columns of B are in *general position*, *i.e.*, each subset of $\ell \leq n$ columns of B is *linearly independent*;
- the *zero structure* of the matrix $C = AB = (c_{ij})$ is known, *i.e.*, a binary matrix $C^0 = \left(c_{ij}^0\right) \in \{0, 1\}^{n \times m}$ is given, and it is known that $c_{ij} = 0$ for each i, j for which $c_{ij}^0 = 0$.

The objective is to obtain as much information as possible about A (which, in the motivating application, describes regulatory interactions among genes and/or proteins), while performing "few" queries (each of which may represent the measuring of a complete pattern of gene expression, done under a different set of experimental conditions). For each query that we perform, we obtain a column B_i, and then the matrix C^0 tells us that certain rows of A have zero inner product with B_i.

As a concrete example, let us take $n = 3$, $m = 5$, and suppose that the known information is given by the matrix:

$$C_0 = \begin{bmatrix} 0 & 1 & 0 & 1 & 1 \\ 1 & 1 & 1 & 0 & 0 \\ 0 & 0 & 1 & 0 & 1 \end{bmatrix}$$

and the two unknown matrices are:

$$A = \begin{bmatrix} -1 & 1 & 3 \\ 2 & -1 & 4 \\ 0 & 0 & -1 \end{bmatrix}, \qquad B = \begin{bmatrix} 4 & 3 & 37 & 1 & 10 \\ 4 & 5 & 52 & 2 & 16 \\ 0 & 0 & -5 & 0 & -1 \end{bmatrix}$$

(the matrix C_0 has zero entries wherever AB has a zero entry). Considering the structure of C_0, we choose to perform four queries, corresponding to the four columns 1, 3, 4, 5 of B, thus obtaining the following data:

$$\begin{bmatrix} 4 & 37 & 1 & 10 \\ 4 & 52 & 2 & 16 \\ 0 & -5 & 0 & -1 \end{bmatrix}. \qquad (10.2.1)$$

What can we say about the unknown matrix A? Let us first attempt to identify its first row, which we call A_1. The first row of the matrix C_0 tells us that the vector A_1 is orthogonal to the first and second columns of (10.2.1) (which are the same as the first and third columns of B). This is the *only* information about A that we have available, and it is not enough information to uniquely determine A_1, because there is an entire line that is orthogonal to the plane spanned by these two columns, However, we can still find *some* nonzero vector in this line, and conclude that A_1 is an unknown multiple of this vector. This nonzero vector may be obtained by simple linear algebra manipulations. For example, we might add a linearly independent column to the two that we had, obtaining a matrix

$$B_1 = \begin{bmatrix} 4 & 37 & 0 \\ 4 & 52 & 0 \\ 0 & -5 & 1 \end{bmatrix},$$

then pick an arbitrary vector v whose first two entries are zero (to reflect the known orthogonality), let us say $v = [0, 0, 1]$, and finally solve $A_1 B = v$, thus estimating

A_1 as vB^{-1}:

$$\hat{A}_1 = [0,0,1]B^{-1} = [0,0,1] \begin{bmatrix} 13/15 & -37/60 & 0 \\ -1/15 & 1/15 & 0 \\ -1/3 & 1/3 & 1 \end{bmatrix} = [-1/3, 1/3, 1].$$

Notice that this differs from the unknown A_1 only by a scaling. Similarly, we may employ the last two columns of (10.2.1) to estimate the second row A_2 of A, again only up to a multiplication by a constant, and we may use the first and third columns of (10.2.1) (which are the same as the first and fourth columns of B) to estimate the last row, A_3.

Notice that there are always intrinsic limits to what can be accomplished: if we multiply each row of A by some nonzero number, then the zero structure of C is unchanged. Thus, as in the example, the best that we can hope for is to identify the rows of A up to scalings (in abstract mathematical terms, as elements of the projective space $\mathbb{P}^{n-1}$). To better understand these geometric constraints, let us reformulate the problem as follows. Let A_i denote the ith row of A. Then the specification of C^0 amounts to the specification of *orthogonality relations* $A_i \cdot B_j = 0$ for each pair i, j for which $c_{ij}^0 = 0$. Suppose that we decide to query the columns of B indexed by $J = \{j_1, \cdots, j_\ell\}$. Then, the information obtained about A may be summarized as $A_i \in \mathcal{H}_{J,i}^\perp$, where "$\perp$" indicates *orthogonal complement*, $\mathcal{H}_{J,i} = \text{span}\{B_j, j \in J_i\}$, and $J_i = \{j \mid j \in J \text{ and } c_{ij}^0 = 0\}$. Suppose now that the set of indices of selected queries J has the property:

$$\text{each set } J_i, \ i = 1, \cdots, n, \text{ has cardinality} \geq n - k, \qquad (10.2.2)$$

for some given integer k. Then, because of the general position assumption, the space $\mathcal{H}_{J,i}$ has dimension $\geq n - k$, and hence the space $\mathcal{H}_{J,i}^\perp$ has dimension at most k.

The case $k = 1$

The most desirable special case is that in which $k = 1$. Then $\dim \mathcal{H}_{J,i}^\perp \leq 1$, hence each A_i is uniquely determined up to a scalar multiple, which is the best that could be theoretically achieved. Often, in fact, finding the sign pattern (such as "$(+, +, -, 0, 0, -, \cdots)$") for each row of A is the main experimental goal (this would correspond, in our motivating application, to determining if the regulatory interactions affecting each given gene or protein are *inhibitory* or *catalytic*). Assuming that the degenerate case $\mathcal{H}_{J,i}^\perp = \{0\}$ does not hold (which would determine $A_i = 0$), once that an arbitrary nonzero element v in the line $\mathcal{H}_{J,i}^\perp$ has been picked, there are only two sign patterns possible for A_i (the pattern of v and that of $-v$). If, in addition, one knows at least one nonzero sign in A_i, then the sign structure of the whole row has been *uniquely* determined (in the motivating biological question, typically one such sign is indeed known; for example, the diagonal elements a_{ii}, i.e. the ith element of each A_i, is known to be negative, as it represents a degradation rate). Thus, we will be interested in this question:

$$\text{find } J \text{ of minimal cardinality such that } |J_i| \geq n - 1, i = 1, \cdots, n.$$
$$\text{(Q1)}$$

If queries have variable unit costs (different experiments have a different associated cost), this problem must be modified to that of minimizing a suitable linear combination of costs, instead of the number of queries.

The general case $k > 1$

More generally, suppose that the queries that we performed satisfy (10.2.2), with $k > 1$ but small k. It is not true anymore that there are only two possible sign patterns for any given A_i, but the number of possibilities is still very small. For simplicity, let us assume that we know that no entry of A_i is zero (if this is not the case, the number of possibilities may increase, but the argument is very similar). We wish to prove that the possible number of signs is much smaller than 2^n. Indeed, suppose that the queries have been performed, and that we then calculate, based on the obtained B_j's, a basis $\{v_1, \cdots, v_k\}$ of $\mathcal{H}_{J,i}^\perp$ (assume $\dim \mathcal{H}_{J,i}^\perp = k$; otherwise pick a smaller k). Thus, the vector A_i is known to have the form $\sum_{r=1}^k \lambda_r v_r$ for some (unknown) real numbers $\lambda_1, \cdots, \lambda_k$. We may assume that $\lambda_1 \neq 0$ (since, if $A_i = \sum_{r=2}^k \lambda_r v_r$, the vector $\varepsilon v_1 + \sum_{r=2}^k \lambda_r v_r$, with small enough ε, has the same sign pattern as A_i, and we are counting the possible sign patterns). If $\lambda_1 > 0$, we may divide by λ_1 and simply count how many sign patterns there are when $\lambda_1 = 1$; we then double this estimate to include the case $\lambda_1 < 0$. Let $v_r = \text{col}(v_{1r}, \cdots, v_{nr})$, for each $r = 1, \cdots, k$. Since no coordinate of A_i is zero, we know that A_i belongs to the set $\mathcal{C} = \mathbb{R}^{k-1} \setminus (L_1 \bigcup \cdots \bigcup L_n)$ where, for each $1 \leq s \leq n$, L_s is the hyperplane in $\mathbb{R}^{k-1}$ consisting of all those vectors $(\lambda_2, \cdots, \lambda_k)$ such that $\sum_{r=2}^k \lambda_r v_{sr} = -v_{s1}$. On each connected component of $\mathcal{C}$, signs patterns are constant. Thus the possible number of sign patterns is upper bounded by the maximum possible number of connected regions determined by n hyperplanes in dimension $k - 1$. A result of L. Schläfli (see [6, 14], and also [17] for a discussion, proof, and relations to Vapnik-Chervonenkis dimension) states that this number is bounded above by $\Phi(n, k - 1)$, provided that $k - 1 \leq n$, where $\Phi(n, d)$ is the number of possible subsets of an n-element set with at most d elements, that is,

$$\Phi(n, d) = \sum_{i=0}^d \binom{n}{i} \leq 2\frac{n^d}{d!} \leq \left(\frac{en}{d}\right)^d.$$

Doubling the estimate to include $\lambda_1 < 0$, we have the upper bound $2\Phi(n, k-1)$. For example, $\Phi(n, 0) = 1$, $\Phi(n, 1) = n + 1$, and $\Phi(n, 2) = \frac{1}{2}(n^2 + n + 2)$. Thus we have an estimate of 2 sign patterns when $k = 1$ (as obtained earlier), $2n + 2$ when $k = 2$, $n^2 + n + 2$ when $k = 3$, and so forth. In general, the number grows only polynomially in n (for fixed k).

These considerations lead us to formulating the generalized problem, for each fixed k: *find J of minimal cardinality such that $|J_i| \geq n - k$ for all $i = 1, \cdots, n$.* Recalling the definition of J_i, we see that $J_i = J \cap T_i$, where $T_i = \{j \mid c_{ij}^0 = 0\}$. Thus, we can reformulate our question purely combinatorially, as a more general version of Question (Q1) as follows. Given sets $T_i \subseteq \{1, \cdots, m\}$, $i = 1, \cdots, n$, and an integer $k < n$, the problem is:

$$\text{find } J \subseteq \{1, \cdots, m\} \text{ of minimal cardinality such that } \left| J \cap T_i \right| \geq n - k,$$

$$1 \leq i \leq n. \tag{Q2}$$

For example, suppose that $k = 1$, and pick the matrix $C^0 \in \{0, 1\}^{n \times n}$ in such a way that the columns of C^0 are the binary vectors representing all the $(n-1)$-element subsets of $\{1, \cdots, n\}$ (so $m = n$); in this case, the set J must equal $\{1, \cdots, m\}$ and hence has cardinality n. On the other hand, also with $k = 1$, if we pick the matrix C^0 in such a way that the columns of C^0 are the binary vectors representing all the 2-element subsets of $\{1, \cdots, n\}$ (so $m = n(n-1)/2$), then J must again be the set of all columns (because, since there are only two zeros in each column, there can only be a total of 2ℓ zeros, $\ell = |J|$, in the submatrix indexed by J, but we also have that $2\ell \geq n(n-1)$, since each of the n rows must have $\geq n - 1$ zeros); thus in this case the minimal cardinality is $n(n-1)/2$.

10.2.2 Set Multicover Formulations

The algorithmic questions posed in the previous section can be cast as variations of a generic combinatorial set multicover problem, defined as follows. Suppose that we are given an universe U, a set of subsets Γ of U and a positive integer k with $|\{u \in \gamma \mid \gamma \in \Gamma\}| \geq k$ for every $u \in U$. Then, our problem is the following integer programming problem:

$$\text{minimize } \sum_{\gamma \in \Gamma} x_\gamma \text{ subject to } \begin{cases} \sum_{u \in \gamma \in \Gamma} x_\gamma \geq k, & \text{for each } u \in U \\ x_\gamma \in \{0, 1\}, & \text{for each } \gamma \in \Gamma \end{cases}$$

Basic versions of question (Q1) and (Q2) in the next section can be cast in a similar formulation; see references [4, 5]. An appropriate *on-line* variation of the set-multicover problem, as outlined in the reference [3], is also appropriate for the reverse engineering problems as will be mentioned in the next section.

10.2.3 Motivations from Systems Biology

The biological motivation stems from an effort by many research groups whose goal is to infer mechanistic relationships underlying the observed behavior of complex molecular networks. We focus our attention here solely on one such approach, originally described in [9, 10], further elaborated upon in [2, 16], and reviewed in [7, 18]. In this approach, the architecture of the network is inferred on the basis of observed global responses (namely, the steady-state concentrations in changes in the phosphorylation states or activities of proteins, mRNA levels, or transcription rates) in response to experimental perturbations (representing the effect of hormones, growth factors, neurotransmitters, or of pharmacological interventions).

In the setup in [9, 10, 16], the time evolution of a vector of state variables $x(t) = (x_1(t), \cdots, x_n(t))$ is described by a system of differential equations:

$$\dot{x} = f(x, p) \iff \begin{aligned} \dot{x}_1 &= f_1(x_1, \cdots, x_n, p_1, \cdots, p_m) \\ &\vdots \\ \dot{x}_n &= f_n(x_1, \cdots, x_n, p_1, \cdots, p_m) \end{aligned}$$

where the dot indicates time derivative and $p = (p_1, \cdots, p_m)$ is a vector of parameters, which can be manipulated but remain constant during any given experiment. The components $x_i(t)$ of the state vector represent quantities that can be in principle measured, such as levels of activity of selected proteins or transcription rates of certain genes. The parameters p_i represent quantities that can be manipulated, perhaps indirectly, such as levels of hormones or of enzymes whose half-lives are long compared to the rate at which the variables evolve. A basic assumption (but see [16] for a time-dependent analysis) is that states converge to steady state values, and these are the values used for network identification. There is a reference value $\bar{p}$ of p, which represents "wild type" (that is, normal) conditions, and a corresponding steady state $\bar{x}$. Mathematically, $f(\bar{x}, \bar{p}) = 0$. We are interested in obtaining information about the Jacobian of the vector field f evaluated at $(\bar{x}, \bar{p})$, or at least about the signs of the derivatives $\partial f_i / \partial x_j (\bar{x}, \bar{p})$. For example, if $\partial f_i / \partial x_j > 0$, this means that x_j has a positive (catalytic) effect upon the rate of formation of x_i. The critical assumption, indeed the main point of [9, 10, 16], is that, while we may not know the form of f, we often do know that *certain parameters p_j do not directly affect certain variables x_i*. This amounts to *a priori* biological knowledge of specificity of enzymes and similar data. In the current context, this knowledge is summarized by the binary matrix $C^0 = (c_{ij}^0) \in \{0, 1\}^{n \times m}$, where "$c_{ij}^0 = 0$" means that p_j does not appear in the equation for $\dot{x}_i$, that is, $\partial f_i / \partial p_j \equiv 0$.

The experimental protocol allows one to perturb any one of the parameters, let us say the kth one, while leaving the remaining ones constant. (A generalization, to allow for the simultaneous perturbation of more than one parameter, is of course possible.) For the perturbed vector $p \approx \bar{p}$, one then measures the resulting steady state vector $x = \xi(p)$. Experimentally, this may for instance mean that the concentration of a certain chemical represented by p_k is kept are a slightly altered level, compared to the default value $\bar{p}_k$; then, the system is allowed to relax to steady state, after which the complete state x is measured, for example by means of a suitable biological reporting mechanism, such as a microarray used to measure the expression profile of the variables x_i. Mathematically, we suppose that for each vector of parameters p in a neighborhood of $\bar{p}$ there is a unique steady state $\xi(p)$ of the system, where ξ is a differentiable function. For each of the possible m experiments, in which a given p_j is perturbed, we may estimate the n "sensitivities"

$$b_{ij} = \frac{\partial \xi_i}{\partial p_j}(\bar{p}) \approx \frac{1}{\bar{p}_j - p_j}\left(\xi_i(\bar{p} + p_j e_j) - \xi_i(\bar{p})\right), \quad i = 1, \cdots, n,$$

where $e_j \in \mathbb{R}^m$ is the jth canonical basis vector. We let B denote the matrix consisting of the b_{ij}'s. (See [9, 10] for a discussion of the fact that division by $\bar{p}_j - p_j$, which is undesirable numerically, is not in fact necessary.) Finally, we let A be the Jacobian matrix $\partial f/\partial x$ and let C be the negative of the Jacobian matrix $\partial f/\partial p$. From $f(\xi(p), p) \equiv 0$, taking derivatives with respect to p, and using the chain rule, we get that $C = AB$. This brings us to the problem stated in the previous section; the general position assumption is reasonable, since we are dealing with experimental data.

10.2.4 Online Versions of Questions of the Type (Q1) or (Q2)

The online versions of the questions of the type (Q1) or (Q2) are more suited to the case when one performs an experimental protocol which is slightly different from the one described in Section 10.2.1 and described below:

– Let $J_i \subseteq \{j \mid c_{ij} = 1\}$ be the indices of the sets chosen in our set-multicover. Then, each $j \in J_i$ is associated with an experiment of the following type:

   – Change (perturb) only the parameter p_j.

   – For the perturbed vector $p \approx \bar{p}$, we measure the resulting steady state value $x_i = \xi_i(p)$. Experimentally, this may for instance mean that the concentration of a certain chemical represented by p_j is kept are a slightly altered level, compared to the default value $\bar{p}_j$; then, the system is allowed to relax to steady state, after which the steady state

x_i is measured, for example by means of a suitable biological reporting mechanism, such as a fluorescent proteins*. Mathematically, we suppose that for each vector of parameters p in a neighborhood of $\bar{p}$ there is a unique steady state $\xi_i(p)$ of x_i, where ξ_i is a differentiable function.

- Estimate the corresponding "sensitivity"

$$b_{ij} = \frac{\partial \xi_i}{\partial p_j}(\bar{p}) \approx \frac{1}{\bar{p}_j - p_j} \left(\xi_i(\bar{p} + p_j e_j) - \xi_i(\bar{p}) \right)$$

(where $e_j \in \mathbb{R}^m$ is the jth canonical basis vector).

The cost of doing these experiments is amortized against the weights of the sets, the unweighted case being the simplest case when we just wish to minimize the number of experiments.

These considerations motivate us to look at the online versions of questions (Q1) and (Q2) which can be abstracted as an online set multicover problem as follows. we have an universe V of elements, a family $\mathcal{S}$ of subsets of V with a positive real cost c_S for every $S \in \mathcal{S}$, and a "coverage factor" (positive integer) k. A subset $\{i_0, i_1, \cdots\} \subseteq V$ of elements are presented online in an arbitrary order. When each element i_p is presented, we are also told the collection of all (at least k) sets $\mathcal{S}_{i_p} \subseteq \mathcal{S}$ and their costs in which i_p belongs and we need to select additional sets from $\mathcal{S}_{i_p}$ if necessary such that our collection of selected sets contains *at least* k sets that contain the element i_p. The goal is to *minimize* the *total cost* of the selected sets.

10.3 Algorithms and Computational Complexities

References [9, 10, 18] survey biological motivations for Jacobian estimation under the assumptions given above, prove various results, and provide simulations of realistic biological systems in which the technique successfully recovers the Jacobian. In the next two subsections, we discuss most recent algorithmic developments for both the offline version and the online version of the problems.

10.3.1 Offline Version

In [4, 5] we investigated the algorithmic complexity of Question (Q2) and provided *randomized* approximation algorithms with *expected* performance ratios of about 2 for $k = 1$. This was obtained in two steps. In the first step, we showed an equivalence of this problem with the set-multicover formulation outlined in

---

*Fluorescent proteins can be used to know the rate at which a certain gene transcribes in a cell under a set of conditions.

Section 10.2.2. We then considered a randomized approximation algorithm for this problem in the following manner via the "linear programming with nontrivial rounding" approach:

- Let $c = \begin{cases} \ln a, & \text{if } k = 1, \\ \ln(a/(k-1)), & \text{if } a/(k-1) \geq e^2 \text{ and } k > 1, \\ 2, & \text{if } \frac{1}{4} < a/(k-1) < e^2 \text{ and } k > 1, \\ 1 + \sqrt{\frac{a}{k}}, & \text{otherwise,} \end{cases}$

- Find a solution vector $\mathbf{x}^* \in \mathbb{R}^{|U|}$ to the LP relaxation of the formulation in Section 10.2.2 via algorithms such as [8]. Let x_j^* denote the jth component of this solution vector.
- Form a family of sets $\mathcal{C}_0 = \{\gamma \colon c x_\gamma^* \geq 1\}$.
- Form a family of sets $\mathcal{C}_1 \subseteq \mathcal{S} - \mathcal{C}_0$ by selecting a set $\gamma \in \Gamma \backslash \mathcal{C}_0$ with probability $c x_\gamma^*$.
- Form a family of sets $\mathcal{C}_2$ by greedy choices: if an $u \in U$ belongs to fewer than k sets in $\mathcal{C}_0 \cup \mathcal{C}_1$, choose any of the remaining sets that contains u.
- Return $\mathcal{C} = \mathcal{C}_0 \cup \mathcal{C}_1 \cup \mathcal{C}_2$ as the solution.

Then, we were able to prove the following result on the performance of this algorithm.

Theorem 10.3.1. *The expected performance ratio of our algorithm is given by*

$1 + \ln a,$ if $k = 1$,

$\left(1 + e^{-(k-1)/5}\right) \ln(a/(k-1)),$ if $a/(k-1) \geq e^2 \approx 7.39$
                                                                                                                 and $k > 1$,

$\min\{2 + 2e^{-(k-1)/5}, \ 2 + \left(e^{-2} + e^{-9/8}\right)\frac{a}{k}\}$
$\approx \min\{2 + 2e^{-(k-1)/5}, \ 2 + 0.46\frac{a}{k}\},$ if $\frac{1}{4} < a/(k-1) < e^2$ and $k > 1$,

$1 + 2\sqrt{\frac{a}{k}},$ if $a/(k-1) \leq \frac{1}{4}$ and $k > 1$.
                                                                                                                                            $\square$

10.3.2 Online Version

In [3] we describe a new randomized algorithm for the online multicover problem based on a randomized version of the *winnowing approach* of [12]. The winnowing algorithm has two scaling factors: a multiplicative scaling factor μ/cs that depends on the particular set S containing i and another additive scaling factor $|\mathcal{S}_i|^{-1}$ that depends on the number of sets that contain i. These scaling factors quantify the appropriate level of "promotion" in the winnowing approach.

// definition //
D1   **for** $(i \in V)$
D2 $\mathcal{S}_i \leftarrow \{s \in \mathcal{S} : i \in S\}$

// initialization //
I1 $\mathcal{T} \leftarrow \varnothing$ // $\mathcal{T}$ is our collection of selected sets //
I2   **for** $(S \in \mathcal{S})$
I3 $\alpha p[S] \leftarrow 0$ // accumulated probability of each set //

// after receiving an element i //
A1 $deficit \leftarrow k - |\mathcal{S}_i \cap \mathcal{T}|$ // k is the coverage factor //
A2   **if** $deficit = 0$ // we need $deficit$ more sets for i //
A3   finish the processing of i
A4 $\mathcal{A} \leftarrow \varnothing$
A5   **repeat** $deficit$ **times**
A6 $S \leftarrow$ least cost set from $\mathcal{S}_i - \mathcal{T} - \mathcal{A}$
A7   insert S to $\mathcal{A}$
A8 $\mu \leftarrow c_S$ // μ is the cost of the last set added to $\mathcal{A}$ //
A9   **for** $(S \in \mathcal{S}_i - \mathcal{T})$
A10 $p[S] \leftarrow \min \left\{ \frac{\mu}{c_S} \left(\alpha p[S] + |\mathcal{S}_i|^{-1} \right), 1 \right\}$
                                // probability for this step //
A11 $\alpha p[S] \leftarrow \alpha p[S] + p[S]$ // accumulated probability //
A12   **with** probability $p[S]$
A13   insert S to $\mathcal{T}$ // randomized selection //
A14 $deficit \leftarrow k - |\mathcal{S}_i \cap \mathcal{T}|$
A15   **repeat** $deficit$ **times**   // greedy selection //
A16   insert a least cost set from $\mathcal{S}_i - \mathcal{T}$ to $\mathcal{T}$

Figure 10.3.1: Algorithm A-Universal.

The algorithm shown in Figure 10.3.1 generalizes and improves some earlier results in [1]. We proved the following performance bounds for this algorithm.

Theorem 10.3.2. *The expected performance ratio of Algorithm A-Universal is at most* $\log_2 m \ln d$ *plus lower order terms, where* d *is the maximum number of elements in any set and* m *is the number of sets.* □

We also discussed in [3] lower bounds on competitive ratios for *deterministic algorithms* for general k based on the approaches in [1].

10.4 Conclusions

Obviously, much research remains to be done regarding the algorithmic and computational complexity of Questions (Q1) and (Q2) and their generalizations, extensions, and specific applications to gene and protein networks. For example:

- Can we design randomized algorithms with expected performance ratios better than the ones in Theorem 10.3.1, especially for $k \in [\omega(n), o(n)]$? It seems that a different rounding strategy with a considerably more nontrivial probabilistic analysis may be necessary in order to achieve this goal.
- The set system Γ may have a structure depending on the biological nature of the dependence of the variables p_j on the variable x_i's. This requires a new integer programming formulation in which, for example, "forbidden" (either mutual or as a group) combination of sets may arise (analogously to what is done in [11], for a different problem in reverse engineering). For example, a basic version of the problem that one might consider involves a given set $S \subseteq 2^U$ of forbidden combinations and adding the constraint $\sum_{\gamma \in s} x_\gamma \leq 1$ for every $s \in S$. Interestingly, the computational complexity of the problem changes substantially with these additional constraints.
- How do we design *deterministic* algorithms to derandomize such algorithms efficiently to provide deterministic algorithms? The greedy strategy is shown not to work effectively in [4, 5], hence another strategy may be necessary. A direct derandomization of the randomized algorithm, via standard techniques such as the method of conditional probabilities or the two-point sampling techniques [13], does not seem to generate a computationally efficient deterministic procedure.

Bibliography

[1] N. Alon, B. Awerbuch, Y. Azar, N. Buchbinder, and J. Naor, "The online set cover problem," *Proc. 35th Annual ACM Symposium on the Theory of Computing*, San Diego, CA, June 2003, pp. 100–105.

[2] M. Andrec, B. N. Kholodenko, R. M. Levy, and E. D. Sontag, "Inference of signaling and gene regulatory networks by steady-state perturbation experiments: Structure and accuracy," *J. Theoretical Biology*, vol. 232, pp. 427–441, 2005.

[3] P. Berman and B. DasGupta, "Approximating the online set multicover problems via randomized winnowing," *Proc. 9th Workshop on Algorithms and Data Structures*, Waterloo, Canada, Aug. 2005, pp. 110–121. (see also LNCS 3608, F. Dehne, A. Lopez-Ortiz, and J. R. Sack, Editors, Springer-Verlag, 2005.)

[4] P. Berman, B. DasGupta, and E. Sontag, "Randomized approximation algorithms for set multicover problems with applications to reverse engineering of protein and gene networks," to appear in *Discrete Applied Mathematics* (special issue on computational molecular biology).

[5] P. Berman, B. DasGupta, and E. Sontag, "Randomized approximation algorithms for set multicover problems with applications to reverse engineering of protein and gene networks," *Proc. 7th International Workshop on Approximation Algorithms for Combinatorial Optimization Problems*, Cambridge, MA, Aug. 2004, pp. 39–50. (see also LNCS 3122, K. Jansen, S. Khanna, J. D. P. Rolim, and D. Ron, Editors, Springer Verlag, 2004).

[6] T. Cover, "Geometrical and statistical properties of systems of linear inequalities with applications in pattern recognition," *IEEE Trans. Electronic Computers*, vol. EC-14, pp. 326–334, 1965.

[7] E. J. Crampin, S. Schnell, and P. E. McSharry, "Mathematical and computational techniques to deduce complex biochemical reaction mechanisms," *Progress in Biophysics & Molecular Biology*, vol. 86, pp. 77–112, 2004.

[8] N. Karmarkar, "A new polynomial-time algorithm for linear programming," *Combinatorica*, vol. 4, pp. 373–395, 1984.

[9] B. N. Kholodenko, A. Kiyatkin, F. Bruggeman, E. D. Sontag, H. Westerhoff, and J. Hoek, "Untangling the wires: A novel strategy to trace functional interactions in signaling and gene networks," *Proceedings of the National Academy of Sciences USA*, vol. 99, pp. 12841–12846, 2002.

[10] B. N. Kholodenko and E. D. Sontag, "Determination of functional network structure from local parameter dependence data," *Proc. arXiv physics/0205003*, May 2002.

[11] X. Lin, C. A. Floudas, Y. Wang, and J. R. Broach, "Theoretical and computational studies of the glucose signaling pathways in yeast using global gene expression data," *Biotechnol Bioeng.*, vol. 84, pp. 864–886, 2003.

[12] N. Littlestone, "Learning quickly when irrelevant attributes abound: A new linear-threshold algorithm," *Machine Learning*, vol. 2, pp. 285–318, 1988.

[13] R. Motwani and P. Raghavan, *Randomized Algorithms*, New York: Cambridge University Press, 1995.

[14] L. Schläfli, "Theorie der Vielfachen Kontinuitat (1852)," *Gesammelte Mathematische Abhandlungen*, vol. 1, pp. 177–392, 1950.

[15] S. Shen-Orr, R. Milo, S. Mangan, and U. Alon, "Network motifs in the transcriptional regulation network of escherichia coli," *Nature Genetics*, vol. 31, pp 64–68, 2002.

[16] E. D. Sontag, A. Kiyatkin, and B. N. Kholodenko, "Inferring dynamic architecture of cellular networks using time series of gene expression, protein and metabolite data," *Bioinformatics*, vol. 20, pp. 1877–1886, 2004.

[17] E. D. Sontag, "VC dimension of neural networks," in *Neural Networks and Machine Learning*, C. M. Bishop, ed., Berlin: Springer-Verlag, pp. 69–95, 1998.

[18] J. Stark, R. Callard, and M. Hubank, "From the top down: Towards a predictive biology of signalling networks," *Trends Biotechnol.*, vol. 21, pp. 290–293, 2003.

Chapter 11

Advances in Fingerprint Recognition Algorithms with Application

Jie Tian, Xinjian Chen, Yangyang Zhang, and Xin Yang

Abstract: Great improvement has been achieved in the on-line fingerprint sensing technology, automatic fingerprint recognition algorithms and fingerprint applications. Various fingerprint recognition techniques, including fingerprint acquisition, classification, enhancement and matching, have been developed. This chapter overviews recent advances in fingerprint recognition techniques and summarizes algorithms proposed for every step with special focuses on the enhancement of low-quality fingerprints and the matching of distorted fingerprint images. Both issues are believed to be significant and challenging tasks. We also discuss the common evaluation for fingerprint recognition algorithms: Fingerprint Verification Competition 2004 (FVC2004) and Fingerprint Vendor Technology Evaluation 2003 (FpVTE2003), based on which we can measure the performance of recognition algorithms objectively and uniformly. In addition, fingerprint recognition has been increasingly used to realize personal identification in civilian's daily life, such as fingerprint mobile phones, ID cards, fingerprint hard disks and so on. In this chapter, we introduce an application case: the fingerprint mobile phone which utilizes fingerprint recognition to ensure information security. It is composed of two parts: one is a front-end fingerprint capture subsystem, and the other is a back-end fingerprint recognition system based on smart phones.

317

11.1   Introduction

Fingerprint has been used for individual identification for a long time. Fingerprints are patterns formed by ridges and valleys flowing on the skin of fingertips. The fingerprint satisfies all the demanded properties, such as universality, permanence and distinctiveness. First, the fingerprint exists prevalently, that is, each person has unique characteristics. Second, the fingerprint can be maintained invariantly for matching until death. Third, the fingerprint has the unique feature details [1] (however, this property is not an established fact but an empirical observation). Based on the minutiae-coordinate model, we assume that the minutiae are distributed randomly and discover that the possibility of making mistakes with different fingerprints is small enough for applications.

Various fingerprint recognition techniques, including fingerprint acquisition, classification, enhancement and matching are developed and advanced rapidly. Depending on the application purpose, we classify the recognition into two categories: identification mode and verification mode [1].

Since fingerprint recognition is usually processed in a huge database, it is considered to be highly necessary to investigate an automatic fingerprint identification system (AFIS) for large-scale recognition [2]. AFIS is a pattern recognition system, typically including acquiring fingerprints from individual, enhancing images, extracting features, comparing features to that in the database. Figure 11.1.1 shows the typical structure of the recognition system. Currently, most AFISs utilize the minutiae-coordinate model for individual identification or verification.

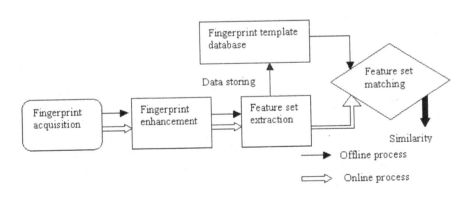

Figure 11.1.1: A typical structure of the recognition system.

Fingerprint recognition has been long accepted for identifying criminals in law enforcement, and is being increasingly used to deal with personal identification in civilian's daily life, such as fingerprint mobile phones, ID card, fingerprint hard disks and so on. In this chapter we describes particularly the mobile phones which utilize fingerprint recognitions to ensure information security. More and more sensitive information are transmitted through or stored in mobile phones nowadays, and information security of mobile phones is an important area of investigation. In the vast majority of the current systems, the only security measure to prevent unauthorized use of the mobile phones is the digital Personal Identification Number (PIN). As the amount of sensitive information stored in mobile phones increases, the need for better security increases as well. Personal identification using biometrics, i.e., personal physiological or behavioral traits, is one of the most promising "real" applications of pattern recognition [3]. It is becoming evident that this technology has the potential to meet the high demand for automated user verification, and to contribute to the security of information systems and network services including mobile phones. Fingerprints have been widely used as a means of personal verification for its time-invariant and uniqueness among people. Fingerprint sensing is often deemed to be the most "practical" biometric option for mobile phones because of low cost and accuracy.

Embedded systems based on biometric authentication is applied as the platform of personal identification. For computing systems, it must meet the real-time requirement and the performance goals [4]. In the early 1980's, embedded computing systems had very limited computing abilities; their primary function was to manage input/output devices [5]. Current embedded systems are much more powerful. Many new methods have been developed in the field of the hardware and software component designs. Engels *et al.* [6] presented a method solving the hardware-software partitioning problem in embedded system design. Their approach is based on transforming the hardware-software partitioning into a deterministic scheduling with rejection which minimizes a function of the completion time of tasks. Sung *et al.* [7] provided a test data selection technique using a fault injection method for hardware and software interaction. El-Kharashi *et al.* [8] proposed a flow for hardware/software co-design including coverification, profiling, partitioning, and co-synthesis.

This chapter overviews the recent progress in fingerprint recognition and summarizes the algorithms proposed for every step, specially focuses on the enhancement of low-quality fingerprints and the matching of distorted images. Both issues are believed to be significant and challenging tasks. We also discuss the standard evaluation of fingerprint recognition algorithms, based on which we can measure the performance of recognition algorithms objectively and uniformly. First, we describe and compare the algorithms of enhancement, especially canvassing the disposer of low-quality fingerprints. We also discuss the most prevalent feature extraction method based on minutiae patterns. Second, we enumerate different methods for fingerprints matching and present the advancement in dealing with distorted fingerprints. Third, two international common evaluation methods for fingerprint recognition algorithms from FVC2004 and FpVTE2003 are introduced. Finally, we introduces an application case: the fingerprint mobile phone which utilizes fingerprint recognition to ensure information security. It is composed of two parts. One is a front-end fingerprint capture subsystem and the other is a back-end fingerprint recognition system based on smart phones. The fingerprint capture subsystem is an external module which contains two parts: an ARM-Core processor LPC2106 and an Atmel Finger Sensor AT77C101B. The LPC2106 processor controls the AT77C101B sensor to capture the fingerprint image.

This chapter is organized as follows. Section 11.2 provides an overview for the recent progress in fingerprint recognition, summarizes the algorithms proposed for every step, and discusses the evaluation benchmark for fingerprint recognition algorithms. Section 11.3 introduces an application case: The fingerprint mobile phone. Finally, Section 11.4 concludes the chapter.

11.2 Advances in Fingerprint Recognition Algorithms

11.2.1 Fingerprint Enhancement

AFIS is based on a comparison of the features in details of ridge and valley structure. Among all the features, terminations and bifurcations, which are usually called minutiae, are the most prominent structures used in AFIS. However, in practice, it is very difficult to reliably extract the minutiae from the input image. Automatic and accurate extraction of minutiae from digital fingerprint images highly relies on the quality of images. Several factors, such as scars, non-uniform contact with the fingerprint sensors, environmental condition during the capturing process, etc., can dramatically degrade the quality of fingerprint images. The main objective of fingerprint enhancement algorithms is to improve the clarity of ridge structures and reduce the noise present in the image. Therefore, it is necessary to apply the enhancement techniques prior to minutiae extraction to obtain a more reliable estimation of minutiae locations; and it is one of the most significant steps in AFIS [9].

In most fingerprint enhancement algorithms, the sequence of main stages is described in Figure 11.2.1. Certainly, the embodied steps of different algorithms will not be the same. For instance, Maio *et al.* [10] introduced an approach which works with gray level images based on direct ridge following, overlapping the steps of binarization and thinning.

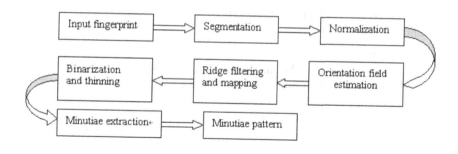

Figure 11.2.1: Complete fingerprint enhancement process.

11.2.2 Typical Fingerprint Enhancement Algorithms

(1) Segmentation

Segmentation is an important step of image preprocessing, which separates the available fingerprint field from the foreground and the noisy region. Effective segmentation not only simplifies the subsequent processing, but also improves the reliability of minutiae extraction considerably.

The general segmentation algorithms are based on variance threshold. The available fingerprint area exhibits a very high variance value, whereas other regions have a relatively low variance. Hence, a variance threshold is used to complete the segmentation. Bazen and Gerez [11] proposed a segmentation algorithm based on pixels features, using the criterion of Rosenblatt's perception to classify the pixels. Chen *et al.* [12] developed an algorithm by utilizing three block features: the clusters degree, the mean information, and the variance for the segmentation of fingerprints.

There is another type of method [4] based on the orientation information of fingerprint images. The method relies on the reliability of the orientation field, and it is not sensitive to the gray contrast. However, it is almost impossible to get accurate orientation graph in regions with discontinuous ridges or around the pore and delta.

The frequency domain based method [13] showed that the surface wave model does not hold in the foreground and noisy regions, and there is very little energy existing in the Fourier spectrum. However, this algorithm cannot treat the areas consisting of unequal texture resulting form the distortion of fingerprint images.

Recently, merging certain features to gain the segmentation's propriety is evidently feasible. Because of different characters, the process has many dubious components. A series of algorithms have been put forward to solve the problem, such as fingerprint segmentation based on D-S Evidence Theory [14], application of URN model in image segmentation [15], the Markov model based segmentation algorithm [16], and so on.

(2) *Normalization*

Due to imperfections in the fingerprint image capture process such as non-uniform ink intensity or non-uniform contact with the capture device, a fingerprint image may display distorted levels of variation in grey-level values along ridges and valleys. Thus, normalization is performed to decrease the dynamic range of the gray scale between ridges and valleys of the image [16], which facilitates the subsequent enhancement steps. Lin *et al.* [17] standardized image intensity values by adjusting the range of grey-level values to lie within a desired range of values. The normalization factor is calculated according to the mean and the variance of the image.

(3) *Orientation field estimation*

For fingerprint images, the orientation field is a matrix representing ridge orientation as a directional vector for each pixel. It reflects the fundamental information existing in fingerprint images, and it is a significant component to measure the fingerprint quality. There are several approaches to calculate it. The widely employed gradient-based approach [17] is based on the fact that the orientation vector is orthogonal to the gradient. In this method, the image is divided into small blocks, and the orientation vector of each block is estimated by averaging the vectors orthogonal to the gradient of all pixels. The template comparison based approach [18] disperses the orientation into finite directions, utilizes special template to calculate the presumed orientation for ridges in the block. Compared to the gradient based approach, this approach obtains result more quickly at the expense of precision.

Due to the noise and the corrupt field existing in fingerprint images, it is necessary to post process the orientation field. Given that the ridge orientation varies slowly in a local neighborhood, the orientation image is then smoothed by using a low-pass filter to reduce the effect of outliers [19]. Figure 11.2.2 gives an original fingerprint image and its orientation field based on the gradient.

(4) *Ridge filtering and mapping*

Fingerprint images usually contain random noise and coarse ridges because of imperfect finger conditions and the environment. It is critical to remove the noise and smooth the ridges by a filter to map the ridges accurately. Existing algorithms can be broadly classified as spatial domain based [17, 20–24] and Fourier domain based [25–27] according to their concrete implementation methods. The former usually employs local ridge properties which consist of ridge frequencies and directions, while the latter mainly utilizes global ridge properties.

Spatial domain based methods handle the band filtering procedure by directly

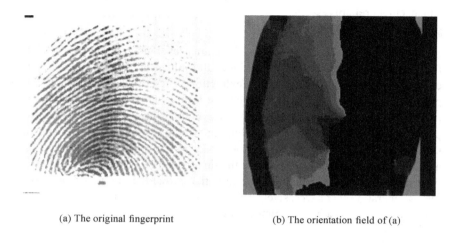

(a) The original fingerprint   (b) The orientation field of (a)

Figure 11.2.2: An original fingerprint image and its orientation field.

convolving the filter operator with the digital fingerprint image. The representative method introduced by Lin *et al.* [17] employed a Gabor filter which has frequency-selective and orientation-selective properties. However, it is very time consuming for applications. A series of modified algorithms based on Gabor filter were proposed to accelerate the process. Vutipong *et al.* [20] implemented a new set of separable Gabor filters for fingerprint enhancement. It is approximately 2.6 times faster than the conventional Gabor filtering.

Fourier domain based methods refer to those coping with the filtering procedure by directly modifying the frequency spectrum of the original image. Since the ridges and valleys present almost equidistantly, the energy of fingerprint images concentrates around a certain frequency. The directional filtering method proposed by Sherlock *et al.* [25] is a sort of Fourier domain based methods that first filter a raw image by a few directional filters, which can transmit the spectrum of a certain direction and attenuate the spectra of other directions, and then form the enhanced image by appropriately combining these filtered images. Compared to the Fourier enhancement, spatial domain based methods such as the Gabor filtering are complicated and computationally complex.

In addition, there are other algorithms [28] that improve the clarity and continuity of ridge structures based on multiresolution analysis of global texture and local orientation by wavelet transform. There are also several dyadic scale-space methods [18, 29] which decompose the fingerprint image into a series of images in different scales, and then analyze and organize the image characters to realize enhancement reliably.

(5) *Binarization and thinning*

Binarization is a process that converts a gray-level image into a binary image. It improves the contrast between the ridges and valleys in a fingerprint image, and consequently facilitates the minutiae extraction. The critical issue is to choose a proper threshold to binarize the image. The local adaptive threshold based algorithms have been employed to generate the binary image. It includes choosing proper threshold referring to the local image window around each pixel and classifying the pixel as the foreground or the background. He *et al.* [24] developed a method to fast and directly binarize the image with its orientation field. As the final step typically performed prior to minutiae extraction, thinning is a morphological operation that successively erodes away the original ridges until they are one pixel wide. A standard thinning algorithm [31] is performed by using two subiterations.

(6) *Minutiae extraction*

In this stage, the local features of minutiae from the thinned image are extracted to obtain fingerprint biometric patterns [19]. The majority of minutiae extraction algorithms are based upon the skeletons of fingerprint images. They are relatively computationally expensive and can produce artifacts such as spurs and bridges. Chikkerur *et al.* [32] proposed a method for feature extraction based on chain coded contour following. Maio *et al.* [10] extracted the minutiae directly from the gray scale image. The main idea of their method is to follow the ridge lines on the gray scale image by "sailing" according to the fingerprint directional image. Jiang *et al.* [33] improved the flexibility of the algorithm based on ridge following.

The extraction process may result in errors which generate the false minutiae and meanwhile miss the genuine minutiae. Therefore, it is necessary to adopt a post processing method to modify the false features. Luo *et al.* [23] utilized human knowledge on fingerprints to post process the condition of ridge break, bridge, blur, and scar, modified spurious minutiae extraction and gained satisfactory performance. The method proposed by Chikkerur *et al.* [32] is mainly based on the following rules. 1) Merging the minutiae that are within a certain distance of each other and have similar angles; 2) removing all points at the border of the interest region; 3) discarding the minutiae whose direction is not consistent with the local ridge orientation; and so forth. Then, we obtain minutiae patterns of input fingerprint images. Figure 11.2.3 presents the extracted minutiae of a thinned fingerprint image after post-processing.

11.2.3 Low Quality Fingerprint Image Enhancement

The quality of fingerprint image degrades due to impression, skin, reader, etc.,

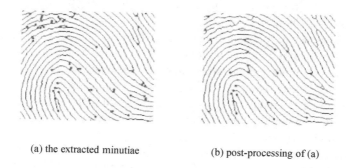

(a) the extracted minutiae   (b) post-processing of (a)

Figure 11.2.3: The extracted minutiae of a thinned fingerprint image with its post-processing result.

(a)   (b)   (c)   (d)

Figure 11.2.4: Four examples of low quality fingerprints. (a) Too dry; (b) too wet; (c) with many scars; (d) molted.

during image capture. Figure 11.2.4 shows four examples of low quality fingerprints. Generally, the adaptability of the AFIS relies on the capability to enhance poor quality fingerprint images. Such enhancement is so important that it seriously affects the performance of the recognition system. It is one of the most crucial and difficult tasks for fingerprint recognition.

Image quality analysis is a critical component of a fingerprint livescan workstation. AFIS rejects a certain percentage of submitted fingerprint images because they fail to satisfy the image quality criteria. Failure to extract minutiae points is usually attributed to poor ridge flow, poor contrast and brightness in the image.

Shen *et al.* [34] proposed a Gabor-feature based method to determine the quality of the fingerprint images.

Many standard and special image enhancement techniques have been developed for poor quality images. Shi *et al.* [35] proposed a new feature Eccentric Moment to locate blurry boundaries using the new block feature of clarified image for segmentation. Zhou *et al.* [36] proposed a model-based algorithm which is more accurate and robust to dispose the degraded fingerprints. They compute a coarse orientation field by traditional methods, and approximate the real orientation with smooth curves.

To enhance the poor quality prints efficiently, we must incorporate a robust ridge filter with respect to the quality of input fingerprint images. Lin *et al.* [17] assumed that parallel ridges and valleys exhibit some ideal sinusoidal-shaped plane waves associated with some noises, which cannot treat poor quality images. Yang *et al.* [31] specified parameters deliberately through some principles instead of experience, preserved fingerprint image structure and achieved image enhancement consistency. This algorithm solved the problem that false estimation of local ridge direction will lead to a poor enhancement. Zhu *et al.* [37] followed Lin's algorithm, but used a circle support filter and tuned the filter's frequency and size differently. This scheme rapidly enhanced the fingerprint image and effectively overcame the blocky effect. Khan *et al.* [38] proposed a method using decimation-free directional filter bank (DFB) structure to improve poor quality fingerprints.

These methods performed local estimation and contextual filtering in a dispersed manner, which often resulted in not only blocky artifacts but also poor estimation of local image characteristics. Another type of mechanism based on non-linear diffusion was also proposed to solve the problem. Xie *et al.* [24] adopted an image structure tensor merging both the coherence enhancement diffusion [39] for processing flow-like pattern and the forward and backward enhancement diffusion [40] for sharpening ridges. These algorithms utilized global features of the ridge flow direction to restore the disconnection caused by the poor quality of images and received good performance.

Compared to the uncertainty of local ridge information, global features can be preserved accurately in the attained fingerprint images. Therefore, many Fourier domain based ridge filters have been presented for the low-quality fingerprint images. Willis *et al.* [26] proposed a Fourier domain based method that boosts up a low quality fingerprint image by multiplying the frequency spectrum by its magnitude. Zhu *et al.* [27] combined the two methods mentioned above by multiplying each filter vector with well designed weights to form a new filter vector. In addition, it applies a top-down iteration technique which makes the method more robust.

11.2.4 Recent Advances in Fingerprint Matching Algorithms

A large number of fingerprint matching approaches have been proposed in the literature. These include methods based on point pattern matching, transform features, structural matching, and graph-based matchers.

Many fingerprint recognition algorithms [41–50] are based on minutiae matching since it is widely believed that the minutiae are the most discriminating and reliable features [53, 54]. They are essentially "Euclidean" matchers. Figure 11.2.5 shows two fingerprints and the map relations of the corresponding minutiae in these two fingerprints. These matchers assume that there exists a similarity transformation with translation, rotation and scaling of the minutiae between the input and the template fingerprints, and they can tolerate, to a limited extent, both spurious minutiae and missing genuine minutiae. In addition, some of them can be modified to tolerate a small bounded local perturbation of minutiae. But they cannot handle large distortions of the minutiae from their true locations. Ratha *et al.* [55] addressed a method based on point pattern matching. The generalized Hough transform is used to recover the pose transformation between two impressions. Chang *et al.* [37] proposed a generalized Hough transform-based approach which converts point pattern matching to peaks detecting in the Hough space of transformation parameters. It discretizes the parameter space and accumulates evidence in the space by deriving transformation parameters that relate two point patterns using a substructure or feature matching technique. However, if there are only a few minutiae points available, it is difficult to accumulate enough evidence in the Hough transform space for a reliable match. Moreover, it is hard to handle large distortions. Ton *et al.* [47] proposed a modified version of the relaxation approach [49] to reduce the matching complexity. However, these algorithms are inherently slow because of their iterative nature and unable to handle large distortions. Some researchers proposed the energy minimization method to point pattern matching [41–46, 48] which establishes the correspondence between a pair of point sets by defining an energy function based on an initial set of possible correspondences and utilizes an appropriate optimization technique such as genetic algorithm, neural network, simulated annealing to find a possible suboptimal match. These methods are very slow and unsuitable for a real-time fingerprint identification system.

Jain *et al.* [56, 57] proposed a novel filter bank-based fingerprint feature representation method. Jiang *et al.* [58] developed a method which relies on a similarity measure defined between local structural features to align two patterns and calculate a matching score between two minutiae lists. Fan *et al.* [59] applied a set of geometric masks to record part of the rich information of the ridge structure. Wahab *et al.* [60] discussed a method using groups of minutiae to define local structural features. The matching is performed based on pairs of corresponding

(a)	(b)	(c)

Figure 11.2.5: Minutiae matching of two fingerprints. (a), (b) Original images; (c) the map relations of the corresponding minutiae in fingerprint images (a) and (b).

structural features that are identified between two fingerprint impressions. However, these methods do not solve the problem of nonlinear distortions.

Some researchers proposed graph-based matchers [61–63], which are essentially a "topological" type of matchers. They allow general transformations, positional errors, missing minutiae, and spurious minutiae. The performance of these algorithms depends heavily upon the availability of ridge features and external alignment information. In a semi-automatic fingerprint identification system, these algorithms can perform well since the minutiae patterns can be aligned and the errors in minutiae and ridge features can be corrected interactively. However, a fully automatic fingerprint matching system may not always guarantee the availability of correct ridge features and external alignment information.

11.2.5 Distorted Fingerprints Matching

How to cope with these nonlinear distortions in the matching process is a challenging task. According to Fingerprint Verification Competition 2004 (FVC2004) [64], the organizers are particularly insisted on distortion, and dry and wet fingerprints. Distortion of fingerprints seriously affects the accuracy of matching. There are two main reasons contributed to the fingerprint distortion. First, the acquisition of a fingerprint is a 3D-2D warping process [51, 52]. The fingerprint captured with different contact centers usually results in different warping mode. Second, distortion will be introduced to fingerprint by the non-orthogonal pressure exerted on the sensor. Figure 11.2.6 displays two examples of large distortion between fingerprints.

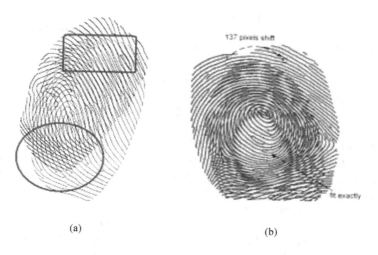

(a)   (b)

Figure 11.2.6: Two examples of large distortion. (a) In the rectangle region, the corresponding minutiae are approximately overlapped. While in the ellipse region, the maximal vertical difference of the corresponding minutiae is above 100 pixels. (b) In the center region, the corresponding minutiae are approximately overlapped. While in the upper region, the maximal horizontal difference of the corresponding minutiae is 137 pixels.

Recently, some algorithms have been presented to deal with the nonlinear distortion in fingerprints explicitly in order to improve the matching performance. Ratha *et al.* [67] proposed a method to measure the forces and torques on the scanner directly. Which prevents capturing with the aid of special hardware when excessive force is applied to the scanner. Dorai *et al.* [68] proposed a method to detect and estimate distortion occurring in fingerprint videos. However, the two methods mentioned above do not work with collected fingerprint images. Cappelli *et al.* [51] proposed a plastic distortion model to cope with the nonlinear deformations characterizing fingerprint images taken from on-line acquisition sensors. This model helps to understand the distortion process. However, it is hard to automatically and reliably estimate parameters due to the insufficient and uncertain information. Lee *et al.* [69] addressed a minutiae-based fingerprints matching algorithm using distance normalization and local alignment to deal with nonlinear distortion. However rich information of the ridge/valley structure is not used, and the matching performance is moderate. To improve the matching accuracy, Senior *et al.* [70] proposed a method to convert a distorted fingerprint image into an equally ridge spaced fingerprint before matching. However, the assumptions of equal ridge spacing is less likely to be true for fingerprints - particularly where ridges break down, such as around minutiae or near the edge of the fingerprint. Watson et al. [61] proposed a method to improve the performance of fingerprint correlation matching by distortion tolerant filters. The improvement was achieved by multiple training fingerprints and a distortion-tolerant MINACE filter. However, the algorithm is difficult to realize on line. Vajna [72] also proposed a method based on triangular matching to cope with the strong deformation of fingerprint images, which graphically demonstrates that large cumulative effects can be resulted from small local distortions. Bazen et al. [73] employed a thin-plate spline model to describe nonlinear distortions between two sets of possible matching minutiae pairs. By normalizing the input fingerprint with respect to the template, this method is able to perform a very tight minutiae matching. Ross et al. [68] used the average deformation computed from fingerprint impressions originating from the same finger based on thin plate spline model to cope with nonlinear distortions. Chen et al. [52] introduced a novel fingerprint verification algorithm based on the determination and inspection of the registration pattern (RP) between two fingerprints. The algorithm first coarsely aligns two fingerprints. Then determines the possible RP by optimally registered each part of the two fingerprints. Next, inspects the possible RP with a genuine RP space. If the RP makes a genuine one, a further fine matching is conducted. Different from the above mentioned methods, Chen et al. [75] proposed an algorithm based on fuzzy theory to deal with nonlinear distortion in fingerprint images. The local topological structure matching was introduced to improve the robustness of global alignment. A similarity computation method based on fuzzy theory, normalized fuzzy similarity measure, was conducted to compute the similarity between the template and input fingerprints. Experimental results indicate that the algorithm works well with the nonlinear distortions. For deformed fingerprints, the algorithm gives considerably higher matching scores compared to conventional matching methods.

11.2.6 Performance Evaluation of the Fingerprint Recognition Algorithm

In the last decade, with the rapid development of fingerprint recognition systems, it is urgent to establish a common benchmark in this field. Participators could evaluate their algorithms on this common benchmark, compare the performance and provide an overview of the state-of-the-art technology in fingerprint recognition. There are two internationally recognized and accredited evaluations: Fingerprint Verification Competition (FVC [64–66]) and Fingerprint Vendor Technology Evaluation (FpVTE [71]).

The FVC2004 [64] (the Third International Fingerprint Verification Competition) were organized by the Biometric System Lab of University of Bologna, the Pattern Recognition and Image Processing Laboratory of Michigan State University and San Jose State University. The aim of FVC2004 is to track recent advances in fingerprint verification, for both academia and industry, and to benchmark the state-of-the-art technology in fingerprint recognition. The FVC2004 results were presented at the International Conference on Biometric Authentication (ICBA, Jan. 2004). The results of this competition give a useful overview of the state-of-the-art technology in this field. The first and second international competitions on fingerprint verification (FVC2000 and FVC2002 [65, 66]) were conducted in 2000 and 2002, respectively. These events received great attention both from academic and industrial biometric communities. The FVC established a common benchmark allowing developers to unambiguously compare their algorithms, and provided the overview of the most recent advancement in fingerprint recognition.

The FVC2004 competition focuses on fingerprint verification software. Participators are required to propose two executable programs for corresponding subcompetitions (open category and light category), which were operated on the same databases [76]. In open category, the enrollment response time was limited to 10 secs and the match time was less than 5 secs for testing. The computing resource for light algorithm was constrained. The limits were as follows: the enrollment time is less than 0.5 secs, the matching time is less than 0.3 secs, the template size is lower than 2KB, and the amount of memory allocated is lower than 4KB. Four fingerprint databases were established for testing, which were collected from three different sensors. Each one has different image sizes and consists of 110 fingers, 8 prints per finger. The algorithm were performed in both genuine mode and impostor mode. For each database, the matching number was 2800 and 4950 times for genuine and impostor matching, respectively. For each algorithm performed in the competition, the statistic performance results were presented by the following performance indicators [73]: 1) Rate of reject to enroll (REJENROLL), rate of reject to match (REJNGRA and REJNIRA); 2) genuine and impostor score histograms; 3) FMR(t) and FNMR(t) and ROC(t); 4) equal error rate (EER), FMR100, FMR1000, ZeroFMR and ZeroFNMR; 5) average match time and average enroll time; 6) maximum memory allocated for enrollment and for match; 7) average and maximum template size.

The Fingerprint Vendor Technology Evaluation (FpVTE) 2003 [77] was an independently administered technology evaluation for fingerprint matching, identification, and verification systems. FpVTE 2003 was conducted by the National Institute of Standards & Technology (NIST) on behalf of the Justice Management Division (JMD) of the U.S. Department of Justice. FpVTE was designed to assess the capability of fingerprint systems to meet requirements for both large-scale and small-scale databases in real world applications. FpVTE 2003 consisted of multiple tests performed with combinations of fingers (e.g., single fingers, two index

Table 11.2.1: The relative surroundings of each test

	LST	MST	SST
Finger-mode	Multi-finger	Single-finger	Single-finger
Fingerprint type	Rolls, Slaps, and Flats (Paper and livescan)	Slaps and Flats (Livescan only)	Flats (Livescan only)
Sample-model	64,000 sets of 1–10 fingerprints	10,000 fingerprints (all right index finger)	1,000 fingerprints (all right index finger)
Time limit	21 days	14 days	14 days

fingers, four to ten fingers) and different types and qualities of operational finger-prints (e.g., flat livescan images from visa applicants, multi-finger slap livescan images from present-day booking or background check systems, or rolled and flat inked fingerprints from legacy criminal databases). Compared to FVC, FpVTE 2003 has many different features, which can reflect algorithms from other aspects. Small to large-scale fingerprint databases are provided for measuring the algorithm's flexibility. FpVTE consisted of three separate segments, which are Large-Scale Test (LST), Medium-Sscale Test (MST), and Small-Scale Test (SST) [76]. Table 11.2.1 illustrates the relative surroundings for each of them.

The performance was measured through the following indicators: 1) match and non-match distribution; 2) ROC(t); 3) slice chart; 4) effect of fingerprint quality; 5) effect of finger -mode, and so on.

11.3 Application to Fingerprint Mobile Phone

The mobile security system is composed of two subsystems. One is front-end fingerprint capture subsystem and the other is back-end fingerprint recognition subsystem based on BIRD smart phone E868 [78]. The structure of the whole system is shown in Figure 11.3.1.

The smart phone device is the platform of mobile security system. All security softwares are implemented on it, including fingerprint recognition software, startup control software, address book management software, etc. The fingerprint capture subsystem is an external module. The main parts of the subsystem are an ARM core processor LPC2106 and an Atmel Finger Sensor AT77C101B. The subsystem is controlled by the LPC2106 processor, and it works in a slave mode. The LPC2106 processor receives the commands from the smart phone via UART

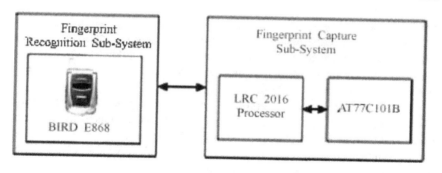

Figure 11.3.1: The mobile security system block diagram.

interface, controls the AT77C101B sensor to capture the fingerprint image, and sends it to the smart phone.

11.3.1 System Hardware Description of Fingerprint Mobile Phone

(1) *ARM Core Processor*

The LPC2106 ARM core processor is manufactured by Philips Company [79]. It is a powerful processor with an ARMTDMI-S core, working at 60 MHz. It has a 128 kilobyte on-chip flash and a 64 kilobyte on-chip static RAM. Moreover, the processor is as compact as 7mm × 7mm in size. It can operate in two low power working modes which make it suitable for mobile applications.

The LPC2106 processor consists of an ARM7TDMI-S CPU, three different types of buses (the ARM7 Local Bus, the AMBA Advanced High Performance Bus (AHB), and the VLSI Peripheral Bus), and twelve peripheral models with different functions.

The ARM7TDMI-S is a general 32-bit microprocessor, which provides high performance and very low power consumption [80]. The ARM architecture is based on Reduced Instruction Set Computer (RISC) principles, the instruction set and related decode mechanism are much simpler than those of Complex Instruction Set Computers. The ARM7TDMI-S processor also employs a unique architectural strategy known as THUMB, which makes it ideally suitable to high-volume applications with memory restrictions, like the mobile security system.

The LPC2106 incorporates a 128K byte flash memory system. This memory may be used for both code and data storage. The flash memory programming may accomplish in several ways: using the serial built-in JTAG interface, employing In System Programming (ISP) and UART0, or by means of In Application Programming (IAP) capabilities. The LPC2106 also provides a 64K byte on-chip static RAM memory for code and/or data storage. The SRAM supports 8-bit, 16-bit, and 32-bit accesses.

Same as the other processors with ARM core, there are 7 exceptions in LPC2106. They are reset, undefined instruction, software interrupt, prefetch abort, data abort, IRQ and FIQ. Except the IRQ, the others have the same processing mode as other ARM core processors.

The LPC2106 processor incorporates an ARM PrimeCell Vectored Interrupt Controller (VIC). It takes 32 interrupt request inputs and assigns them into 3 categories, FIQ, vectored IRQ, and non-vectored IRQ. 16 of the 32 interrupt requests can be assigned to vectored IRQ category. If any of the vectored IRQs are requested, the VIC provides the address of the highest-priority requesting IRQs service routine. It saves the time on reading the interrupt pending register by software. The priority of the vectored IRQs is lower than FIQs and higher than non-vectored IRQs.

The LPC2106 processor is composed of two Universal Asynchronous Receiver and Transmitter (UART) modules. Each UART module contains a baud-rate generator, transmitter, receiver and control unit. UART0 provides only transmit and receive data lines. Otherwise, UART1 provides a full modem control handshake interface. Two UARTs can be operated in interrupt-based or query-based mode, and each channel performs configurable 16-byte FIFOs for receiving.

The LPC2106 processor pins are not connected to a specific peripheral function, but controlled by the GPIO registers. Pins may dynamically be configured as inputs or outputs. Different registers allow setting or clearing outputs simultaneously. The value of output register may be read back as well as the current state of the port pins. Developers can use the GPIO to simulate the function of data/address bus, since the LPC2106 processor do not have the data bus and the address bus.

(2) Atmel Finger Sensor

Atmel's AT77C101B FingerChip IC for fingerprint image capture combines detection and data conversion circuitry in a single rectangular CMOS die [81]. It captures the image of a fingerprint as the finger is swept vertically over the sensor window. It requires no external heat, light or radio source.

The AT77C101B sensor is divided into two sections: sensor array and data conversion (Figure 11.3.2). The sensor array comprises an array of 8 rows by 280 columns, giving 2240 temperature-sensitive pixels and one column selection circuit. The data conversion consists of an analog signal amplifier, two 4-bit Analog-to-Digital Converter (ADC) and a digital signal output circuit.

The 50mm $\times$ 50mm pixel pitch provides a resolution of 500 dpi over 0.4mm $\times$ 4mm image zone. It is adequate to capture a frame of the central portion of a fingerprint at an acceptable image resolution. The resolution also complies with the Image Quality Specification (IQS) from the Integrated Automated Fingerprint

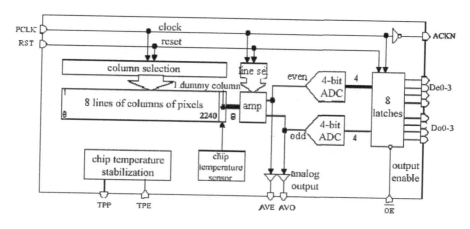

Figure 11.3.2: AT77C101B block diagram (cited from [81]).

Identification System (IAFIS) of Federal Bureau of Investigation (FBI) [82]. The pixel clock can be programmed at up to 2 MHz, with the output of 1780 frames per second. This is adequate for a typical sweeping velocity.

The operating voltage for AT77C101B is 3.3V to 5V, with a power consumption of 20 mW at 3.3V, 1 MHz clock. In normal operation, the sensor is entirely passive, using the thermal energy in the applied fingertip to obtain its measurements. However, if the temperature difference between the finger and the sensor falls too low (less than one degree) a temperature stabilization feature is activated to slightly increase the temperature of the IC and recover the contrast.

There are many benefits in using AT77C101B sensor in the mobile security system. The sensor is small in size and can be embedded in an external mobile module or even in the smart phone. The sensor can work in a low power mode and only consumes less than 10 μA current. The rectangular sensor window can reduce the cost of sensor and the whole mobile system as well.

11.3.2 System Software Description of Fingerprint Mobile Phone

The software for mobile security system mainly contains three algorithms: fingerprint recognition, fingerprint reconstruction, and the energy management. The fingerprint recognition algorithm is described in Section 11.2. The fingerprint reconstruction algorithm is based on the method presented by Shekarforoush et al. [83]. The energy management algorithm is used to save the battery power.

The fingerprint sensor captures a full fingerprint image in three steps: 1) capture 50×70 slices of the fingerprint image; 2) image reconstruction; 3) send the whole fingerprint image to the smart phone, as shown in Figure 11.3.3.

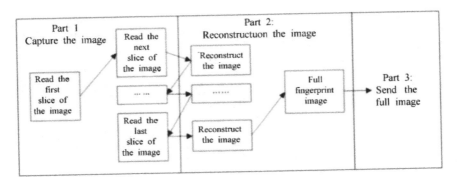

Figure 11.3.3: The fingerprint capture block diagram.

In the first step, when the finger sweeps vertically over AT77C101B's sensor window, the LPC2106 processor will read image slices from it. Each image slice is in a format of 82×80 with 16 gray levels. Since the sensor takes only one image slice each time, it must run many times to obtain a 256×256 fingerprint image during the sweeping.

The rate of the finger swept over the sensor is not always the same at each time. If the rate is slower than a reasonable speed, there will be an overlapping image in the two successive slices. It is an important task to eliminate the overlapped image in image reconstruction period.

A full fingerprint image is 256×256. At least 32 slices is needed to reconstruct a full image. The smallest memory is 137216 bytes which is larger than the LPC2106's memory capacity. Due to limited memory of the LPC2106, the slice capture and image reconstruction are executed alternately as shown in Figure 11.3.3. We use the LPC2106's SRAM as temporary memory for slice image and other temporary data, and use the flash to save the program and the full fingerprint image. The full fingerprint image is sent to the smart phone device by LPC2106 through UART interface at the rate of 230400 bps. The total time for capturing a full fingerprint image is about 1.5 seconds.

The mobile security system is a portable embedded computing system. This battery-powered system needs an efficient energy management to prolong the smart phone operation. Rakhmatov [84], described three approaches for the task scheduling and voltage assignment. Based on the characters of LPC2106 processor, we use a custom method to obtain an optimal trade-off between the system performance and battery consumption.

The fingerprint capture subsystem works in the slave mode. It waits for the command from the application software to the smart phone, and carries out the relevant operations. Most of the operations are capturing a fingerprint image and sending it to the smart phone. We set the LPC2106 processor's UART interrupt flag on during the subsystem initialization. After the initialization, we set the LPC2106 processor and the AT77C101B sensor in sleep mode to save energy. When the smart phone captures a fingerprint image and sends the command, the communication will trigger the working condition for LPC2106 processor.

LPC2106 processor possesses a very good function by turning off selected peripherals. When the LPC2106 processor is in normal mode, the unused peripherals are turned off to save more energy. For example, when the subsystem is capturing a fingerprint image, only part 3 is needed to communicate with the smart phone by using the UART interface. UART interface is turned off during the period of part 1 and part 2. In the same way, AT77C101B sensor is turned off in the period of part 3.

11.3.3 The Prototype of Fingerprint Mobile Phone

Figure 11.3.4 shows a fingerprint mobile phone prototype based on BIRD smart phone E868. The E868 mainly targets the high-end business market. It is capable of supporting up to 65,000-colors, touch-screen and handwriting recognition. The mobile phone provides PDA functions, e-mail, internet, camera, mp3, JAVA technology and much more. The central processing unit of the E868 is a 16-bit embedded processor S1C33. The processor is produced by Epson Company and its working frequency is 13 MHz.

Figure 11.3.4: The prototype of fingerprint mobile phone.

Figure 11.3.5 shows some examples of original captured fingerprints and enhanced fingerprint. The quality of fingerprints in Figure 11.3.5(c) is low as there are scars in the image. The quality of fingerprints in Figure 11.3.5(e) is also low as it is too wet. However, these fingerprints were enhanced well using the proposed algorithm. As shown in Figure 11.3.5, a significant enhancement is seen in captured fingerprints.

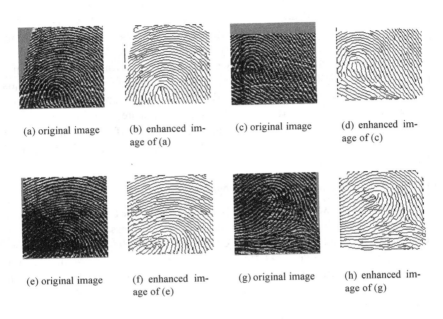

(a) original image   (b) enhanced im-   (c) original image   (d) enhanced im-
                              age of (a)   age of (c)

(e) original image   (f) enhanced im-   (g) original image   (h) enhanced im-
                              age of (e)   age of (g)

Figure 11.3.5: Four examples of enhanced fingerprints.

11.4   Conclusions

This chapter reviewed recent advances in fingerprint recognition systems and summarized algorithms proposed for every step including segmentation, normalization, orientation field estimation, ridge filtering and mapping, binarization and thinning, minutiae extraction and fingerprint matching. We specially focused on the enhancement of low-quality fingerprints and the matching of distorted fingerprint images. Both issues are significant and challenging tasks in fingerprint recognition since they seriously affect the overall performance of the whole recognition system. In this chapter, we also discuss the common evaluation for the fingerprint recognition algorithm by FVC2004 and fpVTE2003, based on which the recognition algorithm can be measured objectively and uniformly. Finally, we described an application case: the fingerprint mobile phone, which utilizes fingerprint recognitions to ensure information security.

Bibliography

[1] A. K. Jain, A. Ross, and S. Prabhakar, "An introduction to biometric recognition," *IEEE Transactions on Circuits and Systems for Video Technology*, vol. 14, no. 1, pp. 4–20, Jan. 2004.

[2] A. K. Hrechak and J. A. McHcgh, "Automated fingerprint recognition using structural matching," *Pattern Recognition*, vol. 23, no. 1, pp. 893–904, Aug. 1990.

[3] K. Uchida, "Fingerprint-based user-friendly interface and pocket-PID for mobile authentication," *Proc. International Conference Pattern Recognition*, Barcelona, Spain, Sept. 2000, pp. 205–209.

[4] W. Wolf and T. Y. Yen, "Embedded computing and hardware-software co-design," *Proc. Microelectronics Communications Technology Producing Quality Products Mobile and Portable Power Emerging Technologies*, San Francisco, CA, Nov. 1995, pp. 450–454.

[5] W. Wolf, "What is embedded computing?" *Computer*, vol. 35, no. 1, pp. 136–137, Jan. 2002.

[6] D. W. Engels and S. Devadas, "A new approach to solving the hardware-software partitioning problem in embedded system design," *Proc. 13th Symposium on Integrated Circuits and Systems Design*, Lansing, MI, Aug. 2000, pp. 275–280.

[7] A. Sung and B. Choi, "An interaction testing technique between hardware and software in embedded systems," *Proc. 9th Asia-Pacific Software Engineering Conference*, Gold Coast, Queensland, Australia, Dec. 2002, pp. 457–464.

[8] M. W. El-Kharashi, M. H. El-Malaki, S. Hammad, A. Salem, and A. Wahdan, "Towards automating hardware/software co-design," *Proc. 4th IEEE International Workshop on System-on-Chip for Real-Time Applications*, Banff, Alberta, Canada, July 2004, pp. 189–192.

[9] D. Maltoni, D. Maio, A. K. Jain, and S. Prabhakar, *Handbook of Fingerprint Recognition*, New York: Springer-Verlag, 2003.

[10] D. Maio and D. Maltoni, "Direct gray scale minutiae detection in fingerprints," *IEEE Transactions on Pattern Analysis and Machine Intelligence*, vol. 19, no. 1, pp. 27–40, Jan. 1997.

[11] A. M. Bazen and S. H. Gerez, "Segmentation of fingerprint images," *Proc. 12th Annual Workshop on Circuits, Systems and Signal Processing*, Veldhoven, Netherlands, Nov. 2001, pp. 276–280.

[12] X. J. Chen, J. Tian, J. G. Cheng, and X. Yang, "Segmentation of fingerprint images using linear classifier," *EURASIP Journal on Applied Signal
Processing*, vol. 4, pp. 480–494, Apr. 2004.

[13] S. Chikkerur, C. H. Wu, and V. Govindaraju, "A systematic approach for
feature extraction in fingerprint images," *Proc. 1st International Conference on Biometric Authentication*, Ft. Lauderdale, FL, Dec. 2004, pp. 344–
350.

[14] L. R. Tang, X. H. Xie, A. N. Cai, and X. Ke, "Fingerprint image segmentation based on D-S evidence theory," *Chinese Journal of Computers*, vol.
26, no. 7, pp. 887–892, 2003.

[15] Y. Y. Ceng, A. N. Cai, and J. A. Su, "Multiple criterion decisions based on
URN model and its application in image segmentation," *ACTA Electronica
Sinica*, vol. 30, no. 7, pp. 1017–1019, 2002.

[16] Y. L. He, J. Tian, and X. P. Zhang, "An algorithm for fingerprint segmentation based on Markov model," *Proc. China Computer Graphics Conference*, Beijing, China, Sept. 2002, pp. 149–156.

[17] L. Hong, Y. Wan, and A. K. Jain, "Fingerprint image enhancement: algorithm and performance evaluation," *IEEE Transactions on Pattern Analysis
and Machine Intelligence*, vol. 20, no. 8, pp. 777–789, 1998.

[18] J. G. Cheng and J. Tian, "Fingerprint enhancement with dyadic scale-
space," *Pattern Recognition Letters*, vol. 25, no. 11, pp. 1273–1284, 2004.

[19] D. S. Zorita, J. O. Garcia, S. C. Llanas, and J. Gonzalez, "Minutiae extraction scheme for fingerprint recognition systems," *Proc. International
Conference Image Processing*, Thessaloniki, Greece, Oct. 2001, pp. 254–
257.

[20] V. Areekul, U. Watchareeruetai, and S. Tantaratana, "Fast separable Gabor
filter for fingerprint enhancement," *Proc. 1st International Conference on
Biometric Authentication*, Hong Kong, China, July 2004, pp. 403–409.

[21] J. W. Yang, L. F. Liu, T. Z. Jiang, and F. Yong, "A modified Gabor filter
design method for fingerprint image enhancement," *Pattern Recognition*,
vol. 24, no. 12, pp. 1805–1817, Aug. 2003.

[22] B. G. Kim, H. J. Kim, and D. J. Park, "New enhancement algorithm for fingerprint images," *Proc. 15th International Conference on Pattern Recognition*, Quebec, Canada, Aug. 2002, pp. 879–882.

[23] X. P. Luo and J. Tian, "Knowledge based fingerprint image enhancement," *Proc. 15th International Conference on Pattern Recognition*, Spain,
Barcelona, Sept. 2000, pp. 783–786.

[24] M. H. Xie and Z. M. Wang, "Fingerprint enhancement based on edge-direct diffusion," *Proc. 3rd International Conference on Image and Graphics*, Hong Kong, China, Dec. 2004, pp. 274–277.

[25] B. Sherlock, D. Monro, and K. Millard, "Fingerprint enhancement by directional Fourier filtering," *Vision, Image and Signal Processing*, vol. 141, no. 2, pp. 87–89, Apr. 1994.

[26] A. J. Willis and L. Myers, "A cost-effective fingerprint recognition system for use with low-quality prints and damaged finger tips," *Pattern Recognition*, vol. 34, no. 2, pp. 255–270, Feb. 2001.

[27] G. C. Zhu and C. Zhang, "A top-down fingerprint image enhancement method based on Fourier analysis," *Proc. 5th Chinese Conference on Biometric Recognition*, Guangzhou, China, Dec. 2004, pp. 439–448.

[28] T. H. Ching, L. Eugene, and Y. C. Wang, "An effective algorithm for fingerprint image enhancement based on wavelet transforms," *Pattern Recognition*, vol. 36, no. 2, pp. 303–312, May 2003.

[29] A. Almansa and T. Lindeberg, "Fingerprint enhancement by shape adaptation of scale-space operators with automatic scale-selection," *IEEE Transactions on Image Processing*, vol. 9, no. 12, pp. 2027–2042, Dec. 2000.

[30] Y. L. He, J. Tian, X. P. Luo, and T. H. Zhang, "Image enhancement and minutia matching in fingerprint verification," *Pattern Recognition Letters*, vol. 24, no. 9-10, pp. 1349–1360, June 2003.

[31] G. Z. Ceng and R. W. Hall, "Parallel thinning with two-subiteration algorithms," *Communications of the ACM*, vol. 32, pp. 359–373, Mar. 1989.

[32] S. Chikkerur, C. Wu, and V. Govindaraju, "A systematic approach for feature extraction in fingerprint images," *Proc. 1st International Conference on Biometric Authentication*, Hong Kong, China, Dec. 2004, pp. 344–350.

[33] X. D. Jiang, W. Y. Yau, and W. Ser, "Detecting the fingerprint minutiae by adaptive tracing the gray-level ridge," *Pattern Recognition*, vol. 34, no. 5, pp. 999–1013, May 2001.

[34] L. L. Shen, A. Kot, and W.M. Koo, "Quality measures of fingerprint images," *Proc. Audio- and Video-Based Biometric Person Authentication*, Halmstad, Sweden, June 2001, 266–271.

[35] Z. C. Shi, Y. C. Wang, J. Qi, and K. Xu, "A new segmentation algorithm for low quality fingerprint image," *Proc. 3rd International Conference on Image and Graphics*, Hong Kong, China, Dec. 2004, pp. 314–317.

[36] J. W. Gu and J. Zhou, "Model-based orientation field estimation for fingerprint recognition," *Proc. IEEE International Conference on Image Processing*, Beijing, China, Sept. 2003, pp. 899–903.

[37] E. Zhu, J. P. Yin, and G. M. Zhang, "Fingerprint enhancement using circular Gabor filter," *Proc. 1st International Conference on Image Analysis and Recognition*, Porto, Portugal, Sept. -Oct. 2004, pp. 750–758.

[38] M. A. U. Khan, M. K. Khan, and M. A. Khan "Fingerprint image enhancement using decimation-free directional filter bank," *Information Technology Journal*, vol. 4, no. 1, pp. 16–20, 2005.

[39] J. Weickert, *A Review of Nonlinear Diffusion Filtering in Scale-Space Theory in Computer Vision*, Berlin: Springer, 1997.

[40] G. Gilboa, Y. Y. Zeevi, and N. A. Sochen, "Forward and backward diffusion processes for adaptive image enhancement de-nosing," *IEEE Transaction on Image Processing*, vol. 11, no. 7, pp. 689–703, July 2002.

[41] N. Ansari, M. H. Chen, and E. S. H. Hou, "A genetic algorithm for point pattern matching," in *Dynamic, Genetic, and Chaotic Programming*, B. Soucek, Ed., New York: John Wiley, 1992, pp. 353–371.

[42] S. Gold and A. Rangarajan, "A graduated assignment algorithm for graph matching," *IEEE Transactions on Pattern Analysis and Machine Intelligence*, vol. 18, no. 4, pp. 377–388, 1996.

[43] S. H. Chang, F. H. Cheng, and W. H. Hsu, "A fast algorithm and point pattern matching: Invariant to translations," *Pattern Recognition*, vol. 30, no. 2, pp. 321–339, Feb. 1997.

[44] S. Sclaro and A. P. Pentland, "Modal matching for correspondence and recognition," *IEEE Transactions on Pattern Analysis and Machine Intelligence*, vol. 17, no. 6, pp. 545–561, June 1995.

[45] G. Scott and H. C. Longuet, " An algorithm for associating the features of two images," *Royal Society of London*, vol. B244, pp. 21–26, 1991.

[46] J. P. P. Starink and E. Backer, "Finding point correspondence using simulated annealing," *Pattern Recognition*, vol. 28, no. 2, pp. 231–240, 1995.

[47] J. Ton and A. K. Jain, "Registering Landsat images by point matching," *IEEE Transactions on Geoscience and Remote Sensing*, vol. 27, no. 5, pp. 642–651, 1989.

[48] V. V. Vinod and S. Ghose, "Point matching using asymmetric neural networks," *Pattern Recognition*, vol. 26, no. 8, pp. 1207–1214, 1993.

[49] A. Ranade and A. Rosenfeld, "Point pattern matching by relaxation," *Pattern Recognition*, vol. 12, no. 2, pp. 269–275, 1983.

[50] X. P. Luo, J. Tian, and Y. Wu, "A minutia matching algorithm in fingerprint verification," *Proc. 15th International Conference on Pattern Recognition*, Barcelona, Spain, Sept. 2000, pp. 833–836.

[51] R. Cappelli, D. Maio, and D. Maltoni, "Modelling plastic distortion in fingerprint images," *Proc. 2nd International Conference on Advances in Pattern Recognition*, Rio de Janeiro, Brazil, Mar. 2001, pp. 369–376.

[52] H. Chen, J. Tian, and X. Yan, "Fingerprint matching with registration pattern inspection," *Journal of Software*, vol. 16, no. 6, pp. 1046–1053, 2005.

[53] B. Jerry, *The Science of Fingerprints: Classification and Uses*. Federal Bureau of Investigation. Washington, D.C., USA: GPO, 1984.

[54] H. C. Lee and R. E. Gaensslen, *Advances in Fingerprint Technology*, New York: Elsevier, 1991.

[55] N. K. Ratha, K. Karu, S. Chen, and A. K. Jain, "A real-time matching system for large fingerprint databases," *IEEE Transactions on Pattern Analysis and Machine Intelligence*, vol. 18, no. 8, pp. 799–813, Aug. 1996.

[56] A. K. Jain, S. Prabhakar, L. Hong, and S. Pankanti, "Filterbank-based fingerprint matching," *IEEE Transactions on Image Processing*, vol. 9, no. 5, pp. 846–859, 2000.

[57] A. Ross, A. K. Jain, and J. Reisman, "A hybrid fingerprint matcher," *Pattern Recognition*, vol. 36, no. 7, pp. 1661–1673, 2003.

[58] X. Jiang and W. Y. Yau, "Fingerprint minutiae matching based on the local and global structures," *Proc. 15th International Conference on Pattern Recognition*, Barcelona, Spain, Sept. 2000, pp. 1038–1041.

[59] K. C. Fan, C. W. Liu, and Y. K. Wang, "A randomized approach with geometric constraints to fingerprint verification," *Pattern Recognition*, vol. 33, no. 11, pp.1793–1803, 2000.

[60] A. Wahab, S. H. Chin, and E. C. Tan, "Novel approach to automated fingerprint recognition," *IEE Proc. Visual Image Signal Processing*, vol. 145, no. 3, pp. 160–166, 1998.

[61] D. Isenor and S. Zaky, "Fingerprint identification using graph matching," *Pattern Recognition*, vol. 19, no. 2, pp. 113–122, 1986.

[62] A. Hrechak and J. McHugh, "Automated fingerprint recognition using structural matching," *Pattern Recognition*, vol. 23, no. 8, pp. 893–904, 1990.

[63] I. Hideki, K. Ryuj, and H. Yu, "A fast automatic fingerprint identification method based on a weighted-mean of binary image," *IEICE Transactions on Fundamentals of Electronic*, vol. E76-A, pp. 1469–1482, 1993.

[64] Biometric Systems Lab, Pattern Recognition and Image Processing Laboratory, Biometric Test Center, http://bias.csr.unibo.it/fvc2004/.

[65] Biometric Systems Lab, Pattern Recognition and Image Processing Laboratory, Biometric Test Center, http://bias.csr.unibo.it/fvc2000/.

[66] Biometric Systems Lab, Pattern Recognition and Image Processing Laboratory, Biometric Test Center, http://bias.csr.unibo.it/fvc2002/.

[67] N. K. Ratha and R. M. Bolle, "Effect of controlled acquisition on fingerprint matching," *Proc. 14th International Conference on Pattern Recognition*, Brisbane, Australia, Aug. 1998, pp. 1659–1661.

[68] C. Dorai, N. Ratha, and R. Bolle, "Detecting dynamic behavior in compressed fingerprint videos: Distortion," *Proc. Conference on Computer Vision and Pattern Recognition*, Hilton Head Island, SC, June 2000, pp. 2320–2326.

[69] D. Lee, K. Choi, and J. Kim, "A robust fingerprint matching algorithm using local alignment," *Proc. 16th International Conference on Pattern Recognition*, Quebec, Canada, Aug. 2002, pp. 803–806.

[70] A. Senior and R. Bolle, "Improved fingerprint matching by distortion removal," *IEICE Transactions on Info and Systems*, vol. E84-D, no. 7, pp. 825–831, 2001.

[71] C. Watson, P. Grother, and D. Cassasent "Distortion-tolerant filter for elastic-distorted fingerprint matching," *Proc. SPIE Optical Pattern Recognition*, Orlando, FL, USA, Mar. 2000, pp. 166–174.

[72] Z. M. K. Vajna, "A fingerprint verification system based on triangular matching and dynamic time warping," *IEEE Transactions on Pattern Analysis and Machine Intelligence*, vol. 22, no. 11, pp. 1266–1276, Nov. 2000.

[73] A. M. Bazen and S. H. Gerez, "Fingerprint matching by thin-plate spline modelling of elastic deformations," *Pattern Recognition*, vol. 36, no. 8, pp. 1859–1867, Aug. 2003.

[74] A. Ross, S. Dass, and A. K. Jain, "A deformable model for fingerprint matching," *Pattern Recognition*, vol. 38, no. 1, pp. 95–103, 2005.

[75] X. J. Chen, J. Tian, and X. Yang, "A new algorithm for distorted fingerprints matching based on normalized fuzzy similarity measure," *IEEE Transactions on Image Processing*, accepted for publication, Jan. 2005.

[76] http://bias.csr.unibo.it/fvc2004/categories.asp [2004-9-14].

[77] FpVTE 2003, http://fpvte.nist.gov [2005-9-14].

[78] BIRD DOEASY E868 Mobile Business Elite Introduction, http://doeasy.net.cn/index-2.htm.

[79] LPC2106/2105/2104 User Manual, Philips Semiconductors.

[80] ARM7TDMI-S (Rev 4) Technical Reference Manual, ARM Limited.

[81] FCD4B14 FingerChip Datasheet, Atmel Corporation.

[82] Electronic Fingerprint Transmission Specification, CJIS-RS-0010 (V7), http://www.fbi.gov/hq/cjisd/iafis.htm.

[83] H. Shekarforoush, J. Zerubia, and M. Berthod. "Extension of phase correlation to subpixel Registration," *IEEE Transactions on Image Processing*, vol. 11, no. 3, pp. 188–200, Mar. 2002.

[84] D. Rakhmatov and S. Vrudhula, "Energy management for battery-powered embedded systems," *ACM Transactions on Embedded Computing Systems*, vol. 2, no. 3, pp. 277–324, Aug. 2003.

Chapter 12

Adaptation and Predictive Control Observed in Neuromuscular Control Systems

Jiping He

Abstract: The neuromuscular control system for posture and movement is a complex nonlinear system with many advanced features for adaptation and robustness in performance. Valuable lessons on how to design a versatile control system can be learned from investigation of functional structures of the system and strategies in adaptation to various tasks and environmental conditions. In this chapter we present two case studies on how the neuromuscular control system develops adaptive and predictive control strategies to achieve effective stability and performance improvement against perturbations. The first case is to maintain the upright posture against a strong yet predictable perturbation. The second case is to reach designated targets amid a strong pulling force perturbation. As results will demonstrate, a feedforward predictive control strategy is often developed to achieve efficient control with satisfactory performance, instead of energy demanding stiffness control.

12.1 Introduction

Postural stability is the fundamental motor task for any activities of daily living [16]. Fear of potential falling has prevented a large population of people from enjoying many activities. Lifting or moving a heavy object, throwing a ball, swinging a racket, or even waving an arm will all create disturbances to body postural equilibrium. These perturbations to the body postural equilibrium can be classified as predictable internal perturbations because they arise from the intended

347

voluntary movement. All these activities require prior adjustment in postural position or muscle tone in order to maintain postural balance to achieve a good performance [6, 24, 26, 27]. Massion [28] defined such anticipatory action as the postural adjustments associated with voluntary movements and occurring prior to the onset of the disturbance due to the intended movement.

The anticipatory action consists of two major components: adjusting body posture and muscle tone, defined as feedforward control, and resetting sensory feedback gains, defined as feedback control [28, 34]. A proper anticipatory action is important to the performance of the intended movement or task in addition to maintaining a stable postural balance. The investigation of anticipatory action for internal perturbations has largely focused on the postural adjustment, even though the neural control system is believed to utilize both feedforward and feedback control strategies.

Another class of perturbations to body balance is unpredictable, external ones, such as randomly applied pull or push, unexpected slip or trip, or sudden moves of a support base. Many experimental paradigms have been developed to investigate the dynamic stability of postural control system after unexpected perturbations. A sudden move of a supporting platform where a subject stands, either translational or tilting, has been shown a very useful perturbation paradigm to illuminate some fundamental principles in postural control [14, 32]. The release of a supporting force when a subject is leaning forward is an effective way of studying stability boundary and postural recovery margin [20, 38]. Applying a force (push or pull) to the body as a perturbation is another popular paradigm [1, 2, 4, 5, 8, 33].

In most of these investigations the perturbation to a posture is applied randomly to prevent subjects from predicting the timing and direction of perturbation. Hence, the perturbation evokes the pre-programmed, automatic response due to the intrinsic mechanical constraints and sensory feedback control [11, 15, 16, 22, 31, 32]. From these perturbation investigations several control strategies to restore postural balance have been characterized based on time sequence and magnitude of muscle activities. The adoption of a specific strategy depends on perturbation types and a person's physical condition [23, 38]. Each strategy has its applicable boundary and is task dependent.

A person's physical condition will deteriorate due to injury, disease or aging, resulting in a longer response time, smaller achievable forces, or a slower rate of force generation. This will affect the person's ability to restore balance after a perturbation if no adaptation in control strategy is developed. When the specific knowledge is acquired about a perturbation, certain anticipatory actions can be taken to improve stability and performance [17, 24, 27, 28, 30]. The adoption of appropriate preparation actions, in anticipation of a predictable perturbation to the body balance, is a very effective control strategy. The anticipation or feedforward action will disrupt the existing balance and often rely on interaction with external forces for final balance, especially in the case of external perturbations. Therefore, an appropriate planning on timing and amplitude of the anticipatory action is crucial for stability, and often the sensory feedback control is relied upon for robust stability. The question is how much the neural control system relies on the anticipatory action for postural control when a perturbation is predictable. Investigation into the stability boundary and proper control strategy for restoring a perturbed upright posture will help to develop effective preventive and rehabilitative measures.

In this research project we set out to understand how the CNS coordinates the feedforward and feedback control in adaptation to a repeated external perturbation to improve performance of stability. The approach and results will be reported in this chapter. The experimental paradigm is described to investigate potential adjustment in body posture and/or muscle tone when a perturbation becomes predictable. A prominent observation in a pattern of anticipatory action is formulated and discussed. The performance improvement of this special strategy is then evaluated. We also discuss some adjustment to the anticipatory control strategy when the predicted perturbation changes.

12.2   Control of Postural Stability

12.2.1 Experimental Design

Subjects and Experimental Protocol

The experimental protocol was approved by the university IRB before the study (HS #03699-96). Ten healthy subjects participated in the study (age 24–41). Prior to the data collection in the experiment, every subject signed the informed consent form approved by the IRB.

To study the adaptation to a perturbation and the subsequent preparation action, we need to evoke noticeable preparation actions in subjects for a specific perturbation. Such preparation or anticipatory actions to counter a perturbation require knowledge of the perturbation and the confidence that the perturbation will occur. Therefore, we designed the experimental protocol that was distinctly different from most postural control study. Instead of applying a perturbation to the postural balance in a pseudo-random fashion to prevent subjects from predicting the timing, the direction and the amplitude of a perturbation, we used an identical perturbation repeatedly in every trial. Furthermore, we also used an audible warning beep in fixed duration (0.7 s) prior to the application of the perturbation to give subjects enough preparation time. This protocol allowed subjects to quickly acquire the knowledge and the effect of the perturbation to the body postural equilibrium.

The perturbation to the standing posture was a short duration impulse force (200 ~ 250 ms) generated by a high pressure (85 ~ 95 psi) pneumatic actuator to pull the subject forward. A quick release mechanism was developed to control the maximum force and duration of the perturbation. The amplitude of the pulling force was adjusted for each subject such that a small step would be necessary to maintain balance without appropriate preparation actions. The force was approximately 20 ~ 25% of the body weight, the specific value depending on physical fitness of each individual subject. A harness was worn by each subject to protect from potential fall. The harness was connected to a cable fixed to the ceiling and was loose enough not to interfere with the subject's normal body sway and stepping motion. Figure 12.2.1 demonstrates the experimental setup and the event sequences.

The objective for the subject was to maintain the posture without stepping and to minimize body sway and bending when perturbed. The instruction to the subject was to adopt whatever strategy that would reduce the impact of the perturbation to the balance. Stepping was allowed if the upright posture could not be maintained, but was strongly discouraged.

The standard procedure of the experiment was to repeat the perturbation trial 35 times for each subject. The experiment would stop when the subject became tired, resulting in less total number of trials for some subjects. To investigate what adjustments the subject may have adopted to improve performance we inserted two trials (11th and 21st trials) in which the pulling force was disabled but the warning beep was still generated. This would prompt the subject to activate the preparation action in anticipation of the perturbation. Because such anticipatory action was taken to counteract the effect of the perturbation, the absence of the perturbation would give the anticipatory action enough time to develop its effect on body posture. These trials without perturbation (11th and 21st trials) are combined with experienced trials in data analysis to establish the anticipation action. The ground reaction force and body movement would provide data to confirm the anticipatory action (see Figures 12.2.2–12.2.4 and 12.2.6 for detail).

To examine the performance improvement with the adapted control we also inserted two trials (16th and 26th trials) in which the perturbation would be applied without the warning beep. The absence of the audible warning beep would deprive subjects the information to initiate the anticipatory postural adjustment. The resulting response should be similar to the ones when the subject still had no a prior knowledge of the perturbation. These trials without the beep were therefore combined with early trials to study the sensory feedback control and to evaluate postural balance performance without anticipation.

The number of these abnormal trials is small enough (4 out of 35) in order to

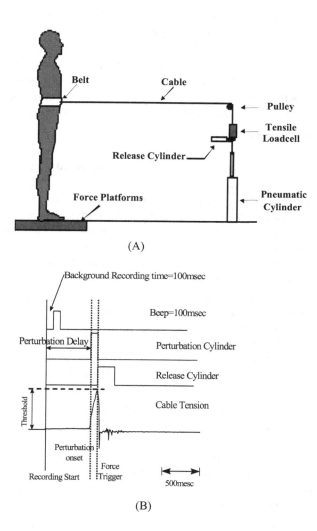

(A)

(B)

Figure 12.2.1: Schematics of experimental setup (A) and the time sequence of events (B). A subject stands upright with arm folded. A perturbation to the stance posture is applied through a cable connected to a pneumatic cylinder. A release cylinder is used to modulate the amplitude and duration of the perturbation based on the load cell signal. The time delay between the cue and perturbation is adjustable.

maintain subjects' confidence in initiating the anticipatory action. Because these trials provide additional information supplement to that available from normal trials these inserted abnormal trials improve, rather than diminish, the statistical power of data analysis. The consideration of subjects fatigue is the reason for limiting the total number of trials. Continuity of trials and consistent perturbation are important factors in the experiment because rest or interruption between trials will influence the response and especially the anticipatory control.

Data Collection and Processing

To evaluate the performance and adaptation process in postural control, motion data (kinematics) with ground reaction forces were recorded simultaneously with muscle activities. A four-camera MacReflex system (AOA/Qualysis West) was used for 3D kinematic data. Two in-ground force plates (Bertec) measured ground reaction forces under each foot. An 8-channel electromyographic (EMG) system (Therapeutic Unlimited, Iowa City, IA) provided means for recording major leg muscle activities. Surface electrodes with pre-amplification were placed on tibialis anterior, soleus, gastrocnemius, vastus medialis, rectus femoris, hamstrings and sometimes countralateral rectus femoris. The kinematics and force plate data were recorded at 120Hz. EMG's were recorded at 840Hz. Each data type was recorded through a separate computer. Kinematics was on a PowerMac 7100 through a series connection with MacReflex system. Ground reaction force was on a 486/66 PC and EMG on a Pentium 75, each connected through a separate 16-channel A/D board by Data Translation. The three systems were synchronized through an external trigger circuit that would also generate the audible cue and activate the perturbation.

The entire duration of the postural response to the perturbation was sub-divided into four periods: preparation or anticipation, early sensory feedback, automatic postural response (APR), and residual response. The first period was defined from 200 ms after the audible beep to 10 ms after the onset of the perturbation, to detect anticipatory activity in preparation for pending perturbation. No sensory feedback response was possible in this period because of transmission delay in neural pathways. The second period was from 10 ms to 50 ms after the onset of the perturbation to measure response activities before the automatic postural response (pre-APR). This period would represent spinal segmental level sensory feedback responses. Any supra-spinal control in response to the perturbation would take a longer time delay. The third period (from 50 ms to 150 ms after the onset of the perturbation) would measure the sensory feedback response or APR. This period would generate the most correction in postural control. The fourth period (to the end of the recording) was assumed to represent post-APR or residual activity in returning the body to the original upright posture.

Kinematics: The kinematic data were represented by (x, y, z)-coordinates of

16 markers. The body was assumed to consist of five rigid segments: trunk (with head and arms), pelvis, femur, tibia and foot of the supporting leg. Three markers were placed on each segment so that the 3-D orientation of the segment was uniquely determined. One additional marker was placed on the heel of stepping leg to provide information on the distance of stepping. The marker positions were pre-processed before the calculation of joint angles. The pre-processing involved elimination of outliers in camera trace [12], low-pass filtering with Woltring's spline algorithm developed in 1980 and assurance of proper segment lengths in determining the final joint trajectory. The center of mass (CoM) was estimated from the marker positions of every segment. Though 3D data were recorded, only the sagittal plane data were analyzed for this investigation.

EMG: Surface EMG data were high-pass filtered with a cut-off frequency of 20Hz. The root-mean-square (RMS) values were recorded with a 5.5 ms window width before digitizing at 840 Hz. The area under the RMS values of EMG was calculated for each of the four separate periods specified above. This area was corrected for the background activity estimated from the average EMG value over a 200 ms time window before the audible cue. The functional significance of these measures is best illustrated in discussion of Figures 12.2.3 and 12.2.4 later in the next section.

Ground Reaction Forces: The force plate data were low-pass filtered and the center of pressure (CoP) under each foot was calculated. The data were then presented in raster format for inspection before averaged according to trial types specified in the procedure section above and repeated in the next paragraph. The feedforward or anticipatory control action was measured by deviations of forces (in Newtons) and CoP traces (in cm) from the baseline data before the onset of the perturbation.

Statistical Analysis: The experimental data were grouped according to trial types for analysis. The training trials without warning cue were used to examine the normal automatic postural responses (APR) and to establish the baseline. The first 3 trials with warning cue and perturbation were assumed as novice and adaptation trials when subjects learned the protocol. The next seven trials were used to establish adaptation and to detect anticipatory action. The inserted single trials (11th and 21st) without pulling force were used to examine the full effect of the anticipatory action on the body balance and the subsequent feedback response in correcting postural deviation caused by this anticipatory action. The inserted single trials (16th and 26th) without warning cue after adaptation were used to detect the effect of anticipatory actions on performance. The trials followed immediately afterwards were grouped to examine the effect of the previous abnormal trials (no perturbation or no cue) on anticipatory action and performance. The remaining trials were grouped together for analysis of anticipatory action and APR ratio.

The performance was evaluated by the magnitude of hip bending or stepping. No stepping and a small angle excursion at the hip were considered a better performance. Comparison of movement trajectories, EMG patterns and ground reaction force traces among different trial types was performed separately on the four periods of the total postural response. The anticipatory adjustment to posture and muscle tone was evaluated by comparing changes in CoP and EMG magnitudes during the first period (prior to the onset of the perturbation). The sensory feedback gain adjustment was evaluated by comparing the total EMG during pre-APR or the second period. A larger or quicker activity in EMG would indicate a pre-adjusted sensory gain. Data from periods 3–4 would represent the effect of anticipatory control action on performance and the normal response pattern. The statistical significance of the differences among data from these groups was determined by the analysis of variation (ANOVA). The single parameter ANOVA was used for every analysis based on the difference in magnitude. A p-value less than 0.05 would be considered as statistically significant.

12.2.2 Adaptation and Predictive Control Strategy

We instructed subjects to use early trials to learn the specific perturbation characteristics. After three trials most subjects became familiar with the timing of the cue with respect to the perturbation and the impact of the perturbation on the postural balance. A consistent control strategy and response pattern emerged.

Center of Mass (CoM) and Center of Pressure (CoP) Trajectory

Figure 12.2.2 shows typical CoM and CoP traces from our study in cm. Solid lines represent data from trials without anticipatory action and dashed lines the trials with anticipatory action. A clear and consistent difference between the two data types is the motion observed in both CoM and CoP traces before the onset of the perturbation. The movement is small but distinctive.

The CoM trace from experienced trials showed a slight backward movement starting from approximately 300 ms (± 100 ms) before the onset of perturbation. Instead of a companion motion of CoP in the same direction, the CoP trace moved opposite of CoM trace during this period, a forward motion in more noticeable magnitude. This suggested voluntary activation of ankle extensors to cause the body to sway backward in anticipation of a forward pulling perturbation. This anticipatory movement was observed from all subjects except one in our study.

Table 12.2.1 lists the statistical analysis of movement magnitude in CoP traces during the first period among different trial types. The magnitude was calculated from the baseline estimated from the quiet standing before the audible cue. The anticipatory CoP motion was absent in novice trials and those immediately following the no pull trials. The motion was significant in all other trials. The opposite directions in initial sway of CoM and CoP indicated some voluntary muscle ac-

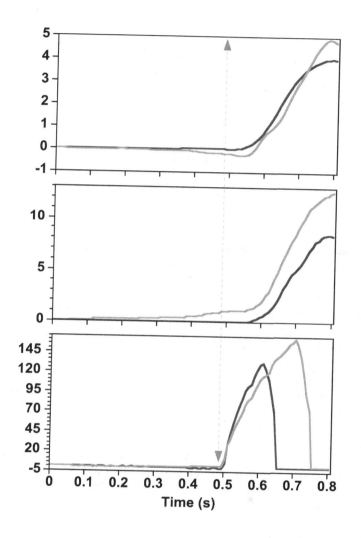

Figure 12.2.2: Typical perturbation responses from an early trial (red, representing novice response) and a later trial (light blue, representing experienced response). In early trials, the subject was passively waiting for the perturbation. In experienced responses, both the center of mass (top plot) and the center of pressure (middle plot) started to move before the onset of perturbation (indicated by the vertical arrow). This slight preparatory activity enabled the subject to maintain the upright postural equilibrium without stepping for a larger perturbation (160 N, bottom plot).

Table 12.2.1: Effect of anticipatory action on CoP

Trial types	Maximum anticipatory CoP movement (cm)	p-value
With anticipation	1.72 ± 1.66	$\ll 0.01$
Without anticipation	0.227 ± 0.61	

tion at the ankle joint. EMG from ankle muscles was analyzed to confirm the observation.

Muscle Activities before and during the Perturbation

Figure 12.2.3 shows the representative EMG activity of soleus muscle from 3 sets of trials. The top plot shows EMG from those trials without pulling force after the cue. The middle plot shows EMG from later ones for experienced responses with anticipatory action. The bottom plot shows EMG from novice responses or the ones immediately following no puling force trials.

No change in EMG from baseline data was observed until perturbed in novice trials and those following no pulling force trials, indicating that subjects were passively waiting for the perturbation and relied on the normal sensory feedback control or the automatic postural response for postural stability. The APR to the perturbation started with a latency of 100ms±10ms, similar as reported in many other postural perturbation studies [31]. The anticipatory activity for the perturbation occurred when subjects learned the pattern and the effect of the perturbation, represented by EMG bursts above the baseline starting before the onset of perturbation, as observed in experienced trials. These small bursts had the effect of moving the CoP and CoM to reduce the impact of the perturbation on the postural equilibrium (Figure 12.2.2). Since the subjects had associated the audible cue with the forthcoming perturbation in a fixed time delay after learning, they would activate the feedforward control after hearing the cue. When we released the pull cable at the time of applying perturbation, the effect of the preparation activity for counterbalancing the perturbation was exaggerated because the expected counter force did not come.

Figure 12.2.4 shows the corresponding average EMG of soleus muscle for all the early and later trials. The solid trace is averaged EMG for the early trials representing novice responses. The dashed trace is for the experienced responses. There is a clear increase in muscle activity for experienced trials before the onset of the perturbation.

To measure muscle activity during various periods in responding to the perturbation, we used the following formula to calculate the EMG activity intensity for

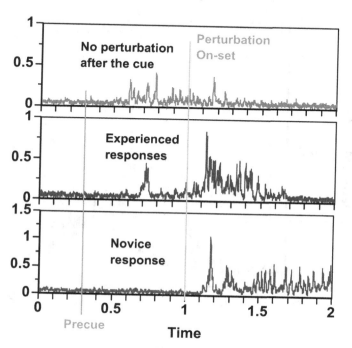

Figure 12.2.3: Typical EMG activity of soleus muscle in response to a perturbation for novice (bottom) and experienced (middle) trials. There were significant activities above the baseline before the onset of the perturbation in the experienced response, indicating the adoption of a feedforward control strategy to improve performance. Top plot gives EMG from one trial in which no actual perturbation occurred after cue. Since the subject was expecting a perturbation after the cue, the preparation activity was clearly activated.

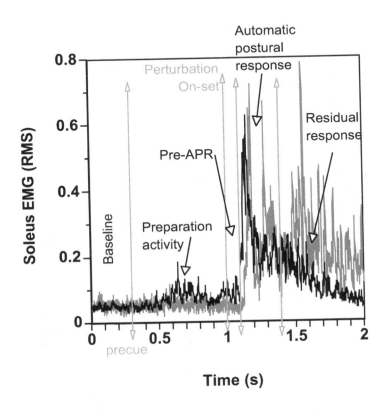

Figure 12.2.4: The averaged EMG activity of soleus muscle responses to the same perturbation from novice trials (solid trace) and experienced trials (dashed trace). Increased muscle activity before perturbation could be observed in the experienced trials. Four periods are defined in this figure for calculation of muscle activity measures: baseline, preparation, preautomatic postural response (APR), and APR. See Table 12.2.2.

Table 12.2.2: Activity intensity of soleus muscle

	Preparation activity	Pre-APR activity	Automatic postural activity	Post-APR activity
Novice responses	1.01 ± 0.01	0.91 ± 0.10	5.76 ± 0.1	4.3 ± 0.84
Experienced responses	1.53 ± 0.26	2.07 ± 0.44	5.22 ± 0.93	2.59 ± 0.77
p-value (0.05)	0.02	0.002	$0.38 > 0.05$	0.01

anticipatory, pre-APR, early sensory feedback, and automatic postural responses. The measures are scaled by the baseline activity to obtain an EMG intensity index (EI) for the perturbation response

$$EI = \frac{1}{t_2 - t_1} \int_{t_1}^{t_2} EMG(t)dt \Big/ EI_{base}.$$

The measure for soleus muscle activity indicated significant activities in both preparation and pre-automatic postural response phases. Table 12.2.2 summarizes the analysis of soleus activity for two types of trials: novice and experienced. In novice response, no preparation activity could be observed before the onset of perturbation. The EMG activity level was the same as the baseline level. No pre-APR activity could be seen either. The EMG activity was even smaller in this period than the baseline level.

Figure 12.2.5 shows EMG activities from other major leg muscles. On the left is the representative of novice response to the perturbation. On the right is the typical experienced response. Among the six major leg muscles monitored (gastrocnemius, vastus medialis or lateralis, rectus femoris, hamstring, tibialis anterior, and countralateral rectus), no significant activities could be detected prior to the automatic postural response. The difference among the novice and experienced responses is mainly in amplitudes of activities of these muscles. The experienced response had generally smaller amplitudes. But the latency and phase relations remained similar. This is consistent with the previous findings that specific knowledge about the perturbation and training has no effect on latency and pattern of automatic posture responses [7].

Anticipatory Activity, Body Momentum, and Performance

When a perturbation is applied to a body posture, it introduces extra momentum to the body's total momentum. An anticipatory action will also introduce extra momentum to the body. If the two momenta have the opposite effect on the

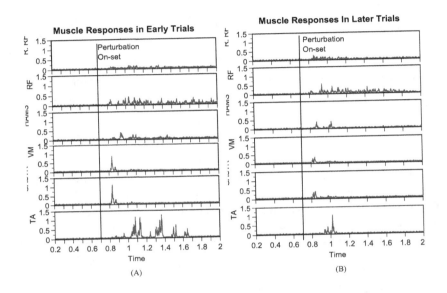

Figure 12.2.5: Typical EMG activities of major leg muscles (soleus is shown in Figures 12.2.3 and 12.2.4) recorded from the postural perturbation experiment. (A) shows muscle responses from the first trial when the perturbation force is smaller than the stepping threshold. (B) shows the muscle responses from a later trial when no stepping occurred although the perturbation force was above the initial stepping threshold.

dynamic balance of the body posture, the anticipatory action will in effect cancel the perturbation induced momentum. A perfect cancellation requires the perfect timing and magnitude of the anticipatory action. An anticipatory action that is too large or initiated in wrong time may have adverse effect on the body postural balance. The body will fall backward if a large anticipatory action is initiated but the perturbation is not applied. To relate the performance of postural control to the anticipatory action we calculated the momentum from the anticipatory action and the maximum excursion of CoP for each trial.

The data are separated into two groups according to trial types. The first two trials, no cue trials, and the trials following the no perturbation ones are in group 1, representing no anticipation. All other trials are in group 2, representing trials with anticipatory actions. The difference in maximum excursion from the two groups are statistically significant ($p < 0.04$). The difference in the anticipatory momentum is also statistically significant ($p < 0.05$). A linear regression to relate the two variables generates a slope of -0.21 ($p < 0.0024$). This negative correlation indicates that the larger the amount of anticipatory action, the better the performance will be (smaller CoP excursions). It also indicates that no subject created a perfect cancellation between the anticipation and the perturbation.

12.2.3 Feedforward vs. Feedback Control

There have been many studies to characterize the sensory feedback control of the posture equilibrium [2, 16]. The sensory feedback control component of postural control is largely automatic responses that are determined by both biomechanical constraints and neural sensory structures [11, 31]. The pattern of automatic response to a perturbation depends on the task, the availability of sensory information, and the type of perturbations [16]. This pattern, defined by muscle responses with ~100ms latency to a specific perturbation, however, is independent of the precuing of the perturbation. That is, the specific knowledge about the timing, direction and amplitude of a perturbation has little effect in the response [7, 19].

Another component of the control structure in maintaining body posture is the feedforward control, or the anticipatory action for the perturbation. The evidence for such action has been widely documented [17, 24, 26, 34]. The feedforward control is highly dependent on the prior experience with the perturbation and its effect on the postural equilibrium.

Feedforward Control and Performance

In this study, the perturbation was consistent and predictable to the subject. From the first few trials, each subject gathered the specific knowledge about the perturbation: the time of arrival after the cue, the amplitude and the effect. They developed a feedforward control strategy in anticipation of the perturbation as demonstrated in the previous section. Most subjects adopted a simple yet effective strategy: a slow backward sway before the arrival of the perturbation (a forward pulling force). This slight backward sway started with a mild increase in EMG activity of soleus muscle as shown in Figures 12.2.3 and 12.2.4.

This small increase in background activity of the soleus muscle had an effect of modulating the leg impedance to the perturbation. At the same time, it introduced a nonzero initial velocity to the body. This nonzero velocity generated a momentum against the impact of the perturbation to the body posture. The effect was improved performance of postural control measured by the maximum excursion of the center of the ground reaction force (CoP).

The feedforward control on reducing the impact of the perturbation on body posture can be discussed in terms of total body momentum. The approach of using total momentum for the formulation of dynamics and control of a multi-linkage mechanical structure has been demonstrated to be very instructive in understanding the stability.

The total (angular and linear) momentum of the body (Mb) at any time t is given by:

$$M_b = \sum_{i=1}^{n} \left[(I_i \omega_i + x_i \times m_i v_i) + m_i v_i \right]$$

where n is the number of body segments considered in a model, m_i, x_i, v_i, I_i, and ω_i are the mass, position, velocity, moment of inertia and angular velocity, respectively, for segment i. At quiet standing, the total body momentum should be zero in ideal situation. Due to the small magnitude body sway, the real body momentum at an arbitrary time t is only close to zero, $|M_b| = |M_{b0}| > \varepsilon \, (\geq 0)$, where ε is a very small positive number and bounded by the stability margin.

When a perturbation is applied to the body, the momentum generated from the perturbation force $F(t)$ is, from the Newtonian mechanics $F = mdv/dt$,

$$M_p = \int_{t_1}^{t_2} F(t)dt.$$

This perturbation momentum will accelerate the body away from the original stable momentum value M_{b0}. Depending on the magnitude of M_p the postural control system needs to determine an appropriate strategy to reduce the body momentum to the nominal body momentum M_{b0} and to restore the upright posture. If the momentum created by the perturbation is too large such that neither ankle strategy and hip strategy alone nor in combination is adequate to restore the postural balance, the subject will have to take a step to prevent from falling.

How is a small, backward sway effective in improving performance? If subjects take a strategy that relies on sensory feedback control only, that is, passively standing upright and waiting for the perturbation, the effect of the perturbation on the total body momentum will be a change almost identical to Mp. This may introduce a large deviation of body momentum that requires a substantial force within a short period to reverse the perturbation momentum.

On the other hand, the small, backward sway prior to the perturbation generates a change in the body angular velocity and introduces a CoM motion. This change in velocity creates a momentum opposite to the perturbation momentum. Therefore, a portion of the perturbation momentum will be consumed by the preparation momentum. This preparation momentum is the body momentum at the onset of the perturbation:

$$M_{\text{prep}} = \sum \left(I_i \omega_i + x_i \times m_i v_i \right) + \sum m_j v_j.$$

Comparing with the trials in which no active motion is exerted by CNS, the effect of perturbation on the body momentum will be reduced by an amount proportional to

$$M_c = M_p - M_{\text{prep}}.$$

Since M_{prep} is much larger than M_{b0}, the net effect of the perturbation to the upright posture of the body is reduced for a much improved performance. The data analysis showed that the larger the M_{prep}, the better the performance will be (smaller CoP excursion). Since the perturbation intensity level was on the threshold of stepping, the feedforward control action maintained the stability without a step while the missing of this small action and a larger sensory feedback gain caused the subject to take a step.

To further confirm that the improved performance was due to the feedforward control, we inserted two trials for each subject (trial 15 and 25) in which the perturbation was applied without precuing, i.e., the beep was turned off (the actuator noise in the pneumatic system can not be considered as a cue because the time constant is very short $\ll 100$ ms). These trials were inserted after the subjects had become familiar with the perturbation and had adopted a consistent control strategy in which they used the feedforward control. Without exception, every subject took a step to recover from the disturbed posture, even though the perturbation level was exactly the same from other trials in which they successfully recovered the posture without a step when the beep was given.

Feedback Control and Stability

Theoretically, the anticipatory action can cancel the perturbation effect on body equilibrium completely. However, we do not observe a complete cancellation in any such experiment. The amount of anticipation action depends on the confidence of the subject in the knowledge of the expected perturbation and the physical fitness in creating large forces quickly. In general we observe a conservative strategy in subjects to maintain body postural stability.

Figure 12.2.6 shows CoP traces from three consecutive trials after the subject had become familiar with the protocol: 1) perturbation with cue, 2) cue only, and 3) perturbation with cue again. These are trials 10, 11, 12 in the first group and 20, 21, and 22 in the second group. The baseline data for a quiet standing is given in the figure for reference of CoP sway amplitude.

In the first of the three trials, the subject activated the feedforward control after receiving the cue. The postural equilibrium was maintained without stepping. In the second trial, the subject was expecting the perturbation after receiving the cue, therefore activated the feedforward control as indicated by the movement of CoP prior to the perturbation. The perturbation did not occur as expected. The subject stopped the mild body sway and returned to equilibrium because the anticipatory action was so mild that the sensory feedback control was adequate to correct the body motion. In the following trial, the condition was exactly the same as the first one. The subject did not activate the feedforward control strategy because of uncertainty associated with the cue from the immediate prior experience. Though the perturbation force was the same as in the first trial, the subject stepped because of the instability caused by the lack of anticipatory action and a higher response intensity due to high sensory feedback gains. The high sensory feedback gain is evidenced by the increased EMG responses given in Figure 12.2.7.

The feedforward control action involves two major components: to modify posture by activating certain muscles prior to the perturbation on-set, and to modify the sensory feedback gain. Pre-lean in platform perturbation and pre-sway in our study are actions of feedforward control. These preparation adjustments of posture have the effect of reducing the impact of the perturbation on body equilibrium by introducing a counteracting momentum. However, the sensory feedback can never be turned off. The feedback control provides the robust stability when perturbations can not be predicted accurately or completely unpredictable, as in the case of trials 11–12 and 21–22 in Figure 12.2.6.

Coordination of Feedforward and Feedback Control

The data presented in the previous section demonstrate that the neural control system for postural stability utilizes both feedforward and feedback control. In summary, the experimental paradigm developed in this study, a combination of a perturbation with cue, a perturbation without cue, or cue without perturbation, provides a new avenue to investigate the effect of feedforward and feedback control action on the performance of postural equilibrium. The presence of cue and the consistency in perturbation characteristics enable subjects to develop a feedforward control strategy to improve the performance of stability control of the postural equilibrium. The strategy consists of an adjustment of initial body align-

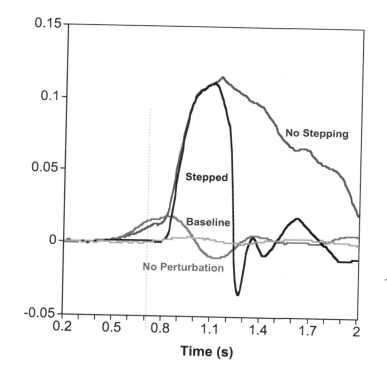

Figure 12.2.6: The effect of combined feedforward and feedback control compared to feedback control alone. When the feedforward control action was missing, the subject had a higher intensity automatic postural response (shown in Figure 12.2.7) that caused the heel off the ground rapidly (reflected in the faster rise of CoP at the beginning of the response). Then a quick stepping occurred at 400 ms after the onset of the perturbation.

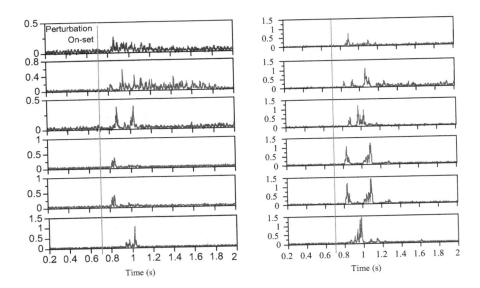

Figure 12.2.7: The effect of feedforward control on EMG activity of major leg muscles during the postural perturbation response. When the subject activated the feedforward control, the amplitude of automatic response was generally smaller (shown on the left column), suggesting a suppression of control gain for the automatic postural response. When the subject did not activate the feedforward control because of uncertainty on perturbation, he utilized a higher control gain. The result was a high response amplitude as shown on the right column (notice the different scales). A step was triggered to recover the equilibrium (the second EMG bursts, also see Figure 12.2.6).

ment and sway velocity and a reduction in sensory feedback gains to respond to the perturbation. When uncertain of the perturbation or the cue, subjects will not activate the feedforward control because pre-leaning and sway may cause instability of the body equilibrium if the perturbation is not as expected. The general reduction in the amplitude of the automatic response to a perturbation after repeated exposure should be the effect of both habituation and adaptation.

Several important questions of postural equilibrium control remain to be investigated. If both habituation and adaptation contribute to the reduction in response amplitude to a perturbation, how much contribution will each one have and under what conditions will one be more dominant than the other? If the specific action of feedforward control depends on perturbation type and effect of perturbation on body equilibrium, is it the prior experience of the perturbation effect or the prior knowledge of exact characteristics of the perturbation that is more important? Since a minimum time latency is required to process the sensory information and to activate the voluntary control action, what will be the effective range of time delay between a cue and a perturbation for a subject to use feedforward control? Will the time delay be age or sensory type dependent?

The ability to control the type and timing of cue to a perturbation will allow us to address these questions. A further refinement of the experimental protocol is under way to provide more flexibility in controlling each parameter of the experiment. A more detailed computer simulation of a postural control model is also under development.

12.3   Adaptive Control Strategy in Arm Reaching Movement

The similar adaptation and development of a predictive control strategy are also observed in non-human primates. This section we present results from investigation of adaptation strategies represented by simultaneous changes in neuronal activities of primary motorcortex and kinematics of arm movement.

One objective of this investigation is to determine whether motorcortex neurons adapt to a transient perturbation applied repeatedly to a reaching movement and how if they do. The results will have a significant application in designing cortically controlled neuroprostheses. Rhesus monkeys were trained to reach 8 targets located at the vertices of a 13-cm cube, as shown in Figure 12.3.1. A transient force perturbation was applied whenever monkey's hand moved 25mm away from the center position. Spike trains of over 40 cortical neurons were recorded from chronically implanted electrode arrays, along with EMG and kinematics from the moving arm. All cells demonstrated a secondary burst of activity following the perturbation. Increased cortical activity following the perturbation suggests the active role of cortical neurons in correcting disturbances in arm movement tra-

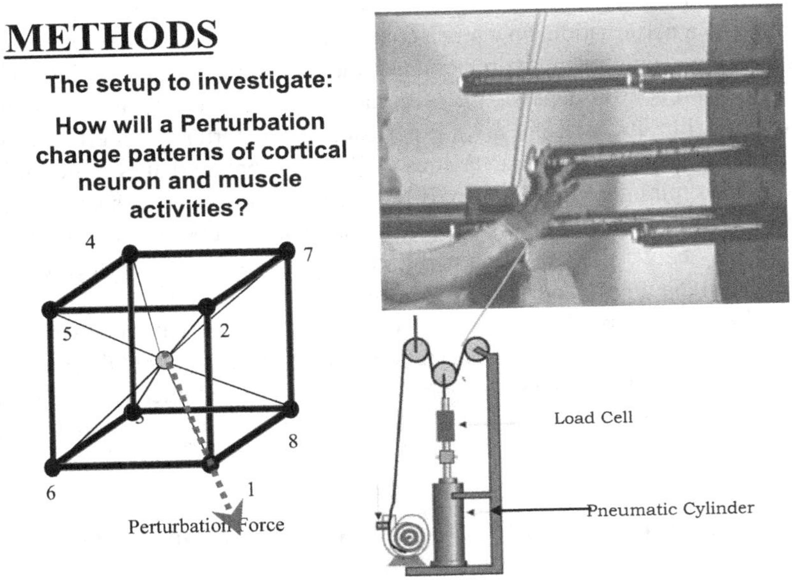
METHODS
The setup to investigate:
How will a Perturbation change patterns of cortical neuron and muscle activities?
4
7
5
2
8
6
1
Perturbation Force
Load Cell
Pneumatic Cylinder

Figure 12.3.1: Experimental workspace for 3-D center-out task. The left diagram shows the cube of 8 targets and the direction of the perturbation force. The right panel shows the apparatus used to deliver the force perturbation. The string from the recoil is routed through 3 pulleys to the subject's wrist. To deliver a perturbation, the recoil is locked and the pneumatic cylinder is actuated to deliver a pulling force to the wrist. A load cell is used to measure tension in the string.

jectory. The adaptation of cortical neuron responses is reflected in the increased firing rate before the perturbation onset. This pattern stabilizes after one week's practice. The adapted pattern disappears quickly after the removal of the perturbation.

The role of primary motor cortex in planning and control of arm movement has been investigated extensively yet no consensus has been reached. By using a single cell recording technique while monkeys perform visually guided center out reaching task Georgopoulos and colleagues (Schwartz et al 1988; Georgopoulos et al 1988) found a relationship between the discharge rate of many motor cortical neurons and the direction of the hand during multi-joint arm movements. The nature of the tuning suggests that a population of neurons is capable of encoding the movement direction of the hand. This hypothesis has been experimentally supported in single-unit studies performed during two- and three-dimensional reaching and drawing movements [36]. In 1998, Moran and Schwartz developed a model of single-cell activity to describe the dynamic, time-varying correlations between cortical activity and arm movement. Their model predicts that cortical cells activities are modulated with speed in a way that depends on the cell's preferred direction, and the cortical activity pattern clearly precedes each increment of the movement in a continuous way throughout the task.

Cortical neuron firing patterns have also been shown to change with tasks and loading conditions [3]. Kalaska and co-workers [21] found that the discharge rate of cortical neurons changed when a constant load was applied to the hand during reaching tasks. In many cells, they observed a continuous gradation of discharge with the direction of load, but this sensitivity was not uniform across all cells. In an isometric task, Georgopoulos [9] observed changes in the tonic firing rates of single neurons when the monkey was required to exert a bias force, but found that the cell's directional tuning was preserved across different directions of an applied bias force.

The key question is whether the primary motorcortex is involved in the control of dynamics of the arm movement. To investigate the dynamic control exerted by the motorcortex we applied a perturbation to the arm during movement. The effect of the perturbation is unexpected at the beginning but becomes predictable after many regular repetitions. We are reporting some preliminary observations from this investigation.

12.3.1 Arm Reaching Movement, Perturbation and Cortical Recordings

The Institutional Animal Care and Use Committee approved the behavioral paradigm, surgical procedures and general animal care. The general guidelines by AAALAC and the Society for Neuroscience were followed.

Three rhesus monkeys (maccaca mullata) were trained to perform the center-out- reaching tasks. The eight targets were located at the vertices of a cubic workspace. The start position was at the center of the 13 cm cube. The center light was illuminated (center-on) to indicate the start of a trial, and the monkey would push and hold the center button until a target was presented (Target on). The monkey reached to the illuminated target and held the button for receiving a liquid reward. The eight targets were presented in a randomized block design with 5 replications of each target, for a total of forty trials (1 set = 40 trials).

The complete study was completed in four phases over a 20-day period. Five days of normal trials were completed to establish the baseline. Perturbation trials began on day 6 and continued for eight days. The first five sets of trials on the first perturbation day were still normal trials and the perturbation started on the sixth set. During these perturbation trials, a transient force (duration of 60 ms) was applied to the monkey's wrist after the hand moved 25 mm away from the center position. On day 13, the perturbation was removed on seventh set. This marked the beginning of the extinction trials, and 2 more days of experiments were performed without the perturbation.

Cortical neuron activities were recorded from 64 micro-wire electrodes that were chronically implanted in the motor cortex areas associated with arm movement. Spike times were recorded for each active channel using the Plexon System. Waveform template match technique was used to identify and distinguish valid cortical neuron spikes. Muscle activities were recorded from seven major arm muscles using special surface electrodes. Kinematics for the moving arm were recorded using the OptoTrak system. Markers were placed at shoulder, elbow, wrist and MPJ of the middle finger. The perturbation force was recorded by a load cell attached to the pneumatic piston creating the short pulling motion. Data from each trial was aligned on the start of the movement.

12.3.2 Development of Predictive Control Strategies

Figure 12.3.2 shows the accumulative histogram of spike counts from a single cortical cell recorded from five repetitions of the reaching movement toward target 1. The perturbation was applied during every movement and the perturbation on-off time was marked in the figure, together with other event markers.

During the initial perturbation set, the startle effect made the monkeys miss targets within the allotted time. However, the initial high failure rate of 45% quickly reduced to less than 5% within a few sets of repetitions. They learned to accommodate the perturbation in reaching the target within the given time. Hand trajectory data demonstrates an improvement in performance as the effect of the perturbation was reduced. After two days of practice to move toward the target under the perturbation, monkeys developed a strategy, in anticipation of the perturbation, that the movement trajectory became deviated away from the

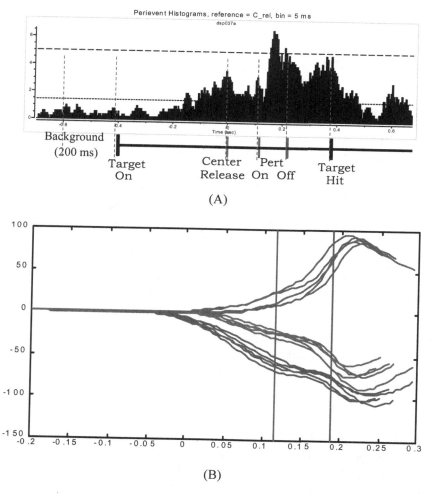

Figure 12.3.2: (A) Histogram of spike frequency of a single cortical cells during a perturbed movement. The hand movement was not completely stopped when the target was hit as indicated by the velocity traces given in (B). This is because the target is a button to be pushed by a finger and the hand movement is recorded.

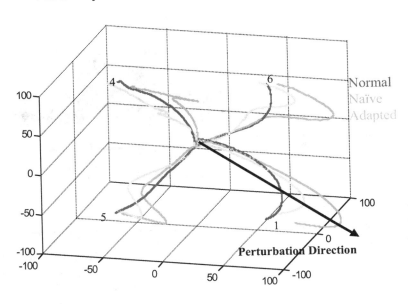

Figure 12.3.3: Average hand paths toward four of the eight targets are shown. The three traces for each target represent the normal, early perturbation and later adapted movement. Notice the deviation in the adapted trajectory before the perturbation effect.

direction of the perturbation as shown in Figure 12.3.3.

Simultaneous recordings from over 40 cortical neurons were made as the monkey performed the task. Some cells exhibited a modulated burst of activity (phasic) that began 100 ms prior to movement onset. Other cells exhibited similar phasic activity, but at latencies following the start of movement. All of these "task-related" cells demonstrated a secondary burst of activity following the perturbation. An analysis of variance (F test, $p < 0.05$) identified a differential response among targets. Perievent and inter-spike-interval histograms showed that the discharge patterns of many of the neurons remained very consistent across the duration of the study. The timing and magnitude of the cortical response to the perturbation was very stable across the two weeks of perturbation trials.

However, a noticeable increase in activity level of many cortical neurons right before the onset of the perturbation can be observed. This is shown in Figure 12.3.4. Histograms of spike frequency from eight different cortical neurons are shown in this figure for three conditions: normal reaching before any perturbation was applied (the left column), first successful reaching motion under perturbation (the middle column), and after eight days of repeated perturbations (right column).

Among these eight neurons, the fourth and fifth seemed well tuned to the movement task as the activity level was proportional to the velocity profile with approximately 100 ms lead time. The top one was not well tuned to the movement since the activity level increased as the movement was terminating. The bottom one was almost silent across the movement duration. Other cells were combinations of these cases.

Despite the difference in these neurons' activity pattern during the movement, they all responded to the perturbation with significant increases in activity level after the perturbation, as shown in the middle column. This response cannot be attributed to the startle effect because it can be observed consistently after many days of exposing to the same perturbation.

After eight days of repeated perturbations for every movement, the perturbation became a fixed feature of the movement task. From Figure 12.3.3 we have shown that the movement trajectory had been adjusted to accommodate the perturbation effect on arm dynamics. This adjustment was deviation in trajectory before the perturbation effect. If the adjustment was intentional as the result of adapting to the perturbation, there should have corresponding changes in motor cortical activity patterns. Comparing the activity patterns shown on the right column to those in the middle, an increase in firing frequency before the perturbation is noticeable. Another change is the general increase in overall activity level of all neurons shown here. These neurons were recorded from the same electrode array whose recording area span less than 1.5 mm^2 in the motor cortex.

Figure 12.3.5 shows the after-effect of the adaptation to the perturbation. Activity patterns shown in this figure are from the same eight neurons at later stages of dealing with the perturbation and after the adaptation to the perturbation became stable, first day after the perturbation was removed, and after three days without being perturbed. When the adapted neuron activity patterns became stable, we removed the perturbation for three consecutive days to examine whether the anticipation actions remain and for how long.

As shown by the activity patterns in the middle column of the figure, the anticipation remained for more than a day after we removed the perturbation. Other than the bursts in response to the perturbation were missing, all the basic patterns were the same as those after adapted to the perturbation. The increased activity

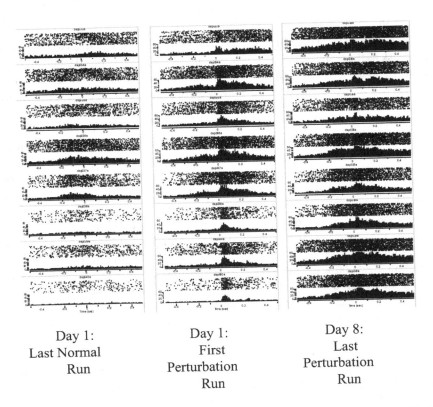

Day 1:
Last Normal
Run

Day 1:
First
Perturbation
Run

Day 8:
Last
Perturbation
Run

Figure 12.3.4: Comparison of cortical neuron activity patterns under normal (left) and in response to a short impulsive pulling perturbation (middle). Shown on the right column are responses after eight days of repeated perturbation trials.

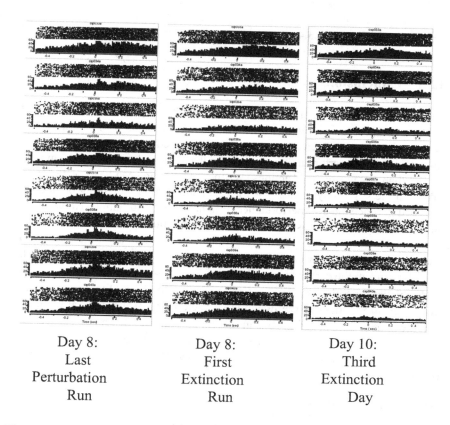

Day 8:	Day 8:	Day 10:
Last	First	Third
Perturbation	Extinction	Extinction
Run	Run	Day

Figure 12.3.5: The neuronal activity patterns after adaptation to the perturbation (left), first day after the perturbation was removed (middle), and three days later (right). Adapted changes remained for a day and the activity patterns returned to the original ones shown in the left column of Figure 12.3.4.

levels were still observable. Since we still activated the piston and only discon-nected the string to the wrist, the anticipation activities in the cortical neurons were also present. Only after a day of without perturbation the anticipation activi-ties disappeared. After two days, the activity patterns from these neurons returned to the original patterns before the perturbation was ever applied. This can be seen by comparing the patterns from the right column in this figure to those from the left column in Figure 12.3.4.

12.4   Conclusions

To develop a neuroprosthetic system controllable directly by cortical commands, we have to understand how the cortical neural system deals with external pertur-bations to a given task, either expected or unexpected. When control a task with uncertain external perturbations, sensory feedback control system maybe more ef-fective than any pre-planned actions. When a perturbation can be expected and its effect on the system dynamics is learned, a control strategy relying on anticipa-tion will produce the best performance in light of the perturbation. However, this strategy will generate disaster if the anticipated perturbation is different. On the other hand, a practical neuroprosthetic system should be able to do versatile tasks and learn to perform new tasks. How will this versatility and learning change the commands from the cortical signal detectors?

In this investigation we designed an experimental protocol that may generate insightful information on the dynamic characteristics of the cortical control of arm movement. The data from this investigation indicates that many cortical neu-rons respond to a positional perturbation and makes anticipative adjustment after learned the effect of the perturbation. The adjustment, however, is temporary in the motor cortex.

Acknowledgements

This work was supported in part by funding from the Whitaker Foundation, the National Science Foundation, PHS NS-37088, and in part by a grant for open research projects (ORP-0302) from the Key Lab of Complex Systems and Intelligence Science, Chinese Academy of Sciences.

Bibliography

[1] N. B. Alexander, N. Shepard, M. J. Gu, and A. Schultz, "Postural control in young and elderly adults when stance is perturbed: Kinematics," *J. Geron-tology: Medical Sciences*, vol. 42, pp. 79–87, 1992.

[2] B. Bussel, R. Katz, E. Pierrot-Deseilligny, C. Bergego, and A. Hayat, "Vestibular and proprioceptive influences on the postural reactions to a sudden body displacement in man," *Prog. Clin. Neurophysiol.*, vol. 8, pp. 310–322, 1980

[3] R. Caminiti, P. B. Johnson, and A. Urbano, "Making arm movements within different parts of space: Dynamic aspects in the primate motor cortex," *J. Neurosci.*, vol. 10, no. 7, pp. 2039–58, 1990.

[4] M. R. Carhart and G. T. Yamaguchi, "Motor control of stepping responses to postural perturbations," *Proc. 19th Ann. Meeting American Soc. Biomechanics*, Stanford, CA, Aug. 1995.

[5] J. M. Chandler, P. W. Duncan, and S. A. Studenski, "Balance performance on the postural stress test: Comparison of young adults, healthy elderly, and fallers," *Physical Therapy*, vol. 70, pp. 410–415, 1990.

[6] P. J. Cordo and L. M. Nashner, "Properties of postural adjustment associated with rapid arm movements," *J. Neurophysiol.*, vol. 47, pp. 287–302, 1982.

[7] H. C. Diener, F. B. Horak, G. Stelmach, B. Guschlbauer, and J. Dichgans, "Direction and amplitude precuing has no effect on automatic posture responses," *Exp. Brain Res.*, vol. 84, pp. 219–223, 1991.

[8] M. C. Do, Y. Breniere, and P. Brenguier, "A biomechanical study of balance recovery during the fall forward," *J. Biomechanics*, vol. 12, pp. 933–939, 1982. vol. 8, no. 8, pp. 2928–37, 1988.

[9] A. P. Georgopoulos, J. Ashe, N. Smyrnis, and M. Taira, "The motor cortex and the coding of force," *Science*, vol. 256, pp. 1692–1695, June 1992.

[10] A. P. Georgopoulos, R. E. Kettner, and A. B. Schwartz, "Primate motor cortex and free arm movements to visual targets in three-dimensional space-II: Coding of the direction of movement by a neuronal population," *J. Neurosci,*

[11] J. He, W. S. Levine, and G. E. Loeb, "Feedback gains for correcting small perturbations to standing posture," *IEEE Trans. Auto. Contr.*, vol. 36, pp. 322–332, 1991.

[12] J. He and C-X. Tian, "A statistical smoothness measure for tracking motion data," *J. Human Movement Sciences*, vol. 17, pp. 189–200, 1998.

[13] H. Hemami and A. Katab "Constrained inverted pendulum model of evaluating upright postural stability," *J. Dyn. Sys. Meas. Control*, vol. 104, pp. 343–349, 1982.

[14] F. B. Horak, *Adaptation of Automatic Postural Responses*, in *Acquisition of Motor Behavior in Vertebrates*, J. Bloedel, T. J. Ebner, and S. P. Wise, Eds., Cambridge, MA: MIT Press, 1996.

[15] F. B. Horak and H. C. Diener, "Cerebellar control of postural scaling and central set in stance," *J. Neurophysiol.*, vol. 72, pp. 479–493, 1994.

[16] F. B. Horak and J. M. Macpherson, "Postural orientation and equilibrium," in *Exercise: Regulation and Integration of Multiple Systems*, vol. 12, L. B. Rowell and J. T. Sheperd, Eds., Oxford University Press, 1996.

[17] F. B. Horak and Moore, "The effect of prior leaning on human postural responses," *Gait & Posture*, vol. 1, pp. 203–210, 1993.

[18] F. B. Horak and L. M. Nashner, "Influence of central set on human postural responses," *J. Neurophysiol.*, vol. 62, pp. 841–853, 1989.

[19] G. A. Horstmann and V. Dietz, "A basic posture control mechanism: The stabilization of the center of gravity," *Electroencepholography and Clinical Neurophysiol.*, vol. 76, pp. 165–176, 1990.

[20] M. Hoshiyama, Y. Suzuki, Y. Kaneoke, Y. Koike, A. Takahashi, and S. Watanabe, "Modulation of human balancing recovery in forward fall by optokinetic stimulation," *Env. Med.*, vol. 36, pp. 191–194, 1992.

[21] J. Kalaska, D. Cohen, M. Hyde, and M. Prud'homme, "A comparison of movement direction-related versus load direction-related activity in primate motor cortex, using a two-dimensional reaching task," *J. Neurosci.*, vol. 9, pp. 2080–2102, 1989.

[22] A. D. Kuo, "An optimal control model for analyzing human postural balance," *IEEE Trans. Biomed. Eng.*, vol. 42, pp. 87–101, 1995.

[23] A. D. Kuo and F. E. Zajac, "A biomechanical analysis of muscle strength as a limiting factor in standing posture," *J. Biomechanics*, vol. 26, pp. 137–150, 1993.

[24] W. Lee, "Anticipatory control of postural and task muscles during rapid arm flexion," *J. Motor Behav.*, vol. 12, pp. 185–196, 1980.

[25] J. M. Macpherson, "Strategies that simplify the control of quadrupedal stance," *J. Neurophysiol.*, vol. 60, pp. 204–231, 1988

[26] B. Maki and R. S. Whitelaw, "Influence of expectation and arousal on center of pressure responses to transient postural perturbations," *J. Vestib. Res.*, vol 3, pp. 25–39, 1993.

[27] C. D. Marsden, P. A. Merton, and H. B. Mortan, "Anticipatory postural responses in the human subject," *J. Physiol.*, vol. 275, pp. 47–48, 1977.

[28] J. Massion, "Movement, posture and equilibrium: Interaction and coordination," *Prog. Neurobiol*, vol. 38, 35–56, 1992.

[29] G. McCollum and T. K. Leen, "Form and exploration of mechanical stability limits in erect stance," *J. Motor Behav.*, vol. 21, pp. 225–244, 1989.

[30] W. E. McIlroy and B. E. Maki, "Changes in early automatic postural responses associated with prior planning an execution of a compensatory step," *Brain Res.*, vol. 631, pp. 203–211, 1993.

[31] L. M. Nashner, "Fixed patterns of rapid postural responses among leg muscles during stance," *Exp. Brain Res.*, vol. 30, pp. 13–24, 1977.

[32] L. M. Nashner and G. McCollum, "The organization of human postural movements: A formal basis and experimental synthesis," *Behavioral and Brain Science*, vol. 8, pp. 135–172, 1985.

[33] A. E. Patla, J. S. Frank, and D. A. Winter, "Balance control in the elderly: Implications for clinical assessment and rehabilitation," *Canadian J. Public Health*, vol. 83, suppl. 2, pp. S29–S33, 1992.

[34] A. Prochazka, "Sensorimotor gain control: A basic strategy of motor systems?" *Prog. Neurobiol.*, vol. 33, pp. 281–307, 1989.

[35] C. L. Riach and K. C. Hayes, "Anticipatory postural control in children," *J. Motor Behav.*, vol. 22, pp. 250–266, 1990.

[36] A. Schwartz, "Motor cortical activity during drawing movements: Single unit activity during sinusoid tracing," *J. Neurophysiol.*, vol. 68, no. 2, pp. 528–41, 1992.

[37] A. B. Schwartz, R.E. Kettner, and A.P. Georgopoulos, "Primate motor cortex and free arm movements to visual targets in three-dimensional space-I: Relations between single cell discharge and direction of movement," *J. Neurosci.*, vol. 8, no. 8, pp. 2913–27, 1988.

[38] A. Schultz, M. J. Gu, N. B. Alexander, T. Boismier, and N. T. Shepard, "Postural control in young and elderly adults when stance is challenged: Clinical versus laboratory measurements," *Ann. Oto. Rhino. Laryngol.*, vol. 102, pp. 508–517, 1993.

Chapter 13
Robust Adaptive Approximation Based Backstepping via Localized Adaptive Bounding

Yuanyuan Zhao and Jay A. Farrell

Abstract: Adaptive approximation methods estimate unmodeled nonlinear functions to improve tracking performance. This chapter investigates methods for the on-line estimation of functions that bound the achievable adaptive approximation accuracy. Whereas the results currently in the literature involved global forgetting, the adaptation laws for the estimated bounds herein involve local forgetting. The importance of local versus global forgetting is motivated in the text and illustrated with an example. Such localized bounds are useful for self-organizing adaptive approximators that adjust the structure of the function approximator during the operation of the system.

13.1 Introduction

Since the first stability results appeared, robust adaptive nonlinear control has been extensively developed to retain closed-loop stability properties in the presence not only of large parametric uncertainty, but also modeling errors such as nonparametric nonlinear model error, additive disturbances, and unmodeled dynamics [3]. On-line approximation methods [1, 2, 4, 7–10, 12–14] are designed to achieve stability and accurate reference input tracking for systems with partially unknown nonlinearities, by implementing approximations to the unknown nonlinear dynamics during the operation of the system.

Nonlinear close-loop systems which incorporate on-line approximators can be analyzed using Lyapunov stability methods. Both the feedback control law and the approximator parameter estimation equations are derived such that the time derivative of a Lyapunov function has some desirable properties (e.g., negative definiteness). The theory for approximation based nonlinear control is provided in, e.g., [2, 4, 7, 8, 10, 12]. The design and analysis of adaptive systems have been extensively addressed, including controller structure selection, automatic adjustment of the control law, and complete proofs of stability. Its application based on the feedback linearization method is developed in, e.g., [7, 8]. On-line approximation based control by backstepping is considered in, e.g., [4, 10, 12].

Since on-line approximation based control can never achieve an exact modeling of unknown nonlinearities, *inherent approximation errors* could arise even if optimal approximator parameters were selected (see Section 13.2). Often, a restrictive assumption is made that a magnitude bound on the inherent approximation error is known. Articles [4, 10, 11] relax the assumption of a known bound on the inherent approximation errors. With a partially known bound, those articles discuss estimation of the bounding parameters and the design of adaptive robust controllers to guarantee global uniform ultimate boundedness.

However, the global features of the leakage modification in [4, 10, 11] cause each parameter estimate θ_k to drift toward certain design parameters when the operating point x leaves a local region S_k that defines the region over which the affect of the parameter θ_k is nonzero (see Section 13.2). Thus, both the approximated function and the bounding function will lose local accuracy on S_k while $x \notin S_k$. Therefore, any knowledge learned from past experience in S_k may not be retained for future use when the state again returns to S_k. This issue of global forgetting was addressed in [15], by deriving localized leakage based adaptation algorithms for both the approximator parameters and bounding parameters. The analysis of [15] focused on the scalar single-input-single-output system:

$$\dot{x} = f(x) + g(x)u$$

with $g(x) = 1$ and $x \in \Re^1$.

In this chapter, we will extend the results of [15] to general, second order, triangular systems. In addition, the existing adaptive bounding approaches utilize leakage methods. The presentation herein considers leakage, a deadzone, and the ϵ-modification. The stability and robustness results presented herein yield a smaller m.s.s. bound on the tracking error than those previously obtained in the literature; in addition, the bounding function and function approximation information are retained as a function of the operating point even as the operating point moves around the operating envelope.

13.2   Problem Formulation

We consider a class of n-order single-input single-output (SISO) systems described as

$$\dot{x}_i = (f_i^o(\bar{x}_i) + f_i(\bar{x}_i)) + g_i^o(\bar{x}_i)x_{i+1}, \ 1 \le i \le n-1 \tag{13.2.1}$$

$$\dot{x}_n = (f_n^o(\bar{x}_n) + f_n(\bar{x}_n)) + (g_n^o(\bar{x}_n) + g_n(\bar{x}_n))u \tag{13.2.2}$$

where $x = [x_1, \cdots, x_n]^T$ is the state vector and u is the control signal. We let $\bar{x}_i = [x_1, \cdots, x_i]^T$ for notational compactness. The functions $f_i^o(\bar{x}_i)$, $g_i^o(\bar{x}_i)$, $i = 1, \cdots, n$, represent the known model at the design stage (i.e., the design model). The functions $f_i(\bar{x}_i)$, $i = 1, \cdots, n$, and $g_n(\bar{x}_n)$ represent nonlinear effects that are unknown at the design stage. Both the known and unknown functions are assumed to be continuous on a known compact set $\mathcal{D}^i$. To ensure controllability, it is necessary to assume that each g_i^o and g_n are bounded away from zero and of known sign. Therefore, without loss of generality, we will invoke the following assumption:

Assumption 13.2.1. For $i = 1, \cdots, n$ each function $g_i^o(\bar{x}_i)$ has a lower bound such that $g_i^o(\bar{x}_i) \ge g_{l_i}(\bar{x}_i) \ge g_l > 0$, $\forall \bar{x}_i \in \mathcal{D}^i$, where $g_{l_i}(\bar{x}_i)$ is a known function and g_l is a known constant. Also, $g_n(\bar{x}_n)$ has lower bound such that $g_n(\bar{x}_n) \ge g_{l_n}(\bar{x}_i) \ge g_l > 0$, $\forall \bar{x}_n \in \mathcal{D}^n$. □

13.2.1 Reference Trajectory and Baseline Controller

When the physical system is designed, the specified operating range is defined for each element of the state vector as a compact set $\mathcal{A}_i$. Let $\mathcal{D}^i = \mathcal{A}_1 \times \mathcal{A}_2 \times \cdots \times \mathcal{A}_i$. The control system should ensure that the state remains in the physical operating region $\mathcal{D}^n$. It is therefore reasonable to assume that the desired state trajectory is in sufficiently inside $\mathcal{D}^n$. These constraints are stated more rigorously in the next paragraph.

There is a desired trajectory $x_d(t)$ with derivatives $x_d^{(i)}$, $i = 1, \cdots, n$, each of which is available and bounded. The trajectory x_d is defined such that the corresponding state trajectory $x(t)$ required to track $x_d(t)$ will lie entirely in $\mathcal{D}^n$. This implies that each region $\mathcal{A}_i$ contains the trajectory x_{ic} (see Section 13.3) that the state x_i is expected to track and that the region $\mathcal{D}^n$ contains all trajectories $x_c = [x_d, x_{2c}, \cdots, x_{nc}]^T$. In fact, we will assume existence of a small constant $\gamma > 0$ such that

$$\gamma \le \min_{x \in \{\Re^n - \mathcal{D}^n\}}(\|x_c(t) - x\|)$$

for any $t \ge 0$. This condition states that the desired trajectory is at least a distance γ from the boundary of $\mathcal{D}^n$. The unknown functions f_i and g_n will be approximated over the region $\mathcal{D}^n$.

Throughout the chapter

$$\tilde{x}_i = x_i - x_{ic}, \ 1 \le i \le n$$

are the tracking errors. The variable $x_{1c} = x_d$. Based on the design model and using backstepping, a baseline tracking controller is

$$x_{2c} = \frac{1}{g_1^o} \left(-k_1 \tilde{x}_1 + \dot{x}_d - f_1^o \right),$$

$$x_{(i+1)c} = \frac{1}{g_i^o} \left(-k_i \tilde{x}_i + \dot{x}_{ic} - f_i^o - g_i^o \tilde{x}_{i-1} \right), \quad i = 2, \cdots, n-1,$$

$$u = \frac{1}{g_n^o} \left(-k_n \tilde{x}_n + \dot{x}_{nc} - f_n^o - g_n^o \tilde{x}_{n-1} \right).$$

Each control gain k_i is a designer specified positive constant that will determine the decay rate for errors due to disturbances and initial conditions. When the unknown nonlinear effects $f_i(\bar{x}_i)$, $i = 1, \cdots, n$, and $g_n(\bar{x}_n)$ are identically zero, then this baseline controller can be proven to be exponentially stable.

In the presence of model error, tracking performance could deteriorate and stability could be lost. The on-line approximation based controller developed in the subsequent sections is designed to maintain stability and tracking performance, even in the presence on unmodeled nonlinear effects. This will be achieved without using large magnitude switching nor high gain control (i.e., the control gains k_i can be the same as specified for the baseline controller). The controller will also include terms to ensure that the region $\mathcal{D}^n$ is an attractive and positively invariant set.

13.2.2 Approximator Definition

For $\bar{x}_i \in \mathcal{D}^i$, we define approximations to the unknown functions $f_i(\bar{x}_i)$ and $g_n(\bar{x}_n)$ as $\hat{f}_i(\bar{x}_i) = \theta_{f_i}^T \Phi_{f_i}(\bar{x}_i)$ and $\hat{g}_n(\bar{x}_n) = \theta_{g_n}^T \Phi_{g_n}(\bar{x}_n)$, where the parameter vectors θ_{f_i} and θ_{g_n} will be estimated on-line. For $\bar{x}_i \notin \mathcal{D}^i$, $\hat{f}_i(\bar{x}_i) = 0$ and $\hat{g}_n(\bar{x}_n) = g_l$. The vector $\Phi_{f_i}(\bar{x}_i)$ is a user specified regressor vector containing the basis functions for the approximation. Denote the support of the k-th basis function of $\Phi_{f_i}(\bar{x}_i)$ vector by $S_{f_i,k} = \{\bar{x}_i \in \mathcal{D}^i \,|\, \Phi_{f_i,k}(\bar{x}_i) \neq 0\}$. Let $\bar{S}_{f_i,k}$ denote the closure of $S_{f_i,k}$. Note that each $\bar{S}_{f_i,k}$ is a compact set. For each i, the $\Phi_{f_i}(\bar{x}_i)$ vector is defined as a set of positive, locally supported* functions $\Phi_{f_i,k}(\bar{x}_i)$ for $k = 1 \cdots, N_i$ such that each set $S_{f_i,k}$ is connected with $\mathcal{D}^i =$

---

*'Locally supported' means that $\rho(S_{f_i,k}) < \mu \ll \rho(\mathcal{D})$, where for set A,

$$\rho(A) = \max_{x,y \in A} \left(\|x - y\| \right).$$

$\bigcup_{k=1}^{N_i} S_{f_i,k}$ where $N_i > 0$ is a finite integer. This ensures that for any $\bar{x}_i \in \mathcal{D}^i$, there exists at least one k such that $\Phi_{f_i,k}(\bar{x}_i) \neq 0$. Therefore, $\{S_{f_i,k}\}_{k=1}^{N_i}$ forms a finite cover for $\mathcal{D}^i$. Similarly, we define the support of the k-th basis function of $\Phi_{g_n}(\bar{x}_n)$ as $S_{g_n,k}$ with closure $\bar{S}_{g_n,k}$. The sets $\bar{S}_{g_n,k}, k = 1 \cdots, N_n$ also form a finite cover of region $\mathcal{D}^n$.

In this chapter, we are not concerned with the selection of particular basis vectors Φ_f or Φ_g. Any basis vectors which satisfy the above assumptions are qualified candidates for the regressor vectors. Splines, certain radial basis functions, etc. satisfy these assumptions.

We define a set of parameters $\theta_{f_i}^*$ that are optimal in the sense:

$$\theta_{f_i}^* = \arg\min_{\theta} \left(\max_{\bar{x}_i \in \mathcal{D}^i} \left| f_i(\bar{x}_i) - \theta^T \Phi_{f_i}(\bar{x}_i) \right| \right) .$$

Since $\mathcal{D}^i$ is compact and each f_i is continuous, the vector $\theta_{f_i}^*$ exists and is well-defined. Note that these optimal parameters are unknown. The vector $\theta_{f_i}^*$ is not used in the implemented control law, but is useful for the analysis that follows. Define the parameter estimation error vector

$$\tilde{\theta}_{f_i} = \theta_{f_i} - \theta_{f_i}^*.$$

Let

$$\delta_{f_i}(\bar{x}_i) = f_i(\bar{x}_i) - (\theta_{f_i}^*)^T \Phi_{f_i}(\bar{x}_i)$$

represent the *inherent* or *residual approximation error*. Note that by the definition of $\theta_{f_i}^*$ above, the maximum value of $\delta_{f_i}(\bar{x}_i)$ on $\mathcal{D}^i$ is bounded. This maximum value can be affected by the choice of the size and type of basis vector $\Phi_{f_i}(\bar{x}_i)$, but for a given choice of basis vector it cannot be decreased by the choice of the parameter vector θ_{f_i}. The upper bound on the magnitude of the residual approximation error only depends on the designer's choice of approximator. The quantities $\theta_{g_n}^*$, $\tilde{\theta}_{g_n}$ and $\delta_{g_n}(\bar{x}_n)$ are defined similarly.

With the above definitions, (13.2.1)–(13.2.2) can be expressed as

$$\dot{x}_i(t) = f_i^o + (\theta_{f_i}^*)^T \Phi_{f_i} + \delta_{f_i} + g_i^o x_{i+1}$$
$$\dot{x}_n(t) = f_n^o + (\theta_{f_n}^*)^T \Phi_{f_n} + \delta_{f_n} + \left(g_n^o + (\theta_{g_n}^*)^T \Phi_{f_n} + \delta_{g_n} \right) u$$

where to simplify the notation the $\bar{x}_i$ dependence of the various functions has not been shown.

13.2.3 Bound Approximation

By the definition of the δ_{f_i} and δ_{g_n}, the magnitude of these inherent approximation error functions are bounded on $\mathcal{D}^i$; however, the bound is not known. Our

control approach will utilize an estimate of these upper bound functions. Therefore, we assume a form for the bounding functions with multiplicative parameters
that will be estimated. To save computational effort, we reuse the same basis
elements; however, the approach easily extends to the case of different basis elements.

By the above discussion, there exists a positive constant vector Ψ_{fi}^*, $i = 1, \cdots, n$, referred as the *optimal bounding parameter*, such that

$$|\delta_{f_i}| \le \left[\psi_{f_i,1}^* \cdots \psi_{f_i,N_i}^*\right] \begin{bmatrix} \phi_{f_i,1} \\ \vdots \\ \phi_{f_i,N_i} \end{bmatrix} = (\Psi_{f_i}^*)^T \Phi_{f_i}, \ \forall \bar{x}_i \in \mathcal{D}^i. \quad (13.2.3)$$

The vector $\Psi_{f_i}^*$ is not unique since any $\bar{\Psi}_{f_i}^* > \Psi_{f_i}^*$ satisfies this assumption. To
avoid confusion, the *optimal bounding parameter* is defined to be the vector with
the smallest 1-norm such that (13.2.3) is satisfied. A vector $\Psi_{g_n}^*$ yielding a bound
on $|\delta_{g_n}|$ is defined similarly. Note that the optimal bounding parameter vectors
$\Psi_{f_i}^*$ and $\Psi_{g_n}^*$ are unknown. They are used only for analytical purpose. The control
law will use approximations $\Psi_{f_i}^T \Phi_{f_i}$ for $|\delta_{f_i}|$ and $\Psi_{g_n}^T \Phi_{g_n}$ for $|\delta_{g_n}|$ to the optimal
bounding functions where the vectors Ψ_{f_i} and Ψ_{g_n} will be estimated on-line. For
the following analysis, we define bounding parameter estimation errors as

$$\tilde{\Psi}_{f_i} = \Psi_{f_i} - \Psi_{f_i}^M \text{ and } \tilde{\Psi}_{g_n} = \Psi_{g_n} - \Psi_{g_n}^M$$

where each element of $\Psi_{f_i}^M$ is defined as $\psi_{f_i,k}^M = \max\{\psi_{f_i,k}^*, \psi_{f_i,k}^0\}$, $k = 1, \cdots, N_i$. The vector $\Psi_{f_i}^0 = [\psi_{f_i,1}^0, \cdots, \psi_{f_i,N_i}^0]^T$ is selected in the design stage.
Similar definitions and discussion apply relative to $\Psi_{g_n}^M$ and $\Psi_{g_n}^o$. With these es-
timated upper bounds, we will select proper terms in the control signal or the
virtual control variables to maintain accurate tracking performance in spite of the
inherent approximation errors.

13.3   Adaptive Backstepping-Based Design

In the reminder of this chapter, we consider the class of second order systems. The
design methodology can also be extended to $n > 2$ order systems by following
the recursive steps of the backstepping procedure described in, e.g. [5].

We define the virtual control signal x_{2c} for the x_1 subsystem as

$$x_{2c} = \frac{1}{g_1^o} \left(-k_1 \tilde{x}_1 + \dot{x}_d - f_1^o - \hat{f}_1 - \beta_{f_1}(x_1, x_{1_c}, \Psi_{f_1}) - v_{f_1}(x_1, x_{1_c})\right).$$

The $\beta_{f_1}(x_1, x_{1_c}, \Psi_{f_1})$ term will be defined later to address the inherent approximation error δ_{f_1}. The $v_{f_1}(x_1, x_{1_c})$ term is designed to return state x_1 to the

approximation region $\mathcal{D}^1$ and keep it there (i.e., to ensure that $\mathcal{D}^1$ is an attractive and invariant set), see Section 13.3. The v_{f_1} term is defined as

$$v_{f_1}(x_1, x_{1_c}) = \begin{cases} 0, & \text{when } x \in \mathcal{D}^2 \\ \bar{b}_{f_1}(x_1), & \text{when } x \notin \mathcal{D}^2 \text{ and } \tilde{x}_1 \geq 0 \\ -\bar{b}_{f_1}(x_1), & \text{when } x \notin \mathcal{D}^2 \text{ and } \tilde{x}_1 \leq 0 \end{cases}$$

where $\bar{b}_{f_1}(x_1)$ is the known upper bound on $|f_1(x_1)|$ valid $\forall\, x_1 \notin \mathcal{D}^1$. The main focus of this chapter is on accurate tracking within $\mathcal{D}^2$ without the use of high gain control or large magnitude switching. After initial condition effects decay, the system should never operated outside of $\mathcal{D}^2$; therefore, we simply assume the existence of the known bound $\bar{b}_{f_1}$ that will be used only outside of $\mathcal{D}^2$. Note that if the bound function $\bar{b}_{f_1}$ is not known, then it could be estimated using the methods suggested in [10, 11].

Given the definition of x_{2c}, the x_1 tracking error dynamics can be written as

$$\dot{\tilde{x}}_1 = f_1^o + \hat{f}_1 + g_1^o x_{2c} - \dot{x}_d - \tilde{\theta}_{f_1}^T \Phi_{f_1} + \delta_{f_1} + g_1^o \tilde{x}_2$$
$$= -k_1 \tilde{x}_1 - \beta_{f_1} - v_{f_1} - \tilde{\theta}_{f_1}^T \Phi_{f_1} + \delta_{f_1} + g_1^o \tilde{x}_2. \qquad (13.3.1)$$

In the definition of the control signal in (13.3.2), x_{2c} will be treated as the command signal for x_2. Therefore, we require the time derivative $\dot{x}_{2c}$:

$$\dot{x}_{2c} = \frac{\partial x_{2c}}{\partial x_1} \dot{x}_1 + \frac{\partial x_{2c}}{\partial \theta_{f_1}} \dot{\theta}_{f_1} + \frac{\partial x_{2c}}{\partial \Psi_{f_1}} \dot{\Psi}_{f_1} + \frac{\partial x_{2c}}{x_d} \dot{x}_d + \frac{\partial x_{2c}}{\dot{x}_d} \ddot{x}_d$$
$$= -\frac{\partial x_{2c}}{\partial x_1}\left(\tilde{\theta}_{f_1}^T \Phi_{f_1}\right) + \frac{\partial x_{2c}}{\partial x_1}\left(f_1^o + \hat{f}_1 + g_1^o x_2 + \delta_{f_1}\right)$$
$$\quad + \frac{\partial x_{2c}}{\partial \theta_{f_1}} \dot{\theta}_{f_1} + \frac{\partial x_{2c}}{\partial \Psi_{f_1}} \dot{\Psi}_{f_1} + \frac{\partial x_{2c}}{x_d} \dot{x}_d + \frac{\partial x_{2c}}{\dot{x}_d} \ddot{x}_d$$
$$= -\frac{\partial x_{2c}}{\partial x_1}\left(\tilde{\theta}_{f_1}^T \Phi_{f_1} - \delta_{f_1}\right) + \hat{\mu}_1$$

where the first two terms on the right cannot be computed analytically and

$$\hat{\mu}_1 = \frac{\partial x_{2c}}{\partial x_1}\left(f_1^o + \hat{f}_1 + g_1^o x_2\right) + \frac{\partial x_{2c}}{\partial \theta_{f_1}} \dot{\theta}_{f_1} + \frac{\partial x_{2c}}{\partial \Psi_{f_1}} \dot{\Psi}_{f_1} + \frac{\partial x_{2c}}{x_d} \dot{x}_d + \frac{\partial x_{2c}}{\dot{x}_d} \ddot{x}_d$$

can be calculated analytically.

We define the control signal u as

$$u = \frac{u_{a2}}{g_2^o + \hat{g}_2 + \beta_{g_2} + v_{g_2}(x)} \qquad (13.3.2)$$

where $u_{a2} = -k_2\tilde{x}_2 + \hat{\mu}_1 - f_2^o - \hat{f}_2 - \bar{\beta}_{f_1}(x_1, \tilde{x}_2, \Psi_{f_1}) - \beta_{f_2}(x_2, \tilde{x}_2, \Psi_{f_2}) - v_{f_2}(x) - g_1^o\tilde{x}_1$. The $\bar{\beta}_{f_1}, \beta_{f_2}$ and β_{g_2} terms will be defined later to accommodate the inherent approximation errors. The terms v_{f_2} and v_{g_2} are defined as

$$v_{f_2}(x) = \begin{cases} 0, & \text{when } x \in \mathcal{D}^2 \\ \bar{b}_{f_2}(x) + \left|\dfrac{\partial x_{2c}}{\partial x_1}\right| \bar{b}_{f_1}(x_1), & \text{when } x \notin \mathcal{D}^2 \text{ and } \tilde{x}_2 \geq 0 \quad (13.3.3) \\ -\bar{b}_{f_2}(x) - \left|\dfrac{\partial x_{2c}}{\partial x_1}\right| \bar{b}_{f_1}(x_1), & \text{when } x \notin \mathcal{D}^2 \text{ and } \tilde{x}_2 \leq 0 \end{cases}$$

$$v_{g_2}(x) = \begin{cases} 0, & \text{when } x \in \mathcal{D}^2 \\ \bar{b}_{g_2}(x), & \text{when } x \notin \mathcal{D}^2 \text{ and } \tilde{x}_2 u_{a2} \geq 0 \quad (13.3.4) \\ g_l, & \text{when } x \notin \mathcal{D}^2 \text{ and } \tilde{x}_2 u_{a2} \leq 0 \end{cases}$$

where $\bar{b}_{f_2}(x), \bar{b}_{g_2}(x)$ are defined as known bounding functions on $|f_2(x)|$ and $g_2(x)$ for $x \notin \mathcal{D}^2$, respectively.

With the above definition of u, the dynamics of $\tilde{x}_2$ are described as

$$\begin{aligned} \dot{\tilde{x}}_2 &= f_2^o + f_2 + (g_2^o + g_2)u - \dot{x}_{2c} \\ &= f_2^o + \hat{f}_2 + (g_2^o + \hat{g}_2 + \beta_{g_2} + v_{g_2})u - \dot{x}_{2c} - \tilde{\theta}_{f_2}^T \Phi_{f_2} + \delta_{f_2} \\ &\quad - \tilde{\theta}_{g_2}^T \Phi_{g_2}u + \delta_{g_2}u - \beta_{g_2}u - v_{g_2}u \\ &= -k_2\tilde{x}_2 + \frac{\partial x_{2c}}{\partial x_1}\left(\tilde{\theta}_{f_1}^T \Phi_{f_1} - \delta_{f_1}\right) - \bar{\beta}_{f_1} - \beta_{f_2} - g_1^o\tilde{x}_1 - v_{f_2} \\ &\quad - \tilde{\theta}_{f_2}^T \Phi_{f_2} + \delta_{f_2} - \tilde{\theta}_{g_2}^T \Phi_{g_2}u + \delta_{g_2}u - \beta_{g_2}u - v_{g_2}u. \quad (13.3.5) \end{aligned}$$

Given the control law defined above, Section 13.3 shows that $\mathcal{D}^2$ is an attractive and positively invariant set. The remainder of the chapter then considers performance within $\mathcal{D}^2$ using various parameter estimation methods.

13.3.1 Analysis for x Outside $\mathcal{D}^2$

The objective of this section is to demonstrate that the definitions of v_{f_1}, v_{f_2} and v_{g_2} given above ensure that all initial conditions will return to and stay within region $\mathcal{D}^2$. For $x \notin \mathcal{D}^2$, parameter adaptation is turned off and the $\Phi_{f_i}, \beta_{f_i}, i = 1, 2$ and Φ_{g_2}, β_{g_2} terms are all zero.

Because Φ_{f_i} and Φ_{g_2} are zero for $x \notin \mathcal{D}^2$, $\delta_{f_i} = f_i(\tilde{x}_i)$ for $i = 1, 2$ and $\delta_{g_2} = g_2(\tilde{x}_2)$, therefore, the tracking error dynamics in (13.3.1) and (13.3.5) simplify to

$$\dot{\tilde{x}}_1 = -k_1\tilde{x}_1 + f_1 - v_{f_1} + g_1^o\tilde{x}_2$$

$$\dot{\tilde{x}}_2 = -k_2\tilde{x}_2 + f_2 - v_{f_2} - g_1^o\tilde{x}_1 + (g_2 - v_{g_2})u - \frac{\partial x_{2c}}{\partial x_1}f_1$$

To analyze performance for $x \notin \mathcal{D}^2$, we define the Lyapunov function

$$\bar{V} = \frac{1}{2}\tilde{x}_1^2 + \frac{1}{2}\tilde{x}_2^2.$$

The derivative of $\bar{V}$ is

$$\frac{d\bar{V}}{dt} = - k_1 \tilde{x}_1^2 - k_2 \tilde{x}_2^2 + \tilde{x}_1(f_1 - v_{f_1})$$

$$+ \tilde{x}_2 \left(f_2 - \frac{\partial x_{2c}}{\partial x_1} f_1 - v_{f_2} \right) + \tilde{x}_2(g_2 - v_{g_2})u.$$

The design of v_{f_i} for $i = 1, 2$ and v_{g_2} will ensure that the last three terms on the right are non-positive. This yields

$$\frac{d\bar{V}}{dt} \leq -k_1 \tilde{x}_1^2 - k_2 \tilde{x}_2^2 < -2\underline{k}\bar{V}$$

Defining $\underline{k} = \min(k_1, k_2)$ and using the Comparison Principle yields,

$$\bar{V}(t) \leq e^{-2\underline{k}t} \bar{V}(0), \text{ for any } t \geq 0$$

$$\|\tilde{x}(t)\| \leq e^{-\underline{k}t} \|\tilde{x}(0)\|.$$

Therefore, for any t larger than $T = \frac{1}{\underline{k}} ln \left(\frac{\|\tilde{x}(0)\|}{\gamma} \right)$, we have that $\|\tilde{x}(t)\| < \gamma$. Because the desired trajectory in $\mathcal{D}^2$ and is at least γ from the boundary of $\mathcal{D}^2$, this implies that x enters $\mathcal{D}^2$ in finite time. Once $x \in \mathcal{D}^2$, the sliding mode term will not allow x to leave $\mathcal{D}^2$.

The reminder of this chapter will only be concerned with the case of $x \in \mathcal{D}^2$, where all $v_{f_i}, i = 1, 2$ and v_{g_2} terms are zero.

13.3.2 Analysis for $x \in \mathcal{D}^2$

When $x \in \mathcal{D}^2$, the adaptive laws for $\theta_{f_i}, \theta_{g_i}, \Psi_{f_i}, \Psi_{g_i}, i = 1, 2$, are defined as

$$\dot{\theta}_{f_1} = P_B \Big\{ \Gamma_{f_1} (\Phi_{f_1} e_1 + Q_{f_1}) \Big\} \tag{13.3.6}$$

$$\dot{\theta}_{f_2} = P_B \Big\{ \Gamma_{f_2} (\Phi_{f_2} e_2 + Q_{f_2}) \Big\} \tag{13.3.7}$$

$$\dot{\theta}_{g_2} = P_{BC} \Big\{ \Gamma_{g_2} (\Phi_{g_2} e_2 u + Q_{g_2}) \Big\} \tag{13.3.8}$$

and

$$\dot{\Psi}_{f_1} = P_B \Big\{ \Gamma_{\Psi f_1} (\Omega_{f_1} \tilde{x}_1 + \bar{\Omega}_{f_1} \tilde{x}_2 + Q_{\Psi f_1}) \Big\} \tag{13.3.9}$$

$$\dot{\Psi}_{f_2} = P_B \Big\{ \Gamma_{\Psi f_2} (\Omega_{f_2} \tilde{x}_2 + Q_{\Psi f_2}) \Big\} \tag{13.3.10}$$

$$\dot{\Psi}_{g_2} = P_{BC} \Big\{ \Gamma_{\Psi g_2} (\Omega_{g_2} \tilde{x}_2 u + Q_{\Psi g_2}) \Big\} \tag{13.3.11}$$

where Γ_{f_i} and $\Gamma_{\Psi f_i}$ for $i = 1, 2$ and Γ_{g_2}, $\Gamma_{\Psi g_2}$, are defined as positive definite matrices representing the learning rates and for notational convenience we defined

$$e_1 = \tilde{x}_1 - \frac{\partial x_{2c}}{\partial x_1} \tilde{x}_2$$

$$e_2 = \tilde{x}_2.$$

The Ω functions are defined as $\Omega_{f_1} = \Phi_{f_1} \omega\left(\frac{\tilde{x}_1}{\epsilon}\right)$, $\bar{\Omega}_{f_1} = \Phi_{f_1} \omega\left(\frac{\left|\frac{\partial x_{2c}}{\partial x_1}\right| \tilde{x}_2}{\epsilon}\right)$,

$\Omega_{f_2} = \Phi_{f_2} \omega\left(\frac{\tilde{x}_2}{\epsilon}\right)$, and $\Omega_{g_2} = \Phi_{g_2} \omega\left(\frac{\tilde{x}_2 u_{a2}}{g_1 \epsilon}\right)$, where the function $\omega(\cdot)$ satisfies Assumption 13.5.1 of Section 13.5. The functions Q_{f_i}, $Q_{\Psi f_i}$, Q_{g_2}, $Q_{\Psi g_2}$ for $i = 1, 2$ are auxiliary terms included to allow analysis of certain robust parameter estimation approaches in later sections. The functions $P_B\{\cdot\}$ and $P_{BC}\{\cdot\}$ implement projection operations that would be distinct for each of (13.3.6)–(13.3.11). The subscript 'B' indicates that the projection operator is designed to ensure parameter boundedness. The subscript 'BC' indicates that the projection operator is designed to ensure parameter boundedness and to ensure that $\hat{g}_2 + \beta_{g_2}$ satisfies the controllability condition of Assumption 13.2.1. An example projection method is defined as (13.5.6) in Section 13.5. The projection operator is analyzed in [6].

Relative to (13.3.9)–(13.3.11) note that quantities of the form $m\omega\ (m/\epsilon)$ are nonnegative; therefore, if the modification terms denoted by a subscripted Q's were zero, then $\dot{\Psi}_{f_i}$ and $\dot{\Psi}_{g_2}$ would monotonically increase when the tracking errors were nonzero.

For any $x \in \mathcal{D}^2$, the stability properties are investigated by considering the following Lyapunov function candidate

$$V = \sum_{i=1}^{2} V_i(\tilde{x}_i, \tilde{\theta}_{f_i}, \tilde{\theta}_{g_i}, \tilde{\Psi}_{f_i}, \tilde{\Psi}_{g_i})$$

where

$$V_1 = \frac{1}{2}\left(\tilde{x}_1^2 + \tilde{\theta}_{f_1}^T \Gamma_{f_1}^{-1} \tilde{\theta}_{f_1} + \tilde{\Psi}_{f_1}^T \Gamma_{\Psi f_1}^{-1} \tilde{\Psi}_{f_1}\right)$$

and

$$V_2 = \frac{1}{2}\left(\tilde{x}_2^2 + \tilde{\theta}_{f_2}^T \Gamma_{f_2}^{-1} \tilde{\theta}_{f_2} + \tilde{\theta}_{g_2}^T \Gamma_{g_2}^{-1} \tilde{\theta}_{g_2} + \tilde{\Psi}_{f_2}^T \Gamma_{\Psi f_2}^{-1} \tilde{\Psi}_{f_2} + \tilde{\Psi}_{g_2}^T \Gamma_{\Psi g_2}^{-1} \tilde{\Psi}_{g_2}\right).$$

The time derivative of the V is $\dot{V} = \sum_{i=1}^{2} \dot{V}_i$, and $\dot{V}_i$ along solutions of (13.3.1) and (13.3.5) are:

$$\dot{V}_1 = \tilde{x}_1 \left[-k_1 \tilde{x}_1 - \beta_{f_1} - \tilde{\theta}_{f_1}^T \Phi_{f_1} + \delta_{f_1} + g_1^o \tilde{x}_2 \right] + \tilde{\theta}_{f_1}^T \Gamma_{f_1}^{-1} \dot{\theta}_{f_1} + \tilde{\Psi}_{f_1}^T \Gamma_{\Psi f_1}^{-1} \dot{\Psi}_{f_1}$$

$$= -k_1 \tilde{x}_1^2 + \tilde{x}_1 g_1^o \tilde{x}_2 + \Delta_1 + \tilde{\theta}_{f_1}^T \Gamma_{f_1}^{-1} \left(\dot{\theta}_{f_1} - \Gamma_{f_1} \Phi_{f_1} \tilde{x}_1 \right).$$

$$\dot{V}_2 = \tilde{x}_2 \left[-k_2 \tilde{x}_2 + \frac{\partial x_{2c}}{\partial x_1} \left(\tilde{\theta}_{f_1}^T \Phi_{f_1} \right) - \frac{\partial x_{2c}}{\partial x_1} \delta_{f_1} \right.$$

$$\left. - \bar{\beta}_{f_1} - \beta_{f_2} - g_1^o \tilde{x}_1 - \tilde{\theta}_{f_2}^T \Phi_{f_2} + \delta_{f_2} - \tilde{\theta}_{g_2}^T \Phi_{g_2} u + \delta_{g_2} u - \beta_{g_2} u \right]$$

$$+ \tilde{\theta}_{f_2}^T \Gamma_{f_2}^{-1} \dot{\theta}_{f_2} + \tilde{\theta}_{g_2}^T \Gamma_{g_2}^{-1} \dot{\theta}_{g_2} + \tilde{\Psi}_{f_2}^T \Gamma_{\Psi f_2}^{-1} \dot{\Psi}_{f_2} + \tilde{\Psi}_{g_2}^T \Gamma_{\Psi g_2}^{-1} \dot{\Psi}_{g_2}$$

$$= -k_2 \tilde{x}_2^2 - \tilde{x}_1 g_1^o \tilde{x}_2 + \Delta_2 + \frac{\partial x_{2c}}{\partial x_1} \tilde{x}_2 \left(\tilde{\theta}_{f_1}^T \Phi_{f_1} \right)$$

$$+ \tilde{\theta}_{f_2}^T \Gamma_{f_2}^{-1} \left(\dot{\theta}_{f_2} - \Gamma_{f_2} \Phi_{f_2} \tilde{x}_2 \right) + \tilde{\theta}_{g_2}^T \Gamma_{g_2}^{-1} \left(\dot{\theta}_{g_2} - \Gamma_{g_2} \Phi_{g_2} \tilde{x}_2 u \right).$$

where

$$\Delta_1 = \tilde{x}_1 \left(\delta_{f_1} - \beta_{f_1} \right) + \tilde{\Psi}_{f_1}^T \Gamma_{\Psi f_1}^{-1} \dot{\Psi}_{f_1},$$

and

$$\Delta_2 = -\tilde{x}_2 \left(\frac{\partial x_{2c}}{\partial x_1} \delta_{f_1} + \bar{\beta}_{f_1} \right) + \tilde{x}_2 \left(\delta_{f_2} - \beta_{f_2} + \delta_{g_2} u - \beta_{g_2} u \right)$$

$$+ \tilde{\Psi}_{f_2}^T \Gamma_{\Psi f_2}^{-1} \dot{\Psi}_{f_2} + \tilde{\Psi}_{g_2}^T \Gamma_{\Psi g_2}^{-1} \dot{\Psi}_{g_2}.$$

Therefore, the derivative $\dot{V}$ is written as

$$\dot{V} = -k_1 \tilde{x}_1^2 - k_2 \tilde{x}_2^2 + \tilde{\theta}_{f_1}^T \Gamma_{f_1}^{-1} \left(\dot{\theta}_{f_1} - \Gamma_{f_1} \Phi_{f_1} \left(\tilde{x}_1 - \frac{\partial x_{2c}}{\partial x_1} \tilde{x}_2 \right) \right)$$

$$+ \tilde{\theta}_{f_2}^T \Gamma_{f_2}^{-1} \left(\dot{\theta}_{f_2} - \Gamma_{f_2} \Phi_{f_2} \tilde{x}_2 \right) + \tilde{\theta}_{g_2}^T \Gamma_{g_2}^{-1} \left(\dot{\theta}_{g_2} - \Gamma_{g_2} \Phi_{g_2} \tilde{x}_2 u \right) + \Delta$$

$$= -k_1 \tilde{x}_1^2 - k_2 \tilde{x}_2^2 + \tilde{\theta}_{f_1}^T \Gamma_{f_1}^{-1} \left(\dot{\theta}_{f_1} - \Gamma_{f_1} \Phi_{f_1} e_1 \right)$$

$$+ \tilde{\theta}_{f_2}^T \Gamma_{f_2}^{-1} \left(\dot{\theta}_{f_2} - \Gamma_{f_2} \Phi_{f_2} e_2 \right) + \tilde{\theta}_{g_2}^T \Gamma_{g_2}^{-1} \left(\dot{\theta}_{g_2} - \Gamma_{g_2} \Phi_{g_2} u e_2 \right) + \Delta$$

$$\tag{13.3.12}$$

where

$$\Delta = \Delta_1 + \Delta_2$$

$$= \tilde{x}_1 (\delta_{f_1} - \beta_{f_1}) - \tilde{x}_2 \left(\frac{\partial x_{2c}}{\partial x_1} \delta_{f_1} + \bar{\beta}_{f_1} \right) + \tilde{x}_2 (\delta_{f_2} - \beta_{f_2} + \delta_{g_2} u - \beta_{g_2} u)$$

$$+ \sum_{i=1}^{2} \left(\tilde{\Psi}_{f_i}^T \Gamma_{\Psi f_i}^{-1} \dot{\Psi}_{f_i} \right) + \tilde{\Psi}_{g_2}^T \Gamma_{\Psi g_2}^{-1} \dot{\Psi}_{g_2} \tag{13.3.13}$$

The above results will be used at various locations in the following sections. First, Section 13.3 will briefly consider the ideal case in which $\Delta = 0$.

13.3.3 Ideal Special Case

In the ideal case where perfect approximation is known to be possible, $\delta_{f_i} = 0$ for $i = 1, 2$ and $\delta_{g_2} = 0$. Therefore, the designer could define $\beta_{f_i} = \beta_{g_2} = 0$ and $\Psi_{f_i} = \Psi_{g_2} = \mathbf{0}$, so that $\Delta = \Delta_i = 0, i = 1, 2$.

We start from (13.3.12) with the adaptation laws defined by (13.3.6)–(13.3.8) without any robust parameter estimation approach (i.e., $Q_{f_1} = Q_{f_2} = Q_{g_2} = 0$). Therefore, when the projection is not active, the derivative of V reduces to

$$\frac{dV}{dt} \leq -k_1 \tilde{x}_1^2 - k_2 \tilde{x}_2^2 \tag{13.3.14}$$

which is negative semi-definite. Note that when the projection is active, (13.3.14) is still preserved [6].

Theorem 13.3.1 (Ideal Case). *For a class of second order systems described by (13.2.1)–(13.2.2) with the adaptive feedback control law of (13.3.2), (13.3.3), (13.3.4) and the parameter adaptation laws of (13.3.6)–(13.3.8) with Q's equal to zero, we have the following stability properties:*

a) For $x(0) \notin \mathcal{D}^2$, $x(t)$ for $t > 0$ converges to region $\mathcal{D}^2$ in finite time.

b) When $x \in \mathcal{D}^2$ and $\delta_{f_i} = \delta_{g_i} = 0$:

4.1) $\tilde{x}_i, x_i, \theta_{f_i}, \theta_{g_2} \in \mathcal{L}_\infty, \quad i=1,2$

4.2) $\tilde{x}_i \in \mathcal{L}_2, i=1,2$;

4.3) $\tilde{x}_i \to 0$ as $t \to \infty, i=1,2$.

Proof. When $x \notin \mathcal{D}^2$, we have already shown in the Section 13.3 that the $v_{f_i}, i = 1, 2$ and v_{g_2} terms we design will return the state to $\mathcal{D}^2$ in finite time.

The proof for the case of $x \in \mathcal{D}^2$ and $\delta_{f_i} = \delta_{g_2} = 0$ is based on the (13.3.14). The negative semi-definiteness of $\frac{dV}{dt}$ implies that the variables $\tilde{x}_i, \theta_{f_i}, i = 1, 2$ and θ_{g_2} are each bounded. Since each term of $\dot{\tilde{x}}_i$ is bounded, $\dot{V}$ can be directly shown to be bounded. Barbalat's lemma implies that $\tilde{x}_i$ each approach zero as t approaches infinity. Finally, since

$$\dot{V} \leq -k_1 \tilde{x}_1^2 - k_2 \tilde{x}_2^2$$

$$V(t) - V(0) \leq -\int_0^t k_1 \tilde{x}_1^2(\tau) + k_2 \tilde{x}_2^2(\tau) d\tau$$

$$V(0) \geq \int_0^t k_1 \tilde{x}_1^2(\tau) + k_2 \tilde{x}_2^2(\tau) d\tau$$

which shows that each $\tilde{x}_i$ is in $\mathcal{L}_2$. $\square$

In more realistic situations, the Δ terms are expected to be small but nonzero on $\mathcal{D}^2$. In such cases, some form of robust parameter adaptation is required to ensure that the parameter estimates do not diverge toward infinity and to ensure that the tracking performance is maintained. In the presence of the nonzero Δ terms, we will not be able to prove convergence of the tracking errors to zero; instead, we will show boundedness of the tracking errors. To be useful to a designer, the bounds should be known functions of the design parameters so that the design is able to select those parameters to make the bounds suitably small.

13.4   Adaptive Bounding Methods

Subsection 13.3 considered the stability results applicable in the ideal case where perfect approximation was possible. In most applications, perfect approximation is not possible; therefore, we are interested in developing bounds on the approximation error and using those bounds in the control law to achieve robustness to the approximation error. In addition, we are interested in analysis of the achievable tracking performance. In this section, we consider the case where $x \in \mathcal{D}^2$ and there are residual approximation errors, i.e.,

$$\delta_{f_i} \neq 0,\, i = 1, 2 \ \text{ or } \ \delta_{g_2} \neq 0.$$

The case of $x \notin \mathcal{D}^2$ was discussed in Section 13.3.

The adaptive bounds (e.g., $\Psi_{f_i}^T \Phi_{f_i}$) are used to define the $\beta_{f_1}, \bar{\beta}_{f_1}, \beta_{f_2}$ and β_{g_2} terms in the control law:

$$\beta_{f_1} = \Psi_{f_1}^T \Omega_{f_1} = \Psi_{f_1}^T \Phi_{f_1}\, \omega\left(\frac{\tilde{x}_1}{\epsilon}\right) \tag{13.4.1}$$

$$\bar{\beta}_{f_1} = \Psi_{f_1}^T \bar{\Omega}_{f_1} = \Psi_{f_1}^T \Phi_{f_1}\, \omega\left(\frac{\left|\frac{\partial x_{2c}}{\partial x_1}\right| \tilde{x}_2}{\epsilon}\right) \tag{13.4.2}$$

$$\beta_{f_2} = \Psi_{f_2}^T \Omega_{f_2} = \Psi_{f_2}^T \Phi_{f_2}\, \omega\left(\frac{\tilde{x}_2}{\epsilon}\right) \tag{13.4.3}$$

$$\beta_{g_2} = \Psi_{g_2}^T \Omega_{g_2} = \Psi_{g_2}^T \Phi_{g_2}\, \omega\left(\frac{\tilde{x}_2 u_{a2}}{g_l \epsilon}\right) \tag{13.4.4}$$

where $\epsilon > 0$ is a small design constant and the function

$$\omega : \Re \mapsto \Re$$

satisfies Assumption 13.5.1 of Section 13.5 which states that

$$0 \leq |m| - m\omega\left(\frac{m}{\epsilon}\right) \leq \eta\epsilon$$

where η is a small constant. Note that each of the above β functions is smooth.

With this definition, starting from (13.3.13) and using (13.4.1)–(13.4.4), we can reduce the expression for Δ:

$$\Delta \leq (\Psi_{f_1}^M)^T \Phi_{f_1} |\tilde{x}_1| - (\Psi_{f_1}^M + \tilde{\Psi}_{f_1})^T \Phi_{f_1} \tilde{x}_1 \omega\left(\frac{\tilde{x}_1}{\epsilon}\right) + (\Psi_{f_1}^M)^T \Phi_{f_1} \left|\tilde{x}_2 \frac{\partial x_{2c}}{\partial x_1}\right|$$

$$- (\Psi_{f_1}^M + \tilde{\Psi}_{f_1})^T \Phi_{f_1} \tilde{x}_2 \left|\frac{\partial x_{2c}}{\partial x_1}\right| \omega\left(\frac{\left|\frac{\partial x_{2c}}{\partial x_1}\right| \tilde{x}_2}{\epsilon}\right) + (\Psi_{f_2}^M)^T \Phi_{f_2} |\tilde{x}_2|$$

$$- (\Psi_{f_2}^M + \tilde{\Psi}_{f_2})^T \Phi_{f_2} \tilde{x}_2 \omega\left(\frac{\tilde{x}_2}{\epsilon}\right) + (\Psi_{g_2}^M)^T \Phi_{g_2} |\tilde{x}_2 u|$$

$$- (\Psi_{g_2}^M + \tilde{\Psi}_{g_2})^T \Phi_{g_2} \tilde{x}_2 u \omega\left(\frac{\tilde{x}_2 u_{a2}}{g_l \epsilon}\right) + \sum_{i=1}^{2} \left(\tilde{\Psi}_{f_i}^T \Gamma_{\Psi f_i}^{-1} \dot{\tilde{\Psi}}_{f_i}\right) + \tilde{\Psi}_{g_2}^T \Gamma_{\Psi g_2}^{-1} \dot{\tilde{\Psi}}_{g_2}$$

$$\leq (\Psi_{f_1}^M)^T \Phi_{f_1} \left(|\tilde{x}_1| - \tilde{x}_1 \,\omega\left(\frac{\tilde{x}_1}{\epsilon}\right)\right)$$

$$+ (\Psi_{f_1}^M)^T \Phi_{f_1} \left(\left|\tilde{x}_2 \frac{\partial x_{2c}}{\partial x_1}\right| - \tilde{x}_2 \left|\frac{\partial x_{2c}}{\partial x_1}\right| \omega\left(\frac{\left|\frac{\partial x_{2c}}{\partial x_1}\right| \tilde{x}_2}{\epsilon}\right)\right)$$

$$+ \tilde{\Psi}_{f_1}^T \Gamma_{\Psi f_1}^{-1} \left[\dot{\tilde{\Psi}}_{f_1} - \Gamma_{\Psi f_1} \Phi_{f_1} \left(\tilde{x}_1 \omega\left(\frac{\tilde{x}_1}{\epsilon}\right) + \tilde{x}_2 \left|\frac{\partial x_{2c}}{\partial x_1}\right| \omega\left(\frac{\left|\frac{\partial x_{2c}}{\partial x_1}\right| \tilde{x}_2}{\epsilon}\right)\right)\right]$$

$$+ (\Psi_{f_2}^M)^T \Phi_{f_2} \left(|\tilde{x}_2| - \tilde{x}_2 \,\omega\left(\frac{\tilde{x}_2}{\epsilon}\right)\right)$$

$$+ \tilde{\Psi}_{f_2}^T \Gamma_{\Psi f_2}^{-1} \left(\dot{\tilde{\Psi}}_{f_2} - \Gamma_{\Psi f_2} \Phi_{f_2} \tilde{x}_2 \omega\left(\frac{\tilde{x}_2}{\epsilon}\right)\right)$$

$$+ (\Psi_{g_2}^M)^T \Phi_{g_2} \left(|\tilde{x}_2 u| - \tilde{x}_2 u \cdot \omega\left(\frac{\tilde{x}_2 u_{a2}}{g_l \epsilon}\right)\right)$$

$$+ \tilde{\Psi}_{g_2}^T \Gamma_{\Psi g_2}^{-1} \left(\dot{\tilde{\Psi}}_{g_2} - \Gamma_{\Psi g_2} \Phi_{g_2} \tilde{x}_2 u \cdot \omega\left(\frac{\tilde{x}_2 u_{a2}}{g_l \epsilon}\right)\right)$$

Assumption 13.5.1 can be extended as shown in Section 13.5 to provide the inequality†

$$\left| \tilde{x}_2 u \right| - \tilde{x}_2 u \cdot \omega \left(\frac{\tilde{x}_2 u_{a2}}{g_l \epsilon} \right) \leq \left| \frac{g_l u}{u_{a2}} \right| \eta \epsilon \leq \eta \epsilon ;$$

therefore,

$$\Delta \leq \eta \epsilon \left(\sum_{i=1}^{2} 2 (\Psi_{f_i}^M)^T \Phi_{f_i} + (\Psi_{g2}^M)^T \Phi_{g2} \right)$$

$$+ \tilde{\Psi}_{f_1}^T \Gamma_{\Psi f_1}^{-1} \left(\dot{\Psi}_{f_1} - \Gamma_{\Psi f_1} (\Omega_{f_1} \tilde{x}_1 + \bar{\Omega}_{f_1} \tilde{x}_2) \right) \tag{13.4.5}$$

$$+ \tilde{\Psi}_{f_2}^T \Gamma_{\Psi f_2}^{-1} \left(\dot{\Psi}_{f_2} - \Gamma_{\Psi f_2} \Omega_{f_2} \tilde{x}_2 \right) + \tilde{\Psi}_{g2}^T \Gamma_{\Psi g2}^{-1} \left(\dot{\Psi}_{g2} - \Gamma_{\Psi g2} \Omega_{g2} \tilde{x}_2 u \right) .$$

The following three subsection will each continue the analysis from this point while considering an alternative localized robust adaptation approach.

13.4.1 Projection with σ-Modification

To implement a localized σ-modification approach for $i = 1, 2$ design parameter vectors $\theta_{f_i}^0$, θ_{g2}^0, $\Psi_{f_i}^0$ and Ψ_{g2}^0 are selected and the Q terms in the parameter adaptation laws are defined as

$$Q_{f_i} = -\sigma_{f_i} R_{f_i} (\theta_{f_i} - \theta_{f_i}^0) \tag{13.4.6}$$

$$Q_{g2} = -\sigma_{g2} R_{g2} (\theta_{g2} - \theta_{g2}^0) \tag{13.4.7}$$

$$Q_{\Psi f_i} = -\sigma_{\Psi f_i} R_{f_i} (\Psi_{f_i} - \Psi_{f_i}^0) \tag{13.4.8}$$

$$Q_{\Psi g2} = -\sigma_{\Psi g2} R_{g2} (\Psi_{g2} - \Psi_{g2}^0). \tag{13.4.9}$$

These terms ensure that the adapted parameter estimates do not drift too far from the design parameter values. The matrices $R_{f_i} = \text{diag}(\Phi_{f_i}) \in \Re^{N \times N}$ and $R_{g2} = \text{diag}(\Phi_{g2}) \in \Re^{N \times N}$, where $\text{diag}(v)$ is the square diagonal matrix with diagonal components equal to the vector v. The standard σ-modification approach does not include the R matrices. The inclusion of the R matrices localizes the effect of the σ-modification for any specific parameter to the region of the operating envelope where that parameter actually affects the approximator.

After substituting the σ-modification terms of (13.4.6)–(13.4.9) into (13.3.6)–(13.3.11), (13.3.12) and (13.4.5), the derivative $\dot{V}$ becomes

$$\dot{V} = - k_1 \tilde{x}_1^2 - k_2 \tilde{x}_2^2 + d_0$$

$$- \sum_{i=1}^{2} \left(\sigma_{f_i} \tilde{\theta}_{f_i}^T R_{f_i} (\theta_{f_i} - \theta_{f_i}^0) + \sigma_{\Psi f_i} \tilde{\Psi}_{f_i}^T R_{f_i} (\Psi_{f_i} - \Psi_{f_i}^0) \right)$$

$$+ \sigma_{g2} \tilde{\theta}_{g2}^T R_{g2} (\theta_{g2} - \theta_{g2}^0) + \sigma_{\Psi g2} \tilde{\Psi}_{g2}^T R_{g2} (\Psi_{g2} - \Psi_{g2}^0) \tag{13.4.10}$$

---

†Note that u_{an} and u_{ad} have the same sign since the denominator of the control equation is ensured to be bounded away from zero such that $\theta_{gn}^T \Phi_{gn} + \beta_{gn} > g_l > 0$.

where

$$d_0 = \eta\epsilon\left(\sum_{i=1}^{2} 2(\Psi_{f_i}^M)^T \Phi_{f_i} + (\Psi_{g2}^M)^T \Phi_{g2}\right) \tag{13.4.11}$$

is a positive bounded scalar variable and is equal to the first two terms on the right of (13.4.5). In (13.4.10), for each term in lines two and three, we apply the inequality

$$-\sigma\tilde{a}^T R(a - a^0) = -\frac{\sigma}{2}\tilde{a}^T R\tilde{a} - \frac{\sigma}{2}(a - a^0)^T R(a - a^0)$$
$$+ \frac{\sigma}{2}(a^* - a^0)^T R(a^* - a^0)$$

with corresponding subscripts on σ and R; the vector a represents $\theta_{f_i}, \Psi_{f_i}, i = 1, 2$, and θ_{g2}, Ψ_{g2}, respectively. Thus, we have

$$\dot{V} = -k_1\tilde{x}_1^2 - k_2\tilde{x}_2^2 + d_0$$
$$+ \frac{1}{2}\sum_{i=1}^{2}\left(\sigma_{f_i}(\theta_{f_i}^* - \theta_{f_i}^0)^T R_{f_i}(\theta_{f_i}^* - \theta_{f_i}^0) + \sigma_{\Psi f_i}(\Psi_{f_i}^M - \Psi_{f_i}^0)^T R_{f_i}(\Psi_{f_i}^M - \Psi_{f_i}^0)\right)$$
$$+ \frac{1}{2}\sigma_{g2}(\theta_{g2}^* - \theta_{g2}^0)^T R_{g2}(\theta_{g2}^* - \theta_{g2}^0) + \frac{1}{2}\sigma_{\Psi g2}(\Psi_{g2}^M - \Psi_{g2}^0)^T R_{g2}(\Psi_{g2}^M - \Psi_{g2}^0)$$
$$\leq -\underline{k}\,\|\tilde{x}\|^2 + d_0 + \rho_1 \tag{13.4.12}$$

where ρ_1 is a bounded, positive scalar variable defined as

$$\rho_1 = \frac{1}{2}\sum_{i=1}^{2}\left(\sigma_{f_i}(\theta_{f_i}^* - \theta_{f_i}^0)^T R_{f_i}(\theta_{f_i}^* - \theta_{f_i}^0) + \sigma_{\Psi fi}(\Psi_{f_i}^M - \Psi_{f_i}^0)^T R_{f_i}(\Psi_{f_i}^M - \Psi_{f_i}^0)\right)$$
$$+ \frac{1}{2}\left(\sigma_{g2}(\theta_{g2}^* - \theta_{g2}^0)^T R_{g2}(\theta_{g2}^* - \theta_{g2}^0) + \sigma_{\Psi g2}(\Psi_{g2}^M - \Psi_{g2}^0)^T R_{g2}(\Psi_{g2}^M - \Psi_{g2}^0)\right). \tag{13.4.13}$$

Therefore, we can summarize these results in the following theorem.

Theorem 13.4.1 (Projection with σ-Modification). *For the second order system described by (13.2.1)–(13.2.2) with the adaptive feedback control law of (13.3.2)–(13.3.4) and the parameter adaptation laws of (13.3.6)–(13.3.11) with the modification terms defined in (13.4.6)–(13.4.9), we have the following stability properties:*

a) $\tilde{x}_i, \tilde{\theta}_{f_i}, \tilde{\theta}_{g2}, \tilde{\Psi}_{f2}, \tilde{\Psi}_{g2} \in \mathcal{L}_\infty, i = 1, 2;$

b) $x_i, \theta_{f_i}, \theta_{g2}, \Psi_{f_i}, \Psi_{g2} \in \mathcal{L}_\infty, i = 1, 2;$

c) $\dot{\tilde{x}}_i$, $\dot{\theta}_{f_i}$, $\dot{\theta}_{g_2}$, $\dot{\Psi}_{f_i}$, $\dot{\Psi}_{g_2} \in \mathcal{L}_\infty$, $i = 1, 2$;

d) $\tilde{x}$ *is small in the mean square sense, satisfying*

$$\int_t^{t+T} \|\tilde{x}(\tau)\|_2^2 d\tau \leq \frac{1}{\underline{k}} V(t) + \frac{1}{\underline{k}} \int_t^{t+T} (d_0(\tau) + \rho_1(\tau)) d\tau. \quad (13.4.14)$$

Proof. Since the case of $x \notin \mathcal{D}^2$ has already been considered, the following only considers the case of $x \in \mathcal{D}^2$.

From (13.4.12), $\dot{V}$ is negative definite whenever $\underline{k}\|\tilde{x}\|^2 > d_0 + \rho_1$. Using standard methods [3] for the analysis of adaptive laws with projection and σ-modification, we have that $\tilde{x}_i$, $\tilde{\theta}_{f_i}$, $\tilde{\theta}_{g_i}$, $\tilde{\Psi}_{f_i}$, $\tilde{\Psi}_{g_i} \in \mathcal{L}_\infty$. This yields directly x_i, θ_{f_i}, θ_{g_i}, Ψ_{f_i}, $\Psi_{g_i} \in \mathcal{L}_\infty$. Together with the boundedness of Φ_{f_i} and Φ_{g_i}, we can show directly that $u \in \mathcal{L}_\infty$ and then $\dot{\tilde{x}}_i$, $\dot{\theta}_{f_i}$, $\dot{\theta}_{g_i}$, $\dot{\Psi}_{f_i}$, $\dot{\Psi}_{g_i} \in \mathcal{L}_\infty$.

For the proof of 4 given by (13.4.14), we integrate both sides of (13.4.12) to obtain

$$V(t + T) - V(t) \leq \int_t^{t+T} (-\underline{k}\|\tilde{x}(\tau)\|^2 + d_0 + \rho_1) d\tau$$

$$\underline{k} \int_t^{t+T} \|\tilde{x}(\tau)\|^2 d\tau \leq V(t) + \int_t^{t+T} (d_0(\tau) + \rho_1(\tau)) d\tau$$

which directly yields (13.4.14). $\qquad\square$

The previously existing approach in the literature [10, 11] used parameter updates with the standard leakage modification, which did not include the term R_{f_i} and the term R_{g_i} as in (13.4.6)–(13.4.9). Consider the adaptive laws with standard σ-modification for θ_{f_i} as an example:

$$\dot{\theta}_{f_i} = \Gamma_{f_i} \left(e_i \Phi_{f_i} - \sigma_{f_i}(\theta_{f_i} - \theta_{f_i}^0) \right), \quad i = 1, 2. \quad (13.4.15)$$

The disadvantage of (13.4.15) is that when either e_i or $\Phi_{f_i,k}$ is zero, $\theta_{f_i,k}$ will converge toward $\theta_{f_i,k}^0$. Remember from Section 13.2 that $\Phi_{f_i,k}$ is only nonzero on $S_{f_i,k}$ which is small relative to $\mathcal{D}^i$. Therefore, $\theta_{f_i,k}$ only affects the f_i approximation on $S_{f_i,k}$. When $\tilde{x}_i \in S_{f_i,k}$, then $\theta_{f_i,k}$ will converge toward the set of values capable of yielding $\underline{k}\|\tilde{x}\|^2 < d_0 + \rho_1$ on $S_{f_i,k}$, but when $\tilde{x}_i \in \mathcal{D}^i - S_{f_i,k}$ it is the case that $\theta_{f_i,k} \to \theta_{f_i,k}^0$. This cause the approximated functions and bounds to lose their local accuracy outside $S_{f_i,k}$. When the state later returns to $S_{f_i,k}$, $\theta_{f_i,k}$ will need to be estimated again to satisfy the tracking bound. This is due to the forgetting caused by σ_{f_i} having a global effect. Similarly, for the other parameter estimates (i.e., θ_{g_2}, Ψ_{f_i}, Ψ_{g_2}), standard leakage terms will result in the problem of global forgetting.

Using the localized adaptive laws proposed in (13.3.6)–(13.3.11) with the modification terms as defined in (13.4.6)–(13.4.9), it is possible to eliminate the problem with global forgetting by localizing the effects of leakage terms to the vicinity of the present operating point. They also decrease the required amount of on-line computation, since at any time instant all parameters associated with zero elements of basis vectors are left unchanged. In addition, due to the inclusion in ρ_1 of R_{f_i} and R_{g_2}, which are local functions of the operating point, the m.s.s. bound can be shown to be significantly smaller than the bound derived from the previously existing approaches.

The σ-modification is an effective method to prevent the estimated parameters from increasing without bound; however, it does allow the estimated parameters to drift (toward $\theta_{f_i}^0$) in a bounded fashion when the $\tilde{x}_i$ is small. Also, the guaranteed $\mathcal{L}_\infty$ property of $\tilde{x}_i$ does not provide a useful bound on $\|\tilde{x}\|$ that the designer can directly influence. Note in particular, that the theorem does not imply that (ultimately) $\underline{k}\|\tilde{x}\|^2 < d_0 + \rho_1$.

13.4.2 Projection with $\tilde{x}$-Modification

As we state in Section 13.4, the main drawback of the fixed σ-modification is that it causes the parameter estimates to drift towards certain design values in the case of small tracking errors. An attempt to remove the drawback is the following localized adaptation laws with $\tilde{x}$-modification.

The localized $\tilde{x}$-modification terms are defined as,

$$Q_{f_i} = -\sigma_{f_i}\|\tilde{x}\|R_{f_i}(\theta_{f_i} - \theta_{f_i}^0) \tag{13.4.16}$$

$$Q_{g_2} = -\sigma_{g_2}\|\tilde{x}\|R_{g_2}(\theta_{g_2} - \theta_{g_2}^0) \tag{13.4.17}$$

$$Q_{\Psi f_i} = -\sigma_{\Psi f_i}\|\tilde{x}\|R_{f_i}(\Psi_{f_i} - \Psi_{f_i}^0) \tag{13.4.18}$$

$$Q_{\Psi g_2} = -\sigma_{\Psi g_2}\|\tilde{x}\|R_{g_2}(\Psi_{g_2} - \Psi_{g_2}^0) \tag{13.4.19}$$

for $i = 1, 2$. For comparison purpose, we choose all design parameters the same as in the σ-modification terms (13.4.6)–(13.4.9).

Substituting (13.3.6)–(13.3.11) with the $\tilde{x}$-modification terms from (13.4.16)–(13.4.19) into (13.3.12) and (13.4.5), the derivative $\dot{V}$ becomes

$$\begin{aligned}
\dot{V} = {}& -k_1\tilde{x}_1^2 - k_2\tilde{x}_2^2 + d_0 \\
& - \|\tilde{x}\|\sum_{i=1}^{2}\left(\sigma_{f_i}\tilde{\theta}_{f_i}^T R_{f_i}(\theta_{f_i} - \theta_{f_i}^0) + \sigma_{\Psi f_i}\tilde{\Psi}_{f_i}^T R_{f_i}(\Psi_{f_i} - \Psi_{f_i}^0)\right) \\
& - \|\tilde{x}\|\sigma_{g_2}\tilde{\theta}_{g_2}^T R_{g_2}(\theta_{g_2} - \theta_{g_2}^0) - \|\tilde{x}\|\sigma_{\Psi g_2}\tilde{\Psi}_{g_2}^T R_{g_2}(\Psi_{g_2} - \Psi_{g_2}^0) \\
\leq {}& -\underline{k}\|\tilde{x}\|_2^2 + d_0 + \|\tilde{x}\|\rho_1 \tag{13.4.20}
\end{aligned}$$

where ρ_1 is defined in (13.4.13).

In (13.4.20), using the inequality

$$pq \leq \alpha^2 p^2 + \frac{1}{4\alpha^2}q^2$$

with $\alpha^2 = \underline{k}/2$, we obtain

$$\dot{V} \leq -\frac{1}{2}\underline{k}\|\tilde{x}\|^2 + d_0 + \rho_2 \tag{13.4.21}$$

where ρ_2 is a positive constant given by

$$\rho_2 = \frac{1}{2\underline{k}}\rho_1^2.$$

Therefore, we can summarize these results in the following theorem.

Theorem 13.4.2 (Projection with $\tilde{x}$ Modification). *For the second order system described by (13.2.1)–(13.2.2) with the adaptive feedback control law of (13.3.2), (13.3.3), (13.3.4) and the parameter adaptation laws of (13.3.6)–(13.3.11) with the modification terms defined in (13.4.16)–(13.4.19), we have the following stability properties:*

a) $\tilde{x}_i, \tilde{\theta}_{f_i}, \tilde{\theta}_{g_i}, \tilde{\Psi}_{f_i}, \tilde{\Psi}_{g_i} \in \mathcal{L}_\infty$, $i = 1, 2$;

b) $x_i, \theta_{f_i}, \theta_{g_i}, \Psi_{f_i}, \Psi_{g_i} \in \mathcal{L}_\infty$, $i = 1, 2$;

c) $\dot{\tilde{x}}_i, \dot{\theta}_{f_i}, \dot{\theta}_{g_i}, \dot{\Psi}_{f_i}, \dot{\Psi}_{g_i} \in \mathcal{L}_\infty$, $i = 1, 2$;

d) $\tilde{x}$ is small in the mean square sense, satisfying

$$\int_t^{t+T} \|\tilde{x}(\tau)\|_2^2 d\tau \leq \frac{2}{\underline{k}}V(t) + \frac{2}{\underline{k}}\int_t^{t+T}(d_0(\tau) + \rho_2(\tau))d\tau. \tag{13.4.22}$$

Proof. Properties 1, 2, 3 of Theorem 13.4.2 are straightforward to show given the form of inequality (13.4.21). The proof is similar to the proof of Theorem 13.4.1 in Section 13.4 and will not be repeated here.

For the proof of 4 given by (13.4.22), we integrate both sides of (13.4.21) to obtain

$$V(t+T) - V(t) \leq \int_t^{t+T}\left(-\frac{1}{2}\underline{k}\|\tilde{x}(\tau)\|^2 + d_0 + \rho_2\right)d\tau,$$

or

$$\frac{1}{2}\underline{k}\int_t^{t+T}\|\tilde{x}(\tau)\|^2 d\tau \leq V(t) + \int_t^{t+T}(d_0(\tau) + \rho_2(\tau))d\tau,$$

which directly yields (13.4.22). $\square$

Similar comments about the localized forgetting apply as were stated in the previous section.

13.4.3 Projection with Deadzone

Another means to remove the issue of parameter drift is to include a deadzone in adaptive laws. Implementation of the deadzone requires knowledge of an assumed bound on certain terms as will be discussed below.

For the deadzone approach, the modification terms in (13.3.6)–(13.3.11) are defined as

$$
Q_{f_i} = \begin{cases} 0, & \text{for } \|\tilde{x}\|_2 > \sqrt{\frac{\bar{\rho}_3 + \mu}{k}}, \\ -\Phi_{f_i} e_i, & \text{otherwise,} \end{cases} \tag{13.4.23}
$$

$$
Q_{g_2} = \begin{cases} 0, & \text{for } \|\tilde{x}\|_2 > \sqrt{\frac{\bar{\rho}_3 + \mu}{k}}, \\ -\Phi_{g_2} e_2 u, & \text{for } \|\tilde{x}\|_2 \leq \sqrt{\frac{\bar{\rho}_3 + \mu}{k}}, \end{cases} \tag{13.4.24}
$$

$$
Q_{\Psi f_i} = \begin{cases} -\sigma_{\Psi f_i} R_{f_i}(\Psi_{f_i} - \Psi_{f_i}^0), & \text{for } \|\tilde{x}\|_2 > \sqrt{\frac{\bar{\rho}_3 + \mu}{k}}, \\ -\Omega_{f_1}\tilde{x}_1 - \bar{\Omega}_{f_1}\tilde{x}_2, & \text{for } \|\tilde{x}\|_2 \leq \sqrt{\frac{\bar{\rho}_3 + \mu}{k}} \text{ and } i = 1, \\ -\Omega_{f_2}\tilde{x}_2, & \text{for } \|\tilde{x}\|_2 \leq \sqrt{\frac{\bar{\rho}_3 + \mu}{k}} \text{ and } i = 2, \end{cases}
$$
$$\tag{13.4.25}$$

$$
Q_{\Psi g_2} = \begin{cases} -\sigma_{\Psi g_2} R_{g_2}(\Psi_{g_2} - \Psi_{g_2}^0), & \text{for } \|\tilde{x}\|_2 > \sqrt{\frac{\bar{\rho}_3 + \mu}{k}}, \\ -\Omega_{g_2}\tilde{x}_2 u, & \text{for } \|\tilde{x}\|_2 \leq \sqrt{\frac{\bar{\rho}_3 + \mu}{k}}, \end{cases} \tag{13.4.26}
$$

for $i = 1, 2$. The constant $\bar{\rho}_3 > 0$ is a known strict upper bound on $(d_0 + \rho_3)$, where d_0 is defined in (13.4.11) and

$$
\rho_3 = \frac{1}{2} \sum_{i=1}^{n} \left(\sigma_{\Psi f i}(\Psi_{f_i}^M - \Psi_{f_i}^0)^T R_{f_i}(\Psi_{f_i}^M - \Psi_{f_i}^0) \right)
$$
$$
+ \frac{1}{2} \sigma_{\Psi g2}(\Psi_{g_2}^M - \Psi_{g_2}^0)^T R_{g_2}(\Psi_{g_2}^M - \Psi_{g_2}^0).
$$

The deadzone is in effect for $\|\tilde{x}\|_2 \leq \sqrt{\frac{\bar{\rho}_3 + \mu}{k}}$ for some positive design constant $\mu > 0$. For $\|\tilde{x}\|_2 > \sqrt{\frac{\bar{\rho}_3 + \mu}{k}}$, the parameter adaptation laws of θ_{f_i}, $i = 1, 2$ and θ_{g_2} do not include any modification terms. When $\|\tilde{x}\|_2 \leq \sqrt{\frac{\bar{\rho}_3 + \mu}{k}}$, all parameter updates stop.

We are now ready to present the applicable stability theorem.

Theorem 13.4.3 (Projection with Deadzone). *Assuming the upper bound* $\bar{\rho}_3 >$ $d_0 + \rho_3 > 0$ *is known, the system described by (13.2.1)–(13.2.2) with the adaptive feedback control law of (13.3.2), (13.3.3), (13.3.4) and the parameter adaptation laws of (13.3.6)–(13.3.11) with the modification terms defined in (13.4.23)–(13.4.26), we have the following stability properties:*

a) $\tilde{x}_i, \tilde{\theta}_{f_i}, \tilde{\theta}_{g_2}, \tilde{\Psi}_{f_i}, \tilde{\Psi}_{g_2} \in \mathcal{L}_\infty, i = 1, 2;$

b) $x_i, \theta_{f_i}, \theta_{g_2}, \Psi_{f_i}, \Psi_{g_2} \in \mathcal{L}_\infty, i = 1, 2;$

c) $\dot{\tilde{x}}_i, \dot{\tilde{\theta}}_{f_i}, \dot{\tilde{\theta}}_{g_2}, \dot{\tilde{\Psi}}_{f_i}, \dot{\tilde{\Psi}}_{g_2} \in \mathcal{L}_\infty, i = 1, 2;$

d) $\tilde{x}$ *is small in the mean square sense, satisfying*

$$\int_t^{t+T} \|\tilde{x}(\tau)\|_2^2 d\tau \leq \frac{1}{\underline{k}} V(t) + \frac{1}{\underline{k}} \int_t^{t+T} (d_0(\tau) + \rho_3(\tau)) d\tau. \quad (13.4.27)$$

e) $\|\tilde{x}\|_2$ *is ultimately bounded by* $\epsilon_d = \sqrt{\frac{\bar{\rho}_3 + \mu}{\underline{k}}}$, *i.e., the total time for* $\|\tilde{x}\|_2 > \epsilon_d$ *is finite.*

Proof. Substituting (13.3.6)–(13.3.11) with the modification terms defined as in Equations (13.4.23)–(13.4.26) in (13.3.12) and (13.4.5), for $x \in \mathcal{D}^2$ and $\|\tilde{x}\|_2 > \sqrt{\frac{\bar{\rho}_3 + \mu}{\underline{k}}}$, $\dot{V}$ is written as

$$\begin{aligned} \dot{V} &\leq -\underline{k}\|\tilde{x}\|_2^2 + d_0 + \rho_3 \\ &\leq -\underline{k}\|\tilde{x}\|_2^2 + \bar{\rho}_3 \leq -\mu < 0. \end{aligned} \quad (13.4.28)$$

Therefore, if $\|\tilde{x}\|_2 > \sqrt{\frac{\bar{\rho}_3 + \mu}{\underline{k}}}$, then V is decreasing. If $\|\tilde{x}\|_2 \leq \sqrt{\frac{\bar{\rho}_3 + \mu}{\underline{k}}}$ then $\tilde{\theta}_{f_i}, \tilde{\theta}_{g_2}, \tilde{\Psi}_{f_i}$ and $\tilde{\Psi}_{g_2}$ are all constant and $\|\tilde{x}\|_2$ is bounded. Thus, $V(t)$ is bounded by the maximum of $V(0)$ or

$$\max_{\|\tilde{x}\|_2 = \sqrt{\frac{\bar{\rho}_3 + \mu}{\underline{k}}}} \left(V(\tilde{x}, \tilde{\theta}_{f_i}(0), \tilde{\theta}_{g_2}(0), \tilde{\Psi}_{f_i}(0), \tilde{\Psi}_{g_2}(0)) \right)$$

which shows that $\tilde{x}_i, \tilde{\theta}_{f_i}, \tilde{\theta}_{g_2}, \tilde{\Psi}_{f_i}, \tilde{\Psi}_{g_2} \in \mathcal{L}_\infty$. Properties 2, 3 can be similarly shown.

For the proof of 4, we integrate (13.4.28) to obtain

$$\underline{k} \int_t^{t+T} \|\tilde{x}(\tau)\|_2^2 d\tau \leq V(t) + \int_t^{t+T} (d_0(\tau) + \rho_3(\tau)) d\tau$$

which yields (13.4.27) and then implies $\tilde{x}$ is small in the mean square sense (m.s.s.).

Next, we will show the Property 5. Assume x starts at t_0 outside the deadzone, enters the deadzone at t_{2i-1}, and leaves it at t_{2i}, for $i \geq 1$. Then, during the interval $t \in [t_{2i-1}, t_{2i}]$, there is no parameter update; $\|\tilde{x}(t_{2i-1})\| = \|\tilde{x}(t_{2i})\|$, thus

$$V(t_{2i-1}) = V(t_{2i}),$$

and outside the deadzone according to (13.4.28)

$$V(t_{2i+1}) - V(t_{2i}) < -\mu(t_{2i+1} - t_{2i}).$$

Therefore, the total time outside the deadzone is

$$T_d = (t_1 - t_0) + \sum_{i \geq 1}(t_{2i+1} - t_{2i}),$$

and

$$T_d < \frac{1}{\mu}\left(V(t_0) - V(t_1) + \sum_{i \geq 1}(V(t_{2i}) - V(t_{2i+1}))\right)$$

$$< \frac{1}{\mu}\left(V(t_0) - V(t_1) + \sum_{i \geq 1}(V(t_{2i-1}) - V(t_{2i+1}))\right)$$

$$< \frac{V(t_0)}{\mu}$$

which is a finite value.   □

Assuming that the design constant $\bar{\rho}_3 > 0$ is a strict upper bound on $(d_0 + \rho_3)$, the ultimate bounds has a useful form that allows the designer to either increase $\bar{k}$ or decrease μ or $\bar{\rho}_3$ to decrease the ultimate bound on the tracking error.

A disadvantage of deadzone modification is that the implementation of the deadzone requires knowledge of $(d_0 + \rho_3)$ or the upper bound on it over the whole region $\mathcal{D}^2$. If $\mathcal{D}^2$ is relatively large, the upper bound $\bar{\rho}_3$ can be conservative, which may result in a large deadzone.

13.4.4 Discussions

We have considered robust adaptation laws with localized forgetting using three different types of modification, i.e., deadzone, σ and $\tilde{x}$ modifications. In [4, 10], the adaptive bounding approach was introduced using the standard leakage approach that has global features. To remove the issue of global forgetting, localized leakage based adaptation was proposed in [15] for simple scalar systems. This chapter has extended those results in various directions: second order systems, nonconstant g_1 unknown g_2, and alternative robust estimation approaches.

Note that the bounds of (13.4.14), (13.4.22), and (13.4.27) although they look similar are subtly different. The $V(t)$ term of the bound of (13.4.27) is nonincreasing, since the Lyapunov function in the case of the deadzone can never increase. The $V(t)$ term in the right hand side of either (13.4.14) or (13.4.22) can increase (in a bounded fashion), since the Lyapunov functions in the leakeage approaches can increase while the tracking error is suitably small.

Due to the fact that $m\,w(m/\epsilon) \geq 0$. If there were no modification terms (i.e., the Q terms were zero), then $\dot{\Psi}_{f_i}$ and $\dot{\Psi}_{g_2}$ would always be non-negative. This implies that for Ψ_{f_i} and Ψ_{g_2} will only increase, possibly leading to high magnitude switching. Therefore, the updates for bounding parameters Ψ_{f_i}, Ψ_{g_2} use the localized σ-modification or $\tilde{x}$-modification.

13.5 Supplementary Information

This section contains certain items that clarify the discussion of the main body of the chapter, but that would also interrupt the flow of that discussion.

13.5.1 A Useful Bound

Assumption 13.5.1. The scalar function $-1 \leq w(z) \leq 1$ satisfies

$$0 \leq |z| - zw\left(\frac{z}{\epsilon}\right) \leq \eta\epsilon \quad \forall z \in \Re, \tag{13.5.1}$$

for any $\epsilon > 0$ for some constant $0 < \epsilon < \infty$. □

Example 13.5.1. Consider $w(z) = \tanh(z)$ as considered in [10]. In this case, the constant η satisfies $\eta = e^{-(\eta+1)}$, i.e., $\eta = 0.2785$. □

Example 13.5.2. Consider

$$w(z) = \text{sat}(z) = \begin{cases} 1, & \text{for } z \geq 1 \\ z, & \text{for } |z| \leq 1 \\ -1, & \text{for } z \leq -1. \end{cases}$$

Then

$$|z| - zw\left(\frac{z}{\epsilon}\right) = \begin{cases} 0, & \text{for } z \geq \epsilon \\ |z| - \frac{z^2}{\epsilon}, & \text{for } |z| \leq \epsilon. \end{cases}$$

Therefore, $|z| - z\,\text{sat}\,(z/\epsilon)$ satisfies Assumption 13.5.1 with $\eta = 0.25$. □

The inequality (13.5.1) in Assumption 13.5.1 can be extended to another form needed in Section 13.4. The bound

$$0 \leq |z| - zw\left(\frac{z_a}{\epsilon}\right) \leq \left|\frac{z}{z_a}\right|\eta\epsilon \tag{13.5.2}$$

will be used to address the case of $z_a \neq z$, but $\text{sign}(z_a) = \text{sign}(z)$.

The inequality (13.5.2) can be shown by multiplying both sides of

$$0 \leq |z_a| - z_a \omega \left(\frac{z_a}{\epsilon}\right) \leq \eta\epsilon$$

by $|z/z_a|$ such that

$$0 \leq |z_a| \left|\frac{z}{z_a}\right| - z_a \left|\frac{z}{z_a}\right| \omega \left(\frac{z_a}{\epsilon}\right) \leq \left|\frac{z}{z_a}\right| \eta\epsilon.$$

Then, we have

$$0 \leq |z| - \text{sign}(z_a)|z_a| \left|\frac{z}{z_a}\right| \omega \left(\frac{z_a}{\epsilon}\right) \leq \left|\frac{z}{z_a}\right| \eta\epsilon$$

$$0 \leq |z| - \text{sign}(z)|z| \omega \left(\frac{z_a}{\epsilon}\right) \leq \left|\frac{z}{z_a}\right| \eta\epsilon$$

which completes the proof.

13.5.2 Adaptation Law with Parameter Projection

The objective of this subsection is to derive convex sets within which the parameter updates of θ_{g2} and Ψ_{g2} can be constrained to ensure that $\hat{g}_2 + \beta_{g2}$ satisfies the controllability condition of Assumption 13.2.1.

For controllability, we must have $\hat{g}_2 + \beta_{g2} > g_l$ which is the same as

$$\theta_{g2}^T \Phi_{g2} + \Psi_{g2}^T \Omega_{g2} > g_l. \tag{13.5.3}$$

From the definition of Ω_{g2} following (13.3.11), we can show

$$-\Psi_{g2}^T \Phi_{g2} \leq \Psi_{g2}^T \Omega_{g2} \leq \Psi_{g2}^T \Phi_{g2},$$

due to the fact that $-1 \leq \omega(\cdot) \leq 1$. Therefore, it is easy to see that inequality (13.5.3) will hold $\forall x \in \mathcal{D}$ if and only if

$$\theta_{g2}^T \Phi_{g2}(x) - \Psi_{g2}^T \Phi_{g2}(x) > g_l, \ \forall x \in \mathcal{D}. \tag{13.5.4}$$

Given that $\sum_{j=1}^N \Phi_{g2,j} \geq 1$, we can easily show that condition (13.5.4) is satisfied if and only if

$$\theta_{g2,j} > g_l, \ 0 \leq \Psi_{g2,j} < \theta_{g2,j} - g_l, \ \forall x \in \mathcal{D} \text{ and } \forall j = 1, \cdots, N \tag{13.5.5}$$

is satisfied. The reason why we prefer to use condition (13.5.5) instead of condition (13.5.4) is that the condition (13.5.5) defines a convex set within which the projection modification can be easily applied.

Then, we use the following parameter projection to constraint the parameter updates

$$P_C\{\dot{\theta}_{g2,j}\} = \begin{cases} \dot{\theta}_{g2,j}, & \text{if } \theta_{g2,j} > g_l \text{ or } \dot{\theta}_{g2,j} > 0, \\ 0, & \text{otherwise}, \end{cases} \tag{13.5.6}$$

$$P_C\{\dot{\Psi}_{g2,j}\} = \begin{cases} \dot{\Psi}_{g2,j}, & \text{if } \left(\left[(\Psi_{g2,j} < \theta_{g2,j} - g_l) \text{ or } \left(\dot{\Psi}_{g2,j} < 0\right)\right]\right. \\ & \left. \text{or } \left[(0 < \Psi_{g2,j}) \text{ or } \left(\dot{\Psi}_{g2,j} > 0\right)\right]\right), \\ 0, & \text{otherwise}. \end{cases}$$
$$\tag{13.5.7}$$

13.6   Numerical Example

We consider for illustrative purpose the second order system given by

$$\dot{x}_1 = \sin(x_1) + x_2 \tag{13.6.1}$$
$$\dot{x}_2 = \sin(x_1)\sin(x_2) + (2 + \sin(x_1)\cos(x_2))u. \tag{13.6.2}$$

Note that (13.6.1)–(13.6.2) are in the form of (13.2.1)–(13.2.2) with $f_1(x_1) = \sin(x_1)$, $f_2(x_1, x_2) = \sin(x_1)\sin(x_2)$, $g_2(x_1, x_2) = 2 + \sin(x_1)\cos(x_2)$. The functions f_1, f_2, and g_2 are all assumed to be unknown during the design stage. Each of these functions will be estimated during operation.

The reference trajectory $x_d(t)$ is generated as the output of a third order, unity DC gain, low pass filter with the transfer function of

$$\frac{x_d(s)}{r(s)} = \frac{64}{(s+4)^3}.$$

By designing a stable low pass prefilter, we attain the continuous, bounded signals of $\dot{x}_d(t)$ and $\ddot{x}_d(t)$ that will be used in the computation of $x_{2c}(t)$ and $u(t)$, respectively. Theoretically, $r(t)$ can be any bounded signal. For the purpose of this simulation $r(t)$ is selected to be the sum of $2\sin(0.04\pi t)$ and a 0.5 Hz square wave with very small magnitude of 0.0005.

The basis functions for the approximation are the biquadratic kernels of the form as

$$\phi_k(x) = \begin{cases} \left(1 - \left(\frac{||x - c_k||}{\mu_k}\right)^2\right)^2, & \text{if } ||x - c_k|| < \mu_k \\ 0, & \text{otherwise}. \end{cases} \tag{13.6.3}$$

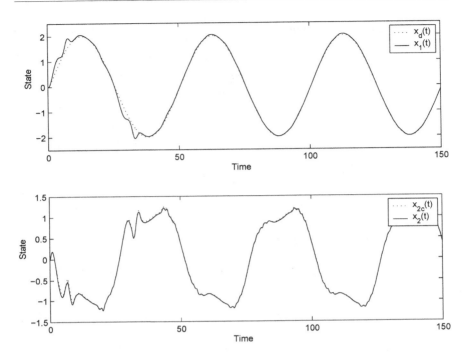

Figure 13.6.1: (a) The state $x_1(t)$ and its reference trajectory $x_d(t)$. (b) The state $x_2(t)$ and its command signal $x_{2c}(t)$.

where c_k is the center location of the k-th basis function and μ_k is a constant which represents the radius of the region of support. For both state x_1 and x_2, the centers are allocated 0.3 units apart with $\mu_k = 0.5$. The operating range for each state $\mathcal{A}_1 = \mathcal{A}_2 = [-3, 3]$.

The control gains are selected to be $k_1 = 1$ and $k_2 = 5$ to ensure that the x_2 dynamic in the inner loop has the faster tracking performance than the x_1 dynamic in the outer loop. The adaptation rate matrices are set to be $\Gamma_{f_1} = 1I_{N_1}$, $\Gamma_{\Psi f_1} = 0.1\Gamma_{f_1}$, $\Gamma_{f_2} = \Gamma_{g_2} = 10I_{N_1 N_2}$ and $\Gamma_{\Psi f_2} = \Gamma_{\Psi g_2} = 0.1\Gamma_{f_2}$ where I_N is the identity matrix in $\Re^N$. In this example, all approximator parameter estimates and bounding parameter estimates are updated based on the localized adaptation laws with σ-modifications. All σ terms are selected to be $1/30$.

Fig. 13.6.1 plots the state outputs $x_1(t)$ and $x_2(t)$ (solid lines) and their command signals (dotted lines). Fig. 13.6.2 shows the tracking errors $\tilde{x}_1$ and $\tilde{x}_2$ that are achieved by using the control law designed in (13.3.2) and parameter adaptation laws given by (13.3.6)–(13.3.11) with σ-modification terms of (13.4.6)–

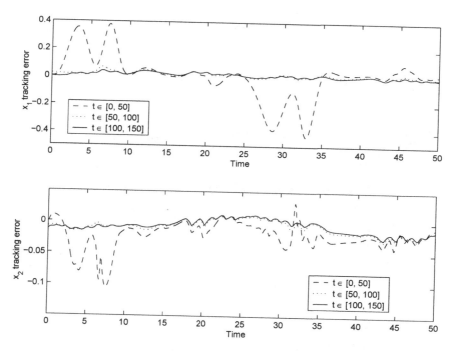

Figure 13.6.2: Tracking errors performance. The dashed lines are the tracking errors in the time interval of $[0, 50]$. The dotted lines are the tracking errors in the time interval of $[50, 100]$. The solid lines are the tracking errors in the time interval of $[100, 150]$.

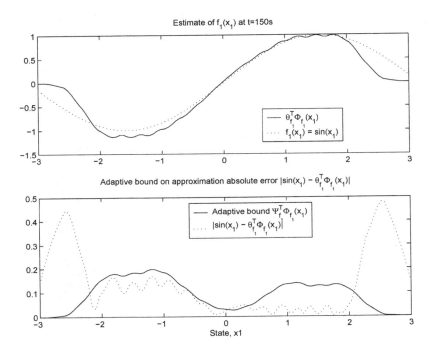

Figure 13.6.3: (a) The actual function $f_1(x_1) = \sin(x_1)$ and the approximation to f_1 after 150s. (b) The approximator absolute error $|\sin(x_1) - \theta_{f_1}^T \Phi_{f_1}(x_1)|$ and the adaptive bound obtained after 150s.

(13.4.9). Since the period of the reference input r is 50s, to facilitate comparison and increase graphical resolution, we have plotted the tracking errors for $t \in [0, 50]$ (dashed line), $t \in [50, 100]$ (dotted line) and $t \in [100, 150]$ (solid line) along the same time axis. It is shown that as experience is accumulated in a given region of state space, tracking performance improves and the improvement is maintained when the region is revisited in the future. It is particularly important to note that the learning is a function of state, not of trajectory as is the case for iterative learning control approaches. For approaches such as those presented herein, performance improvement will extend to different trajectories to the extent that the new trajectories utilize the same regions in state space.

Fig. 13.6.3 provides additional information about the learning abilities of the localized adaptation algorithms. Fig. 13.6.3(a) shows the actual f_1 function compared to the approximation $\hat{f}_1$ at $t = 150$s. It can be clearly seen that $\hat{f}_1$ accurately matches the true function over a portion of the operation range $\mathcal{A}_1$, i.e., $[-2, 2] \subset \mathcal{A}_1$, which contains a large amount of training data. Outside this portion of $\mathcal{A}_1$, $\hat{f}_1$ stays near its initial approximation (i.e., zero). In addition, as shown in Fig. 13.6.3(b), the estimated bound $\Psi_{f_1}^T \Phi_{f_1}$ obtained at $t = 150$s has almost the same shape as approximator absolute errors $|f_1(x_1) - \theta_{f_1}^T \Phi_{f_1}|$ for $x_1 \in [-2, 2]$. Alternatively, for $x_1 \in \mathcal{A}_1 - [-2, 2]$, there is very little learning experience and the corresponding elements of Ψ_{f_1} are unchanged from the initial values (i.e., zero). The f_2, g_2 and corresponding bounding functions are not included as their two dimensional domain complicates their presentation.

13.7 Conclusions

This chapter presents robust adaptive control design methodologies and analysis applicable to uncertain, triangular, second order nonlinear systems. The robust adaptive backstepping design procedure incorporates local learning and localized adaptive bounding functions on the *residual approximation errors*. This is an extension of the algorithms using globalized forgetting in [4, 10, 11] and of those using localized forgetting in [15]. The design methodology can also be extended to $n > 2$ order systems by following the recursive steps of the backstepping procedure described in [5].

The Lyapunov stability analysis is presented herein for three different approaches incorporating projection: σ-modification, deadzone, and $\tilde{x}$-modification. The $\mathcal{L}_\infty$ property of tracking errors $\tilde{x}_i$ and parameter estimation errors, and the m.s.s. boundedness of $\tilde{x}$ is established for all three modification methods. Also, for the deadzone approach, it is shown that the time outside the deadzone is finite.

A numerical simulation is provided to demonstrate the effectiveness of the proposed method.

Acknowledgements

This material is based upon work supported by the National Science Foundation under Grant No. ECS-0322635. Any opinions, findings, and conclusions or recommendations expressed in this material are those of the author(s) and do not necessarily reflect the views of the National Science Foundation.

Bibliography

[1] F.-C. Chen and H. K. Khalil, "Adaptive control of a class of nonlinear discrete-time systems using neural networks," *IEEE Trans. on Automatic Control*, vol. 40, no. 5, pp. 791–801, May 1995.

[2] J. Y. Choi and J. A. Farrell, "Nonlinear adaptive control using networks of piecewise linear approximators," *IEEE Trans. on Neural Networks*, vol. 11, no. 2, pp. 390–401, 2000.

[3] P. A. Ioannou and J. Sun, *Robust Adaptive Control*, Upper Saddle River, NJ: Prentice-Hall, 1996.

[4] J.-P. Jiang and L. Praly, "Design of robust adaptive controllers for nonlinear systems with dynamic uncertainties," *Automatica*, vol. 34, no. 7, pp. 825–840, 1998.

[5] H. Khalil, *Nonlinear Systems*, Upper Saddle River, NJ: Prentice Hall, 2002.

[6] M. Krstic, I. Kanellakopoulos, and P. Kokotovic, *Nonlinear and Adaptive Control Design*, New York: Wiley, 1995.

[7] F. L. Lewis, K. Liu, and A. Yesildirek, "Neual net robot control with guaranteed tracking performance," *IEEE Trans. on Neural Networks*, vol. 6, no. 3, pp. 703–715, 1995.

[8] F. L. Lewis, A. Yesildirek, and K. Liu, "Multilayer neural-net robot controller with guaranteed tracking performance," *IEEE Trans. on Neural Networks*, vol. 7, no. 2, pp. 388–399, 1996.

[9] R. Ordonez and K. M. Passino, "Indirect adaptive control for a class of non-linear systems with a time-varying structure," *International Journal of Control*, vol. 74, no. 7, pp. 701–717, 2001.

[10] M. Polycarpou, "Stable adaptive neural control scheme for nonlinear systems," *IEEE Trans. on Automatic Control*, vol. 41, no. 3, pp. 447–451, 1996.

[11] M. Polycarpou and P. A. Ioannou, "A robust adaptive nonlinear control design," *Automatica*, vol. 32, no. 3, pp. 423–427, 1996.

[12] M. Polycarpou and M. Mears, "Stable adaptive tracking of uncertain systems using nonlinearly parameterized on-line approximators," *International Journal of Control*, vol. 70, no. 3, pp. 363–384, 1998.

[13] R. Sanner and J. Slotine, "Gaussian networks for direct adaptive control," *IEEE Trans. on Neural Networks*, vol. 3, pp. 837–863, 1992.

[14] N. Sureshbabu and J. A. Farrell, "Wavelet based system identification for nonlinear control applications," *IEEE Trans. on Automatic Control*, vol. 44, no. 2, pp. 412–417, 1999.

[15] Y. Zhao, J. A. Farrell, and M. Polycarpou, "Localized adaptive bounds for on-line approximation based control," *Proc. 2004 American Control Conference*, Boston, MA, June 2004, pp. 590–595.

Chapter 14

Dynamically Connected Fuzzy Single Input Rule Modules and Application to Underactuated Systems

Jianqiang Yi, Naoyoshi Yubazaki, and Kaoru Hirota

Abstract: To solve the problems of conventional fuzzy inference, SIRMs (single input rule modules) dynamically connected fuzzy inference model is proposed in this chapter. For each input item, an SIRM is constructed and a dynamic importance degree is defined. The dynamic importance degree consists of a base value insuring the role of the input item throughout a control process, and a dynamic value changing with control situations to adjust the dynamic importance degree. Each dynamic value can easily be tuned based on the local information of current state. The model output is obtained by summarizing the products of the dynamic importance degree and the fuzzy inference result of each SIRM. The SIRMs process the input items dispersedly and the dynamic importance degrees express the control priority orders definitely. The present model is applied to several underactuated systems such as truck-trailer system, ball-beam system, and parallel-type double inverted pendulum system. The controller design approaches are given in detail. The simulation results indicate that the proposed model is effective even for very complex systems.

14.1   Introduction

In the conventional IF-THEN fuzzy inference model [16], all the input items of systems are usually put into the antecedent part of each fuzzy rule. Consequently,

413

both the numbers of fuzzy rules and parameters increase exponentially, and defining fuzzy rules becomes extremely difficult for large-scale systems. Furthermore, each input item in the conventional IF-THEN fuzzy rules has the same weight, making it inadequate for real applications. To solve the problem, SIRMs (single input rule modules) with fixed importance degree connected fuzzy inference model (SIRMs fixed model for short) [38, 39] was proposed, which constructed a SIRM and defined a fixed importance degree for each input item. When applied to function identification, the SIRMs fixed model sharply reduced the number of both the fuzzy rules and parameters. When applied to control systems such as first-order lag plants and second-order lag plants, it also achieved some improvements in control performance compared with the conventional fuzzy inference model. However, the control performance was difficult to be further improved due to the fact that its importance degrees reflecting the influence of the input items on system performance was fixed throughout the whole control process including the rise stage and the settling stage. To obtain even better performance, the importance degrees should be dynamically tuned with continuously changing control situations.

Several researches have studied dynamic adjustment of fuzzy controllers. The multi-unit controller [15] divided state space into several parts, each part is assigned a fuzzy rule module. The time-varying fuzzy control method [19] changed the weight of singleton-type output in the consequent part or switches fuzzy rule modules from one to another according to time schedule. The dynamic fuzzy control method [37] prepared several fuzzy rule modules for different control situations and then used a time series concept to unite all modules. For each of the above methods, however, all the modules must be constructed and the system performance throughout the whole control process must be well understood in advance.

Furthermore, Chen *et al.* [5] described an adaptive fuzzy controller to simplify the membership function construction and the rule development by genetic algorithm. Godjevac and Steele [7] presented an adaptive fuzzy controller for robot navigation, in which the membership functions are tuned by a supervised learning procedure. Margaliot and Langholz [17] designed an adaptive fuzzy controller via fuzzy Lyapunov synthesis, where the fuzzy rules are adapted based on the error between the desired and actual behaviors of the Lyapunov function. These approaches determined automatically the parameters of the membership functions or the fuzzy rules; however, all the parameters were fixed after adaptation. To stabilize an inverted pendulum system [8, 30, 33, 40], for example, the pendulum angular control should have priority over the cart position control when the pendulum is not balanced yet, and the control priority orders should be changed with control situations. Because the conventional fuzzy inference model was adopted and the input items were all treated equally, the adaptive fuzzy controllers have difficulties dealing with such a control problem.

In this chapter, SIRMs dynamically connected fuzzy inference model [34] is proposed. For each input item, a dynamic importance degree is given and a SIRM is constructed. The dynamic importance degree is defined as the sum of a base value insuring the role of the input item throughout a control process, and a dynamic value varying with control situations to control the influence of the input item. The dynamic value can be tuned in real-time just based on the local information of current state. The SIRMs process the input items dispersedly and the dynamic importance degrees express the control priority orders definitely. The model output is obtained by summarizing the products of the dynamic importance degree and the fuzzy inference result of each SIRM. The proposed model has a simple structure and is suitable for complex systems. To show its effectiveness, the proposed model is applied to several underactuated nonlinear systems such truck-trailer system, ball-beam system, and parallel-type double inverted pendulum system. The design approaches of the fuzzy controllers based on the proposed model are given in detail, and control simulation results are shown.

This chapter is organized as follows. Section 14.2 explains the SIRMs dynamically connected fuzzy inference model in detail. Section 14.3 applies the proposed model to the backing-up control of a truck-trailer system. Section 14.4 uses the proposed model to the stabilization control of a ball-beam system. Section 14.5 adopts the proposed model to the stabilization control of a parallel-type double inverted pendulum system. Finally, Section 14.6 provides some concluding remarks.

14.2 SIRMs Dynamically Connected Fuzzy Inference Model

A system with n input items and 1 output item is considered here for convenience. However, this can easily be extended to systems with multiple output items.

Since there are n input items, totally n SIRMs are constructed first as

$$
\left.
\begin{aligned}
\text{SIRM-1:} \quad & \left\{ R_1^j : \ \text{IF } x_1 = A_1^j \ \text{THEN } u_1 = C_1^j \right\}_{j=1}^{m_1} \\
& \vdots \\
\text{SIRM-}i: \quad & \left\{ R_i^j : \ \text{IF } x_i = A_i^j \ \text{THEN } u_i = C_i^j \right\}_{j=1}^{m_i} \\
& \vdots \\
\text{SIRM-}n: \quad & \left\{ R_n^j : \ \text{IF } x_n = A_n^j \ \text{THEN } u_n = C_n^j \right\}_{j=1}^{m_n}
\end{aligned}
\right\}
$$

Each SIRM corresponds to one of the n input items. Here, SIRM-i means the

SIRM corresponding to the ith input item, and R_i^j is the jth rule in the SIRM-i. x_i in the antecedent part of the SIRM-i is the variable corresponding to the ith input item, and u_i in the consequent part of the SIRM-i is an intermediate variable of the output item u. A_i^j and C_i^j are the membership functions of x_i and u_i in the jth rule of the SIRM-i. Furthermore, $i = 1, 2, \cdots, n$ is the index number of the SIRMs, and $j = 1, 2, \cdots, m_i$ is the index number of the rules in the SIRM-i.

If the observed value of the input item x_i ($i = 1, 2, \cdots, n$) is given as x_i^0, the agreement of the antecedent part of the jth rule in the SIRM-i simply becomes $A_i^j(x_i^0)$. Supposing the universe of discourse of the output item u is U, the inference result u_i^0 of the SIRM-i can be conducted by

$$
u_i^0 = \frac{\displaystyle\int_U \left[\bigoplus_{j=1}^{m_i} \left(A_i^j(x_i^0) \otimes C_i^j \right) \right] u \, du}{\displaystyle\int_U \left[\bigoplus_{j=1}^{m_i} \left(A_i^j(x_i^0) \otimes C_i^j \right) \right] du}.
$$

In the above, $\otimes$ means the operator between the agreement and the membership function of the consequent part such as 'min' or 'product'. On the other hand, $\oplus$ means the operator on all the fuzzy rules in the same SIRM, and 'max' or 'sum' is the typical one. Therefore, the well-known min-max-gravity method [18], product-sum-gravity method [18, 20], and simplified inference method [20, 22] can be used here.

Usually, each input item is considered to take an unequal role in control performance. Among the input items, some may contribute significantly to control performance, while the contribution of others is relatively small. Some input items may improve control performance more if their roles are strengthened, while others may not have a positive influence on control performance if emphasized. Hence, assigning larger weights to such input items that positively contribute to the control performance, and assigning at the same time smaller weights to the other input items to restrict their roles, would be in accordance with experts' experience and would improve the total control performance.

In the SIRMs fixed mode [38, 39], an importance degree was introduced for each input item to explicitly indicate the different role of each input item. Because of the introduction of the importance degrees, each input item can be directly managed and can realize a different function with each other. As a result, control performance was improved compared with that by the conventional fuzzy inference model.

However, because the importance degrees in the SIRMs fixed model were fixed throughout the whole control process, the same importance degrees had to be used even at entirely different situations like the rise stage and the settling stage of a first-order plant. Although this led to simple structure for the fuzzy controller, further improvement in control performance becomes difficult. For instance, take the first-order lag plant into consideration. Suppose that the output error and its change are selected as the input items. As well known, the rise time and overshoot are two main indexes to evaluate control performance. The rise time is contradictory to overshoot. If the rise time is to be shortened, then the value of the importance degree of the output error should be increased. However, it will cause large overshoot because of the fixed importance degree. On the other hand, if the importance degree takes a large value at the beginning of the control, and gets smaller gradually when the plant output approaches the desired value, then short rise time and small overshoot can both be achieved. This means that the importance degrees should be dynamically adjusted according to control situations in order to obtain better performance.

Here, a dynamic importance degree is introduced for each input item. Given a base value w_i^0, a dynamic variable Δw_i and a breadth B_i for input item i ($i = 1, 2, \cdots, n$), the dynamic importance degree w_i^D of the input item i is defined by $w_i^D = w_i^0 + B_i \Delta w_i^0$ as the sum of the base value w_i^0 and the dynamic value $B \Delta w_i^0$, where Δw_i^0 is the actual output value of the dynamic variable Δw_i. The base value insures the basic function of the input item throughout a control process. The dynamic variable, varying in $[0.0, \ 1.0]$, reflects control situations and should be dynamically tuned in real time. The breadth then sets an upper limit to the range within which the dynamic value varies. By tuning the dynamic variable according to control situations, the dynamic value will adjust the corresponding dynamic importance degree between $[w_i^0, \ w_i^0 + B_i]$.

After the value of every dynamic importance degree is determined in real-time according to control situations, the SIRMs dynamically connected fuzzy inference model obtains its output by summarizing the products of the dynamic importance degree and the fuzzy inference result of each SIRM for all the input items as $u^0 = \sum_{i=1}^{n} w_i^D u_i^0$.

Since the importance degrees are related linearly with the model output, those input items with larger importance degrees will contribute more significantly to the system, and those with smaller importance degrees will influence less the system. Moreover, since the relation among the input items defined by the importance degrees is linearly mapped to the model output, it is possible to achieve specified control purpose by adjusting intuitively the importance degrees. This resembles the setup of the coefficients in PID control; however, the dynamic importance degrees usually are nonlinear functions of the corresponding input items.

Because the consequent variable of each SIRM corresponds to the same output item and the product of the importance degree and the fuzzy inference result of each SIRM is a part of the model output, the importance degree can be regarded as a part of the scaling factor of the output item. Therefore, dynamically tuning the importance degrees means adjusting the scaling factor of the output item according to control situations. Although several methods [14, 15, 36] of tuning the scaling factors of fuzzy controllers were presented, each of the methods was to repeatedly modify the scaling factors based on a global performance criterion after a certain length of time of control until almost optimal result was obtained. The scaling factors during tuning were different from those after tuning; however, they were in fact fixed during each control trial. On the contrary, since each importance degree in the proposed model is supposed to have a different value with each other, the variable in the consequent part of each SIRM essentially has a different scaling factor in spite that it corresponds to the same output item. In other words, the output item actually holds a different scaling factor in different SIRM. Also, because the importance degrees are adjusted according to control situations, the scaling factor of each SIRM changes automatically with control situations. Moreover, because tuning a dynamic importance degree only needs the local information of current state, veritable real-time tuning can easily be realized.

If the breadths of the importance degrees are all set to zero, the proposed model is essentially equivalent to the SIRMs fixed model. Therefore, the SIRMs fixed model can be considered as a special case of the model proposed here.

To validate the effectiveness of the SIRMs dynamically connected fuzzy inference model, the control problems of several underactuated systems are taken into consideration next.

14.3   Backing-Up Control of Truck-Trailer System

The truck-trailer system [21] considered here is depicted in Figure 14.3.1. The truck is only allowed to move backward at a constant small speed v. Parameters l and L are respectively the truck length and the trailer length. α is the relative angle of the truck with the trailer, β is the angle of the trailer from x-axis, and θ is the steering angle of the front wheels of the truck. All the three angles are measured counterclockwise. (x, y) is the position of the rear center of the trailer. α, β, x, y can be considered as four state variables of the system, and θ is the manipulated variable. Suppose that the system will not become slippery on any wheel, and the sampling period is T. Then, the state variable equations [31] of the truck-trailer system at $k + 1$ sampling step is given in discrete-time format by

$$\alpha_{k+1} = \alpha_k + \frac{vT}{l} \tan \theta_k - \frac{vT}{L} \sin \alpha_k \qquad (14.3.1)$$

$$\beta_{k+1} = \beta_k + \frac{vT}{L} \sin \alpha_k \qquad (14.3.2)$$

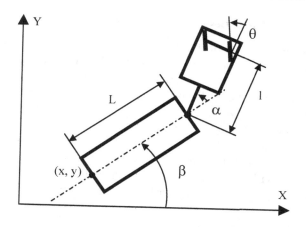

Figure 14.3.1: Truck-trailer system configuration.

$$y_{k+1} = y_k + vT \cos \alpha_k \sin \left(\beta_k + \frac{vT}{2L} \sin \alpha_k \right) \qquad (14.3.3)$$

$$x_{k+1} = x_k + vT \cos \alpha_k \cos \left(\beta_k + \frac{vT}{2L} \sin \alpha_k \right) \qquad (14.3.4)$$

The truck-trailer system is often used as a benchmark for new control approaches. Besides nonlinearity and nonholonomy, the system is also underactuated. Furthermore, when the relative angle of the truck with the trailer becomes right angle, the system falls into the so-called jackknife state, which is difficult to control. Therefore, backing-up control of the truck-trailer system is a challenging problem. Alexander *et al.* [1] applied a three-neuron controller to the truck-trailer system and backed up the system successfully if the initial relative angle was small. Altafini *et al.* [2] designed a feedback control scheme including two switching surfaces and switching logic for the linearized model. Chen and Zhang [4] succeeded in backing up the truck-trailer system along a suboptimal distance trajectory under the condition that the relative angle was less than $45°$. Tanaka and Sano [25] presented a model-based fuzzy controller, which realized backward movement control of the truck-trailer system even from a jackknife state. Yeh and Chen [29] designed a multi-state fuzzy controller for backing up the truck-trailer system, but the fuzzy rules were difficult to set up.

In this section, the SIRMs dynamically connected fuzzy inference model is used for backing-up control of the truck-trailer system.

14.3.1 Fuzzy Controller Design

The backing-up control of the truck-trailer system is to move backward the truck-trailer system from a given initial state and eventually back up the truck-trailer system along the x-axis. Therefore, the relative angle α, the trailer angle β, and the vertical position y all have to be controlled to zero, while it is not necessary to control the horizontal position x. Here, the relative angle α_k, the trailer angle β_k, and the position y_k at k sampling step are selected as the input items x_1, x_2, x_3 after normalized by the scaling factors $45°$, $45°$, 10 m, respectively. As the output item, the steering angle θ_k at k sampling step is then obtained by following fuzzy controller based on the SIRMs dynamically connected fuzzy inference model. Therefore, the fuzzy controller includes three SIRMs and three dynamic importance degrees.

Suppose that the trailer angle and the vertical position both are zero. Note that the moving speed v is negative in the backing-up control. If the relative angle is positively big, it is clear from (14.3.1) that positive steering angle will make the relative angle small. Therefore, we can set up the SIRM for the relative angle in Table 14.3.1. The membership functions NB, ZO, PB are defined in Figure 14.3.2.

Table 14.3.1: SIRM setting for the relative angle

Antecedent x_1	Consequent u_1
NB	−1
ZO	0
PB	1

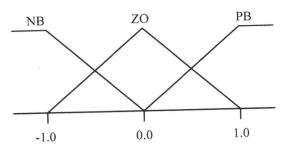

Figure 14.3.2: Membership functions for the SIRMs.

Table 14.3.2: SIRM setting for the trailer angle.

Antecedent x_2	Consequent u_2
NB	1
ZO	0
PB	-1

Table 14.3.3: SIRM setting for the vertical position

Antecedent x_3	Consequent u_3
NB	-1
ZO	0
PB	1

The SIRM for the trailer angle is shown in Table 14.3.2. Assume that the relative angle and the vertical position both are zero. If the trailer angle is positively big, negative steering angle will cause the relative angle to become positive according to (14.3.1). According to (14.3.2), then the trailer angle tends to get small.

Table 14.3.3 shows the SIRM for the vertical position. Suppose that the relative angle and the trailer angle both are zero. If the vertical position is positively big, then from (14.3.1) positive steering angle will make the relative angle negative. From (14.3.2), the trailer angle becomes positive for negative relative angle. Consequently, the vertical position will become small according to (14.3.3).

As pointed out in [1], the relative angle should not be too big. If the relative angle is too big, the backing-up control will become difficult. In particular, if the truck-trailer system falls into a jackknife state, the control may be impossible. Therefore, the relative angle should take control priority over the trailer angle and the vertical position in the backing-up control. Here, the dynamic importance degrees are used to reflect the control priority orders.

The fuzzy rules for the dynamic variable of the relative angle are listed in Table 14.3.4. The absolute value of the normalized relative angle is chosen as the antecedent variable. The membership functions DS, DM, DB are defined in Figure 14.3.3. When the relative angle is big, the inference result of the dynamic variable is also big. As a result, the dynamic importance degree of the relative angle becomes big so that the relative angle has the highest control priority. If the relative angle is small, the relative angle will lose its control priority because the dynamic

Table 14.3.4: Rule setting for the dynamic variable of the relative angle

Antecedent $|x_1|$	Consequent Δw_1
DS	0.0
DM	0.5
DB	1.0

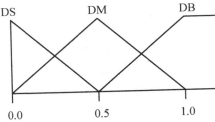

Figure 14.3.3: Membership functions for the dynamic variables.

importance degree gets small and the inference result of the corresponding SIRM is small.

Table 14.3.5 shows the fuzzy rules for the dynamic variable of the trailer angle and the dynamic variable of the vertical position. The absolute value of the normalized relative angle is also used as the antecedent variable of this table. Apparently, the inference results of the two dynamic variables are almost 0 if the relative angle is big enough. The corresponding dynamic importance degrees then approximately equal to their base values, and the trailer angle and the vertical position have lower control priority. If the relative angle is very small, the inference results of the two dynamic variables become almost 1. The corresponding dynamic importance degrees then become relatively big. Consequently, the trailer angle or the positive position will take control priority.

Table 14.3.5: Rule setting for the dynamic variables of the trailer angle and the vertical position

Antecedent $|x_1|$	Consequent Δw_2, Δw_3
DS	1.0
DM	0.5
DB	0.0

Furthermore, the control parameters, i.e., the base value and the breadth, are tuned by the random optimization search method [3]. Suppose that the truck length is $l = 2.8$ m, the trailer length is $L = 5.5$ m, the moving speed is $v = -1.0$ m/sec, and the maximum steering angle is $45°$. In each trial of the random search, sampling period and total control time are fixed to 1 sec and 250 secs. The initial state (the initial values of the relative angle, the trailer angle, the vertical position, and the horizontal position) is set to $(10°, 0°, 10$ m, 80 m$)$. The base values and the breadths are all set to 0 initially. The random search is carried out for $10,000$ trials along such a direction that the total summation of the absolute values of the input items and the output item normalized by the maximum steering angle at each sampling step from the beginning to the end of each trial is reduced.

After the random search, the base values of the dynamic importance degrees become $71.3979, 33.0414, 22.6380$, and their breadths become $19.0578, 8.0592, 7.1929$, respectively. It is found then that the two control parameters of the relative angle are the largest, and the two control parameters of the trailer angle are larger than the two control parameters of the vertical position. Thus, the control parameters together with Tables 14.3.1–14.3.5 guarantees that the relative angle takes control priority over the other two when the relative angle is big. It is also guaranteed that the trailer angle has control priority over the vertical position when the relative angle is small and the trailer angle is big.

14.3.2 Control Simulation Results

To verify the effectiveness of the proposed fuzzy controller, simulations of the backing up control of the truck-trailer system are performed. In the following control simulations, the sampling period is set to 2 secs.

Figure 14.3.4 shows a control result, where the initial state of the truck-trailer system is $(0°, 0°, 20$ m, 80 m$)$. The truck and the trailer are sketched every two sampling steps. Since the initial vertical position is positively big, at first the vertical position takes control priority and the steering angle becomes positive. Accordingly, the relative angle gets negative and the trailer angle gets positive. Then, the truck-trailer system is smoothly backed up to the x-axis and has such a state as $(0.02°, 0.06°, 0.01$ m, -0.64 m$)$ at 100 secs.

In Figure 14.3.5, the control result starting from $(0°, 180°, 20$ m, 60 m$)$ is illustrated. Because the initial trailer angle is big, the trailer angle has control priority over the vertical position and the steering angle takes negative value at the beginning. Then, the relative angle becomes positive and the trailer angle decreases. Since the steering angle is rather small, the truck-trailer system changes its direction slowly. At 100 secs, the truck-trailer system reaches to $(-1.65°, -1.74°, -0.15$ m, 46.40 m$)$.

The control result for the initial state $(-45°, 90°, -10$ m, 80 m$)$ is depicted in

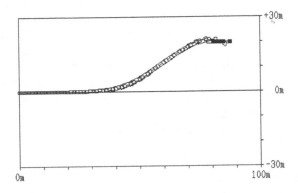

Figure 14.3.4: Control simulation result from $(0°, 0°, 20$ m, 80 m$)$.

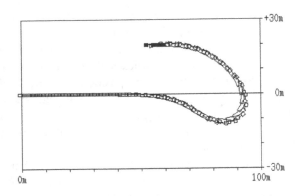

Figure 14.3.5: Control simulation result from $(0°, 180°, 20$ m, 60 m$)$.

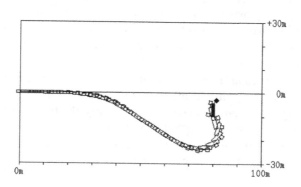

Figure 14.3.6: Control simulation result from $(-45°, 90°, -10$ m, 80 m$)$.

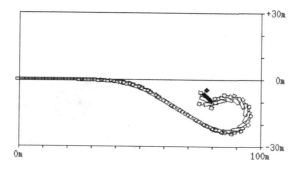

Figure 14.3.7: Control simulation result from $(-90°, 135°, -10$ m, 80 m$)$.

Figure 14.3.6. Both the initial relative angle and the initial trailer angle are in the saturation domain of the membership functions. Because the control parameters of the relative angle are larger than those of the trailer angle, the relative angle takes control priority over the trailer angle and the fuzzy controller first generates negative steering angle. As a result, the relative angle changes to positive and the trailer angle increases a little bit. The state of the truck-trailer system at 100 secs becomes $(0.00°, -0.07°, -0.02$ m, 2.38 m$)$.

The backing-up control of the truck-trailer system starting from a jackknife state $(-90°, 135°, -10$ m, 80 m$)$ is demonstrated in Figure 14.3.7. Because the relative angle is negative and has control priority, the fuzzy controller outputs negative value to the steering angle at first. Then the relative angle increases positively from $-90°$, and the truck-trailer system gets rid of the jackknife situation. At 100 secs, the truck-trailer system is controlled to $(-2.34°, -2.72°, -0.25$ m, 32.43 m$)$.

14.4   Stabilization Control of Ball-Beam System

Figure 14.4.1 shows the ball-beam system [9]. The beam can rotate in a vertical plane around its center, and the ball can roll along the beam. The ball position x is measured from the beam center, and is positive if the ball is on the right side of the beam. The beam angle θ from the horizontal line is positive if the beam rotates counterclockwise. The torque τ applied to the beam center is positive when it is counterclockwise.

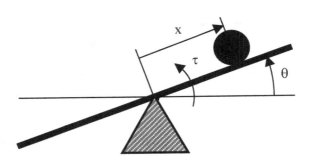

Figure 14.4.1: Ball-beam system configuration.

Suppose that no friction exists in the system, and the ball mass and the beam mass both are uniformly distributed. Then the state variable equation of the ball-beam system is given by [32]

$$
\begin{bmatrix} \dot{x}_1 \\ \dot{x}_2 \\ \dot{x}_3 \\ \dot{x}_4 \end{bmatrix} = \begin{bmatrix} x_2 \\ 0.7143(x_1 x_4^2 - 9.81 \sin x_3) \\ x_4 \\ 0 \end{bmatrix} + \begin{bmatrix} 0 \\ 0 \\ 0 \\ 1 \end{bmatrix} u \qquad (14.4.1)
$$

where the state variables x_1, x_2, x_3, x_4 correspond respectively to the ball position x, ball velocity $\dot{x}$, beam angle θ, and beam angular velocity $\dot{\theta}$. The manipulated variable u is related with the torque by

$$
\tau = 2M x_1 x_2 x_4 + 9.81 M x_1 \cos x_3 + (J_1 + J_2 + M x_1^2)u \qquad (14.4.2)
$$

where M is the ball mass, J_1 is the ball inertia moment, and J_2 is the beam inertia moment.

From (14.4.1), the state variable equation is apparently a nonlinear system, and the manipulated variable equals to the beam angular acceleration. Because the system parameters like the ball mass, the beam mass and the beam length do not appear directly in (14.4.1), they have no influence on the state variable equation. From (14.4.2), however, the system parameters affect the magnitude of the torque to be applied to the beam. If the magnitude of the torque has no upper limit, the system response from the same initial state will be identical for all the ball-beam systems with different system parameters. Because the magnitude of the torque is limited in real applications, the maximum of the manipulated variable should not be too big.

To control the ball-beam system, Joo and Lee [11] applied a hierarchical fuzzy control scheme to the ball-beam system to reduce the number of fuzzy rules. Their scheme used the Takagi-Sugeno (T-S) fuzzy model and many parameters had to be tuned. Lin and Lin [13] presented a reinforcement learning approach to tune fuzzy rules in a network, but the control result of the ball-beam system using the fuzzy rules showed continuous vibration. Shen *et al.* [23] took the position of the ball as a disturbance to the angle of the beam, and designed a decoupled fuzzy controller for the stabilization problem. Taniguchi [26] and Zhang [41] also adopted the T-S model for both the system and the controller. Although the ball-beam system was controlled well, complicated calculation had to be done. Wang [27] described a stable fuzzy control scheme and realized the control of the ball-beam system by 625 fuzzy rules. However, the control scheme was based on linear system and needed to solve two-boundary differential equations. Wang *et al.* [28] presented a feedforward neural network controller, which was tuned by genetic algorithm, to realize the stabilization control.

Table 14.4.1: SIRM setting for the ball

Antecedent x_i $(i = 1, 2)$	Consequent u_i $(i = 1, 2)$
NB	−1
ZO	0
PB	1

Table 14.4.2: SIRM setting for the beam

Antecedent x_1 $(i = 3, 4)$	Consequent u_1 $(i = 3, 4)$
NB	1
ZO	0
PB	−1

Here, the SIRMs dynamically connected fuzzy inference model is adopted for stabilization control of the ball-beam system.

14.4.1 Fuzzy Controller Design

Stabilization control of the ball-beam system is to obtain a proper manipulated variable so that the ball converges to the beam center from arbitrary initial states. Since this requires that the beam is balanced horizontally, the four state variables have to be controlled all to zero. Here, the four state variables after normalization are chosen as the input items x_i $(i = 1, 2, 3, 4)$, and the manipulated variable u without normalization is directly used as the output item. The scaling factors of the four input items are fixed to 1.0 m, 1.0 m/sec, 45.0°, and 100.0°/sec, respectively. The stabilization fuzzy controller for the ball-beam system based on the SIRMs dynamically connected fuzzy inference model is built as follows.

Table 14.4.1 shows the SIRMs for the two input items of the ball, i.e., the ball position and the ball velocity. Equally distributed membership functions NB, ZO, PB are defined the same as in Figure 14.3.2. Suppose that the beam is balanced horizontally. If the ball position is positive, positive torque will make the beam rotate counterclockwise such that the ball rolls toward negative direction. If the ball position is negative, negative torque will make the beam rotate clockwise such that the ball rolls toward positive direction. As a result, the ball moves toward beam center.

Table 14.4.3: Rule setting for the dynamic variables of the ball

Consequent $\Delta w_1, \Delta w_2$		Antecedent $|x_1|$		
		DS	DM	DB
Antecedent $|x_3|$	DS	0.0	0.5	1.0
	DM	0.0	0.0	0.5
	DB	0.0	0.0	0.0

Table 14.4.4: Rule setting for the dynamic variables of the beam

Antecedent $|x_3|$	Consequent $\Delta w_3, \Delta w_4$
DS	0.0
DM	0.5
DB	1.0

Table 14.4.2 gives the SIRMs for the two input items of the beam, i.e., the beam angle and the beam angular velocity. If the beam has a positive angle, the ball will roll toward negative direction. To make the ball converge to the beam center, the beam angle should be reduced. To realize this, negative torque is applied to the beam so that the beam rotates clockwise. If the beam angle becomes negative, then applying positive torque makes the beam rotate counterclockwise. Consequently, the beam tends to get balanced horizontally.

As known from intuition, when the beam angle is big, the ball will move rapidly and the stabilization control of the whole system will become difficult if the beam is not balanced immediately. In this case, emphasizing the ball position control according to Table 14.4.1 may cause the beam angle to get even bigger. To stabilize the whole system efficiently, therefore, the beam angular control should take priority over the ball position control if the beam is not balanced yet. After the beam is almost balanced, the ball position control is permitted.

Table 14.4.3 shows the fuzzy rules for the dynamic variables of the ball position and the ball velocity. The absolute values of the normalized ball position and the normalized beam angle are selected as the antecedent variables. The membership functions DS, DM, DB are defined in Figure 14.3.3. If the beam angle is big, the two dynamic variables will have small values and the corresponding dynamic importance degrees will be small. However, if the beam angle is small and the ball position is big, the two dynamic variables will take big values and the corresponding dynamic importance degrees will become big.

Table 14.4.4 shows the fuzzy rules for the dynamic variables of the beam angle and the beam angular velocity. Only the absolute value of the normalized beam angle is chosen as the antecedent variable. If the beam angle is big, the two dynamic variables will be big and the corresponding dynamic importance degrees will become significant. If the beam angle is small, the two dynamic variables will be small and the corresponding dynamic importance degrees will equal almost to their base values.

The random optimization search method [3] is also used to tune automatically the control parameters, i.e., the base value and the breadth. In each trial, sampling period and total control time are, respectively, fixed to 0.01 secs and 25.0 secs. The ball initial position is set to 0.25 m, while the initial values of the other state variables are all set to 0. The base values and the breadths are all initially set to 0. The random optimization search is run for 5000 trials along such a direction that the total summation of the absolute values of all the state variables at each sampling time from the control beginning to the end of each trial is reduced. After the random search, the four base values of the dynamic importance degrees respectively become 4.6552, 6.9418, 23.3023, 15.3286, and the four breadths of the dynamic importance degrees are 6.9432, 7.0228, 5.1883, 6.4724, respectively.

It is clear from the obtained control parameters that the sum of the base value and the breadth of the beam angle or the beam angular velocity is much larger than that of either of the ball position and the ball velocity. This together with Table 14.4.3 and Table 14.4.4 guarantees that the beam angular control takes priority over the ball position control when the beam angle is big. If the beam is almost balanced, the outputs of the two SIRMs of the beam become almost zero. Then if the ball does not converge to the beam center yet, the contribution from the ball position control in the manipulated variable will exceed that from the beam angular control. As a result, the ball position control becomes possible. By automatically switching the ball position control and the beam angular control, the stabilization control is realized.

14.4.2 Control Simulation Results

Here, several control simulations are performed to demonstrate the performance of the fuzzy controller.

Figure 14.4.2 shows the time responses of the ball position and the beam angle of the ball-beam system. The ball initial position and the beam initial angle are 0.5 m and 30.0°, while the other two initial values are zero. Because the beam initial angle is big, negative manipulated variable is first generated to rotate the beam clockwise. To prevent the ball from rolling left out of the beam, the beam further rotates clockwise and the beam angle becomes negative. Then the beam rotates back and the ball moves back toward the beam center. After 4.35 secs from the beginning of control, the four state variables converge respectively to 0.01 m,

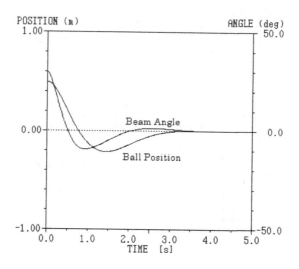

Figure 14.4.2: Stabilization control of Example 1.

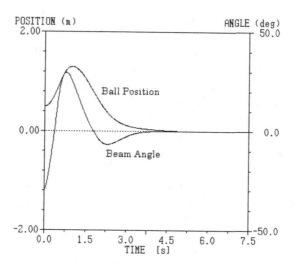

Figure 14.4.3: Stabilization control of Example 2.

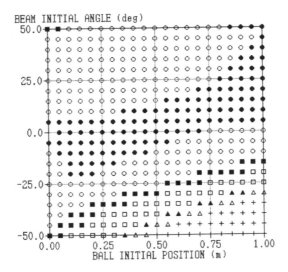

Figure 14.4.4: Stabilization control domain.

0.01 m/sec, 0.1° and 0.1°/sec. The maximum of the manipulated variable is about 15.0.

Figure 14.4.3 shows another control result, where the ball initial position and the beam initial angle are 0.5 m and −30.0°. Apparently, the initial state is more severe because the ball has a tendency to roll right further at the beginning. From Figure 14.4.3, it can be seen that the ball moves outside to about 1.4 m even though the beam is initially rotated counterclockwise. Then i similar to Figure 14.4.2, the beam is rotated clockwise and the ball rolls toward negative direction. At 5.93 secs, the four state variables become to 0.01 m, 0.01 m/sec, 0.1° and 0.1°/sec. The maximum of the manipulated variable is about 20.0.

Figure 14.4.4 shows the stabilization time for the fuzzy controller to stabilize the ball-beam system from different initial states. The ball initial position is shown every 0.05 m from 0.00 m to 1.00 m, and the beam initial angle is selected every 5.0° from −45° to 45°. In the figure, •, ○, ■, □, ▲, △, + denote, respectively, the cases where the whole system is stabilized within 4.0 secs, 5.0 secs, 6.0 secs, 7.0 secs, 8.0 secs, 9.0 secs, and over 9.0 secs for the given initial state. It can be seen from Figure 14.4.4 that the fuzzy controller can stabilize the ball-beam system within 9.0 secs for almost all the initial states.

14.5   Stabilization Control of Parallel-Type Double Inverted Pendulum System

In this section, a more complex underactuated system, the parallel-type double inverted pendulum system, is studied.

As shown in Figure 14.5.1, the parallel-type double inverted pendulum system consists of a horizontal rail, a cart moving on the rail, a longer pendulum 1 hinged on the right side of the cart, a shorter pendulum 2 hinged on the left side of the cart, and a driving unit. In the same vertical plane with the rail, the two pendulums can rotate freely around their own pivots.

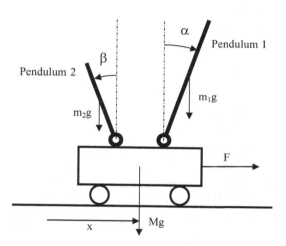

Figure 14.5.1: Parallel-type double inverted pendulum system configuration.

Here, the parameters M, m_1, m_2 (kg) are the masses of the cart, the longer pendulum 1, and the shorter pendulum 2, respectively. The parameter $g = 9.8$ m/sec^2 is the gravity acceleration. Suppose that the mass of each pendulum is distributed uniformly. The length from the gravity center of the longer pendulum 1 to its pivot is given as l_1 (m), which equals to half the length of the longer pendulum 1. The length from the gravity center of the shorter pendulum 2 to its pivot is given as l_2 (m), which equals to half the length of the shorter pendulum 2.

The position of the cart from the rail is denoted as x, and is positive when the cart locates on the right side of the origin. The angles of the longer pendulum 1 and the shorter pendulum 2 from their upright positions are denoted as α and β, and clockwise direction is positive. The driving force applied horizontally to the cart is denoted by F (N), and right direction is positive. Suppose that no friction exists in the pendulum system. Then the dynamic equation [35] of such a parallel-type double inverted pendulum system can be obtained by Lagrange's equation of motion as

$$\begin{cases} a_{11}\ddot{x} + a_{12}\ddot{\alpha} + a_{13}\ddot{\beta} = b_1 \\ a_{21}\ddot{x} + a_{22}\ddot{\alpha} = b_2 \\ a_{31}\ddot{x} + a_{33}\ddot{\beta} = b_3 \end{cases} \tag{14.5.1}$$

where the coefficients are given by

$$\begin{cases} a_{11} = M + m_1 + m_2 \\ a_{12} = m_1 l_1 \cos\alpha \\ a_{13} = m_2 l_2 \cos\beta \\ a_{21} = a_{12} \\ a_{22} = \frac{4}{3} m_1 l_1^2 \\ a_{31} = a_{13} \\ a_{33} = \frac{4}{3} m_2 l_2^2 \end{cases} \tag{14.5.2}$$

$$\begin{cases} b_1 = F + m_1 l_1 \dot{\alpha}^2 \sin\alpha + m_2 l_2 \dot{\beta}^2 \sin\beta \\ b_2 = m_1 l_1 g \sin\alpha \\ b_3 = m_2 l_2 g \sin\beta \end{cases} \tag{14.5.3}$$

For such a parallel-type double inverted pendulum system, stabilization control is impossible if the two pendulums have the same natural frequency. Since the natural frequency of a pendulum depends on the length of the pendulum, it is necessary for the two pendulums of a parallel-type double inverted pendulum system to have different lengths. Furthermore, since the cart moving affects the two pendulums directly, the shorter pendulum with a higher natural frequency tends to respond intensively and is liable to fall down. Therefore, it is known that stabilization control of a parallel-type double inverted pendulum system is the most difficult among the inverted pendulum family [12].

Till now, several approaches have been studied for the stabilization control of parallel type double inverted pendulum system. Based on the singleton-type reasoning method and genetic algorithm, Fujita and Mizumoto [6] constructed a 4-input 1-output fuzzy controller only for balancing the two pendulums of a parallel-type double inverted pendulum system. Since the position control of the cart was not taken into consideration, a limitless rail was necessary to keep the two pendulums upright. Hori et al. [10] studied the stabilization problem by using the T-S model. Kawatani and Yamaguchi [12] first linearized the nonlinear mathematical model of a parallel-type double inverted pendulum system, and then designed a stabilization controller by a state feedback gain and full state observer. Although the controller worked well for small initial angles, stabilization was not guaranteed when the initial angles of the two pendulums become a little bigger. Sugie and Okada [24] derived the linearized mathematical model of a circular parallel-type double inverted pendulum system, and created a stabilization controller based on H_∞ loop shaping design procedure. Besides complicated mathematical analysis, however, stabilization results by the controller showed a lasting vibration with amplitude of about $3.0°$. Zheng *et al.* [42] first trained each single inverted pendulum independently by reinforcement learning approach, and then used a hierarchical scheme to obtain synthetic stabilization controller for the parallel-type double inverted pendulum system.

In this section, a new fuzzy controller for stabilizing a parallel-type double inverted pendulum system is presented based on the SIRMs dynamically connected fuzzy inference model.

14.5.1 Fuzzy Controller Design

Without losing generality, the rail origin is selected as the desired position of the cart. Then, the stabilization control of the parallel-type double inverted pendulum system is to balance the two pendulums upright and move the cart to the rail origin in short time.

The angle α and angular velocity $\dot{\alpha}$ of the longer pendulum 1, the angle β and angular velocity $\dot{\beta}$ of the shorter pendulum 2, the position x and velocity $\dot{x}$ of the cart normalized by their own scaling factors are selected in this order as the input items ($i = 1, 2, \cdots, 6$). The driving force F normalized by its scaling factor is chosen as the output item f. If the six input items all converge to zero, then the stabilization control is apparently achieved. Here, a new fuzzy controller with the six input items and the output item for stabilizing the parallel-type double inverted pendulum system is constructed based on the SIRMs dynamically connected fuzzy inference model.

In case of positive big values of the angle and angular velocity of the longer pendulum 1, if positive driving force is added to move the cart toward right direction, the longer pendulum 1 will rotate counterclockwise toward upright position. Because of its higher natural frequency, however, the shorter pendulum 2 will rotate counterclockwise quickly, causing the angle and angular velocity of the shorter pendulum 2 to become negative. If the angle of the shorter pendulum 2 is negative, the cart has to be moved toward left direction by negative driving force as will be discussed below. If negative driving force is applied, however, the longer pendulum 1 will change its rotation direction to clockwise and its angle and angular velocity will become positive again. As a result, the longer pendulum 1 will repeat rotating clockwise and counterclockwise. Therefore, in order to move the longer pendulum 1 in the upright position in this case, it is necessary to move the cart toward left direction first by negative driving force. Although the longer pendulum 1 will then fall down clockwise further, the shorter pendulum 2 will

Table 14.5.1: SIRM setting for the longer pendulum 1

Antecedent x_i ($i = 1, 2$)	Consequent f_i ($i = 1, 2$)
NB	1
ZO	0
PB	-1

also rotate clockwise faster enough for its angle and angular velocity to become positively larger than those of the longer pendulum 1. Then by moving the cart toward right direction, the two pendulums rotate counterclockwise with synchronization kept and are balanced both to their upright positions. Similarly, if the angle and angular velocity of the longer pendulum 1 are negative big, the cart has to be moved toward right position first by positive driving force. Therefore, the SIRMs of the two input items x_1 and x_2 corresponding to the angle and angular velocity of the longer pendulum 1 can be set as in Table 14.5.1.

When the angle and angular velocity of the shorter pendulum 2 are positive big, the shorter pendulum 2 will fall down clockwise increasingly at its angular velocity if no control action is done at once. If the cart is moved toward left direction, the shorter pendulum 2 will rotate clockwise with acceleration so that the angle of the shorter pendulum 2 gets even bigger and moving upright the shorter pendulum 2 becomes impossible. To balance the shorter pendulum 2 upright in this case, therefore, it is necessary to move the cart toward right direction by positive driving force so that the shorter pendulum 2 will rotate counterclockwise. In the same way, if the angle and angular velocity of the shorter pendulum 2 are negative big, the cart has to be moved toward left direction by negative driving force. Because the longer pendulum 1 has a lower natural frequency, it will rotate at an angular velocity basically smaller than that of the shorter pendulum 2 during this period. The shorter pendulum 2 rotates faster because of its higher natural frequency and becomes synchronized with the longer pendulum 1. Then the two pendulums rotate toward the same direction and get balanced upright. Therefore, the SIRMs of the two input items x_3 and x_4 corresponding to the angle and angular velocity of the shorter pendulum 2 are established as shown in Table 14.5.2.

Suppose that the two pendulums are already balanced upright. In case of positive values of the position and velocity of the cart, if the cart is moved toward right direction by positive driving force, the shorter pendulum 2 will rotate counterclockwise faster because of its higher natural frequency. Consequently, the angle and angular velocity of the shorter pendulum 2 become negative and are bigger than those of the longer pendulum 1 in magnitude. For negative values of the an-

Table 14.5.2: SIRM setting for the shorter pendulum 2

Antecedent x_i ($i = 3, 4$)	Consequent f_i ($i = 3, 4$)
NB	-1
ZO	0
PB	1

Table 14.5.3: SIRM setting for the cart

Antecedent x_i ($i = 5, 6$)	Consequent f_i ($i = 5, 6$)
NB	1
ZO	0
PB	-1

gle and angular velocity of the shorter pendulum 2, the two SIRMs of Table 14.5.2 generate negative driving force to move the cart toward left direction so that the two pendulums are rotated clockwise. When the angle and the angular velocity of the shorter pendulum 2 become positively bigger than those of the long pendulum 1, the two SIRMs of Table 14.5.2 output positive driving force to move the cart toward right direction. As a result, the cart moves right and left alternately, and is likely to move outside gradually. Therefore, if the position and velocity of the cart are positive, negative driving force should first be applied to the cart. After the angle and angular velocity of the short pendulum 2 become positive, positive driving force from the two SIRMs of Table 14.5.2 makes the shorter pendulum 2 rotate counterclockwise. When the angle and angular velocity of the short pendulum 2 turn negative, negative driving force from the two SIRMs of Table 14.5.2 balances the two pendulums toward their upright positions and puts the cart back to the rail origin. On the contrary, if the position and velocity of the cart are negative, positive driving force should be applied first to the cart. Therefore, the SIRMs of the two input items x_5 and x_6 corresponding to the position and velocity of the cart can be given in Table 14.5.3.

As shown in Tables 14.5.1–14.5.3, all the SIRMs infer the output item of the fuzzy controller. The angular control of the shorter pendulum 2 by Table 14.5.2 rotates the shorter pendulum 2 directly toward upright position. However, the angular control of the longer pendulum 1 by Table 14.5.1 makes the longer pendulum

1 fall down further and the position control of the cart by Table 14.5.3 moves the cart away from the rail origin. By this setting of Table 14.5.1 and Table 14.5.3, the shorter pendulum 2 will incline to the same side as the longer pendulum 1 and has a larger angle because of its higher natural frequency. As the shorter pendulum 2 is balanced upright by using Table 14.5.2, then, the longer pendulum 1 is also balanced upright and the cart is moved to the rail origin. From this point of view, the angular control of the longer pendulum 1 and the position control of the cart are realized indirectly.

To make the indirect controls of the longer pendulum 1 and the cart also feasible, the angular control of the longer pendulum 1 and the position control of the cart should be discriminated from the angular control of the shorter pendulum 2. As is well known, the longer pendulum 1 has a bigger momentum, while the shorter pendulum 2 has a higher natural frequency. When the angle of the longer pendulum 1 is big, the longer pendulum 1 will fall down further and make it more difficult to balance the long pendulum 1 upright because of its bigger momentum if the angular control of the longer pendulum 1 is not done immediately. When the angle of the shorter pendulum 2 is big, because of its higher response characteristic, it is relatively easy to move up the shorter pendulum 2 again if relevant control action is executed. Therefore, if the angle of the longer pendulum 1 is big, the angular control of the longer pendulum 1 should be done first with the highest priority so that the shorter pendulum 2 inclines to the same side as the longer pendulum 1. If the angle of the shorter pendulum 2 is big and the two pendulums locate on different sides, then the angular control of the shorter pendulum 2 should be executed so that the two pendulums get inclined to the same side. By performing the angular control of the shorter pendulum 2, the two pendulums are balanced upright. After the two pendulums are almost balanced, the position control of the cart can be started.

In the stabilization control of the parallel-type double inverted pendulum system, therefore, the angular control of the longer pendulum 1 should take the highest priority when its angle is big. When the angle of the shorter pendulum 2 becomes big, the angular control of the shorter pendulum 2 should have the highest priority. The position control of the cart should have the lowest priority before the two pendulums are balanced upright. To make the stabilization control effective, the priority orders of the angular control of the longer pendulum 1, the angular control of the shorter pendulum 2, and the position control of the cart should be reflected in the calculation of the output value of the output item. Equal control priorities will apparently cause contradictions among the angular control of the longer pendulum 1, the angular control of the shorter pendulum 2, and the position control of the cart.

For the longer pendulum 1, the fuzzy rules for the dynamic variables Δw_1 and

Table 14.5.4:   Rule setting for the two dynamic variables of the longer pendulum 1

Antecedent $|x_1|$	Consequent $\Delta w_1, \Delta w_2$
DS	0.0
DM	0.5
DB	1.0

Table 14.5.5:   Rule setting for the two dynamic variables of the shorter pendulum 2

Antecedent $|x_3|$	Consequent $\Delta w_3, \Delta w_4$
DS	0.0
DM	0.5
DB	1.0

Δw_2 of the dynamic importance degrees w_1^D and w_2^D of the input items x_1 and x_2 can be established in Table 14.5.4 by selecting the absolute value of the input item x_1 as the only antecedent variable. By this setting, when the absolute value of the input item x_1 corresponding to the angle of the longer pendulum 1 is big, the inference results of the two dynamic variables will both become big. Therefore, the values of the two dynamic importance degrees will increase so that the angular control of the longer pendulum 1 is emphasized. If the absolute value of the angle of the longer pendulum 1 is close to zero, the inference results of the two dynamic variables will both become almost zero, and the two dynamic importance degrees will approach to their base values. As a result, the influence strength of the angular control of the longer pendulum 1 is weakened.

Similarly for the shorter pendulum 2, the fuzzy rules for the dynamic variables Δw_3 and Δw_4 of the dynamic importance degrees w_3^D and w_4^D of the input items x_3 and x_4 can be set as in Table 14.5.5 by selecting the absolute value of the input item x_3 as the sole antecedent variable. When the absolute value of the input item x_3 corresponding to the angle of the shorter pendulum 2 is big, the inference results of the two dynamic variables will also get big so that the two dynamic importance degrees of the shorter pendulum 2 will increase. When the absolute value of the angle of the shorter pendulum 2 is small, the inference results of the two dynamic variables will also get small so that the two dynamic importance degrees of the shorter pendulum 2 will decrease. Consequently, the control priority order of the shorter pendulum 2 is adjusted according to the situation of its angle.

If the position control of the cart is started before the two pendulums are upright

Table 14.5.6: Rule setting for the two dynamic variables of the cart

Consequent $\Delta w_5, \Delta w_6$		Antecedent $|x_1|$		
		DS	DM	DB
Antecedent $|x_3|$	DS	1.0	0.5	0.0
	DM	0.5	0.0	0.0
	DB	0.0	0.0	0.0

yet, the states of the two pendulums may be destroyed. Therefore, the dynamic variables Δw_5 and Δw_6 of the dynamic importance degrees w_5^D and w_6^D of the input items x_5 and x_6 corresponding to the position and velocity of the cart can be described by the fuzzy rules in Table 14.5.6. Here, the absolute values of the input items x_1 and x_3 are used as the antecedent variables. The real number output of the consequent part is set to 0.0 in those fuzzy rules for $|x_1| = $ DB or $|x_3| = $ DB. By this setting, when the two pendulums are almost stood up, both inference results of the two dynamic variables will become big. As a result, the values of the two dynamic importance degrees of the cart will increase relatively, making the position control of the cart become possible. When one of the two pendulums is still not balanced upright, both the inference results of the dynamic variables will be small. As result, the values of the two dynamic importance degrees of the cart will decrease so that the position control of the cart has low priority order.

To tune automatically the control parameters, the random optimization search [3] is adopted again. Here, the mass and half length of the longer pendulum 1, the mass and half length of the shorter pendulum 2, and the car mass are selected as $m_1 = 0.3$ kg, $l_1 = 0.6$ m, $m_2 = 0.1$ kg, $l_2 = 0.2$ m, and $M = 1.0$ kg. The scaling factors of the input items are set to $15.0°$, $100.0°/\text{sec}$, $15.0°$, $100.0°/\text{sec}$, 2.4 m, 1.0 m/sec, respectively. The scaling factor of the output item is defined as 10.0 times the total mass of the two pendulums and the cart. In each search step, the sampling period and total control time are fixed to 0.01 secs and 25.0 secs. The initial angle of the longer pendulum 1 is set to $5.0°$, while the initial values of the other state variables are set to zeros. All of the base values and the breadths of the dynamic importance degrees are initially set to zeros. The random optimization search is run for $40,000$ steps along such a direction, that the total summation of the absolute values of all state variables and the driving force at each sampling time from the beginning to the end of the total control time is reduced. The base values and the breadths after the random optimization search are shown in Table 14.5.7. As it can be seen from Table 14.5.7, the sum of the base value and the breadth of either input item of the longer pendulum 1 is larger than that of

Table 14.5.7: Control parameters of the dynamic importance degrees

Input item	Base value	Breadth
x_1	2.3694	0.5278
x_2	3.5874	0.0578
x_3	1.9398	0.5298
x_4	1.4012	1.2148
x_5	0.3281	0.0000
x_6	0.0328	0.2690

any other input items. The sum of the base value and the breadth of either input item of the shorter pendulum 2 is much larger than that of either input item of the cart. Apparently, the control parameters reflect the necessary control priority orders very well.

14.5.2 Control Simulation Results

To verify the effectiveness of the proposed stabilization fuzzy controller, several control simulations are done. Figures 14.5.2, 14.5.3, and 14.5.4 show the control results when the initial angles α and β of the two pendulums are respectively set to 5.0° and 0.0° (Example 1), 5.0° and 5.0° (Example 2), 5.0° and 10.0° (Example 3), while the initial values of the other state variables are all fixed to zeros. The initial state in Figure 14.5.2 corresponds to that used in the random optimization search. The left axis and the right axis represent the angle of the pendulums and the position of the cart. The values in $P(0.30, 0.60, 0.10, 0.20, 1.00, 0.01)$ mean in this order the mass and half length of the longer pendulum 1, the mass and half length of the shorter pendulum 2, the mass of cart, and the sampling period. The values in $S(5.0, 0.0, 10.0, 0.0, 0.0, 0.0)$ denote the initial values of the angle and angular velocity of the longer pendulum 1, the angle and angular velocity of the shorter pendulum 2, the position and velocity of the cart, respectively.

In Figure 14.5.2, the angular control of the longer pendulum 1 takes the priority over the other two controls at the beginning of control action because the initial angle of the longer pendulum 1 is bigger than the initial angle of the shorter pendulum 2. Negative driving force is first generated based on the SIRM of the angle of the longer pendulum 1, and moves the cart to the left. Although the longer pendulum 1 falls down further from 5.0° to 11.0° as a result, the shorter pendulum 2 becomes inclined to about 17.0°. Because the contribution of the shorter pendulum 2 exceeds that of the longer pendulum 1 in the driving force from that point, the angular control of the shorter pendulum 2 becomes the main objective.

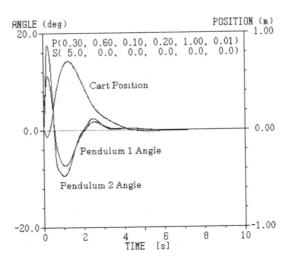

Figure 14.5.2: Control simulation of Example 1.

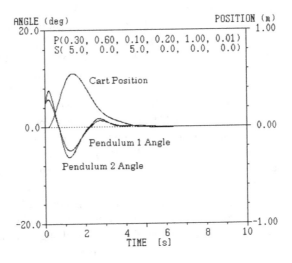

Figure 14.5.3: Control simulation of Example 2.

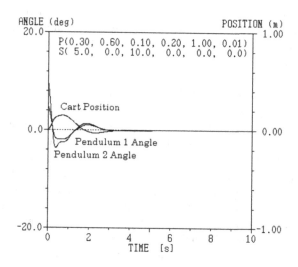

Figure 14.5.4: Control simulation of Example 3.

Consequently, the cart is moved to the right by positive driving force of a maximum about 25.0 N, so that the two pendulums get synchronized to rotate in the same direction. Finally, the two pendulums are gradually balanced to their upright positions and the cart is returned to the rail origin. If the time interval from control action's beginning to such a position that all the state variables converge to $0.1°$, $0.1°/\text{sec}$, $0.1°$, $0.1°/\text{sec}$, 0.01 m, 0.01 m/sec is defined as complete stabilization time, the complete stabilization time of Figure 14.5.2 is about 6.23 secs.

In Figure 14.5.3, although the initial angles of the two pendulums are the same, the angular control of the longer pendulum 1 is done first at the beginning of control action because the two dynamic importance degrees of the longer pendulum 1 are larger. Since negative driving force is generated for positive angle of the longer pendulum 1, the cart is moved a little to the left (although the distance moved is too short to illustrate). Consequently, the angle of the longer pendulum 1 increases to about $6.0°$, and the angle of the shorter pendulum 2 increases to about $8.0°$. From that point, the angular control of the shorter pendulum 2 comes to take the highest priority because the shorter pendulum 2 contributes more to the driving force than the longer pendulum 1. Therefore, positive driving force is applied to move the cart right, and the two pendulums begin to rotate with synchronization. By switching the three controls with control situations smoothly, the pendulum system is finally completely stabilized in about 6.12 secs. The maximum driving force is only about 3.0 N in this case.

As the initial state in Figure 14.5.4, the two pendulums incline to the same side and the initial angle of the shorter pendulum 2 is twice as large as that of the longer pendulum 1. According to the SIRM setting of the two angles, the inference result of the SIRM corresponding to the initial angle of the shorter pendulum 2 becomes also twice as large as that corresponding to the initial angle of the longer pendulum 1. From Tables 14.5.4, 14.5.5, and 14.5.7, the value of the dynamic importance degree corresponding to the initial angle of the shorter pendulum 2 is almost equal to that corresponding to the initial angle of the longer pendulum 1. Therefore, at the beginning of control action the angular control of the shorter pendulum 2 is strengthened first so that the cart is moved to the right by positive driving force. As a result, the two pendulums keep synchronized from the beginning of control action and converge gradually to their upright positions. In this case, the complete stabilization time is about 5.29 secs and the maximum driving force is less than 10.0 N.

Figure 14.5.5 shows the stabilization domain of the initial angles of the two pendulums, for which the proposed fuzzy controller can stabilize the pendulum system. Here, the horizontal axis and the vertical axis stand for the initial angles of the longer pendulum 1 and the shorter angle 2. The initial angles of the two pendulums both are selected every $1.0°$ from $-20.0°$ to $+20.0°$, and the initial values of the other state variables are all fixed to zeros. The symbols •, ○, ■, □ mean that the complete stabilization time is within 4.0 secs, 6.0 secs, 8.0 secs, and 10.0 secs, respectively. Furthermore, the failure limits of the angles of the two pendulums and the position of the cart are set to $[-20.0°, +20.0°]$, $[-20.0°, +20.0°]$, and $[-2.4 \text{ m}, +2.4 \text{ m}]$, respectively. If any of the failure limits is reached during a simulation, the simulation is regarded as a failure.

As it can be seen from Figure 14.5.5, for all sets of the initial angles of the two pendulums in the stabilization domain, the proposed fuzzy controller can completely stabilize the pendulum system in 10.0 secs. For several sets of the initial angles, the fuzzy controller can even stabilize the pendulum system in 4.0 secs. For the longer pendulum 1 with an initial angle between $[-5.0°, +5.0°]$, if the shorter pendulum 2 initially inclines further for up to $5.0°$ beyond the angle of longer pendulum 1, the pendulum system can be stabilized completely in 6.0 secs. For the longer pendulum 1 with an initial angle between $[-10.0°, +10.0°]$, if the initial angle of the shorter pendulum 2 is selected in a range of totally about $10.0°$ around the initial angle of the longer pendulum 1, the stabilization control of the pendulum system is possible.

It is also found that the stabilization domain in Figure 14.5.5 as a whole leans

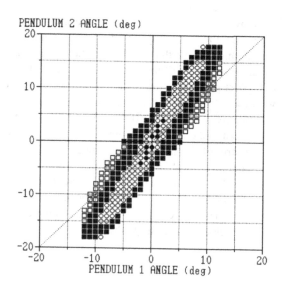

Figure 14.5.5: Stabilization domain.

a little upon the 45° line, i.e., toward the vertical axis of the angle of the shorter pendulum 2. The domain under the line means that the initial angle of the longer pendulum 1 is larger than the initial angle of the shorter pendulum 2. For a set of the initial angles in the domain under the 45° line, the proposed fuzzy controller will first do the angular control of the longer pendulum 1. As a result, the shorter pendulum 2 gets synchronized with the longer pendulum 1 and the angle of the shorter pendulum 2 becomes bigger than that of the longer pendulum 1. In this case, since the driving force gets large with the increase of the difference between the initial angles of the two pendulums, the shorter pendulum 2 will swing intensively and fall down because of its higher response characteristic. On the other hand, the domain above the 45° line means that the initial angle of the shorter pendulum 2 is larger than the initial angle of the longer pendulum 1. For a set of initial angles in the domain above the 45° line, if the initial angle of the shorter pendulum 2 is rather close to that of the longer pendulum 1, the fuzzy controller will still give the highest priority to the angle control of the longer pendulum 1. Because the driving force is small in this case, however, the angle of the shorter pendulum 2 will not increase too much. If the initial angle of the shorter pendulum 2 is much larger than that of the longer pendulum 1, the fuzzy controller will first do the angular control of the shorter pendulum 2 directly at the beginning of control action. Consequently, the shorter pendulum 2 will rotate toward its upright position rather than rotating outside, and the two pendulums keep synchronized from the beginning of control action. In this case, the stabilization control is relatively easy to be realized. Therefore, the stabilization domain concentrates mainly in the part above the 45° line in the first quadrant.

14.6   Conclusions

The SIRMs dynamically connected fuzzy inference model is newly proposed for complex systems. Each input item is assigned with a SIRM and a dynamic importance degree. The dynamic importance degree is defined as the sum of a base value insuring the role of the input item throughout a control process, and a dynamic value varying with control situations to adjust the influence of the input item. Because the dynamic importance degrees are allowed to change with control situations, the proposed model is situation variant. By using the SIRMs and the dynamic importance degrees, switching of the control priority orders of the input items is realized automatically according to control situations. Based on the SIRMs dynamically connected fuzzy inference model, the fuzzy controllers for the backing-up control of the truck-trailer system, the stabilization control of the ball-beam system, and the stabilization control of the parallel-type double inverted pendulum system are designed in detail. These underactuated control plants are difficult to control, especially for the conventional fuzzy model. The SIRMs dynamically connected fuzzy inference model, however, is applicable to such complex systems and needs less fuzzy rules compared with conventional fuzzy inference models. Furthermore, each fuzzy controller based on the SIRMs dynamically connected fuzzy inference model has a simple structure that is easy to understand, and can perform the computation in parallel. For the truck-trailer system, the presented fuzzy controller has three input items, and can even back up the truck-trailer system successfully from a jackknife state. For the ball-beam system, the presented fuzzy controller deals with four input items, and can stabilize the ball-beam system within 9.0 secs for almost all the initial states. For the parallel-type double inverted pendulum system, the presented fuzzy controller handles six input items, and can completely stabilize the system in 10.0 secs for those initial states locating in the stabilization domain. This is the first result for a fuzzy controller to realize complete stabilization control of a parallel-type double inverted pendulum system.

Bibliography

[1] J. R. Alexander, J. P. Coughlin, and J. T. Cox, "A three-neuron computer II," *Proc. IEEE Conference on Systems, Man and Cybernetics*, Beijing, China, Oct. 1996, vol. 1, pp. 384–389.

[2] C. Altafini, A. Speranzon, and B. Wahlberg, "A feedback control scheme for reversing a truck and trailer vehicle," *IEEE Transactions on Robotics and Automation*, vol. 17, no. 6, pp. 915–922, 2001.

[3] N. Baba, "A new approach for finding the global minimum of error function of neural networks," *Neural Networks*, vol. 2, pp. 367–373, 1989.

[4] G. Chen and D. Zhang, "Backing up a truck-trailer with suboptimal distance trajectories," *Proc. IEEE Conference on Fuzzy Systems*, New Orleans, LA, Sept. 1996, vol. 2, pp. 1439–1445.

[5] C. Chen, B. Liu, and J. Tsao, "Adaptive fuzzy logic controller blending the concepts of linguistic hedges and genetic algorithms," *Proc. IEEE Conference on Fuzzy Systems*, Seoul, South Korea, Aug. 1999, vol. 3, pp. 1299–1304.

[6] K. Fujita and M. Mizumoto, "Fuzzy controls of parallel inverted-pendulum under fuzzy singleton-type reasoning method using genetic algorithm," *Proc. 11th Japan Fuzzy System Symposium*, Okinawa, Japan, July 1995, pp. 379–382.

[7] J. Godjevac and N. Steele, "Adaptive fuzzy controller for robot navigation," *Proc. IEEE Conference on Fuzzy System*, New Orleans, LA, Sept. 1996, vol. 1, pp. 136–142.

[8] R. F. Harrison, "Asymptotically optimal stabilising quadratic control of an inverted pendulum," *IEE Proceedings-Control Theory and Applications*, vol. 150, no. 1, pp. 7–16, 2003.

[9] J. Hauser, S. Sastry, and P. Kokotovic, "Nonlinear control via approximate input-output linearization: the ball and beam Example," *IEEE Transactions on Automatic Control*, vol. 37, no. 3, pp. 392–398, 1992.

[10] S. Hori, K. Tanaka, and H. O. Wang, "Multi-objective structure design for mechanical systems," *Proc. IEEE Conference on Fuzzy Systems*, Honolulu, Hawaii, May 2002, vol. 1, pp. 116–121.

[11] M. G. Joo and J. S. Lee, "A class of hierarchical fuzzy systems with constraints on the fuzzy rules," *IEEE Transactions on Fuzzy Systems*, vol. 13, no. 2, pp. 194–203, 2005.

[12] R. Kawatani and T. Yamaguchi, "Analysis and stabilization of a parallel-type double inverted pendulum system," *Transactions of the Society of Instrument and Control Engineers*, vol. 29, no. 5, pp. 572–580, 1993.

[13] C. Lin and C. Lin, "Reinforcement learning for ART-based fuzzy adaptive learning control networks," *Proc. Joint Fourth IEEE Conference on Fuzzy Systems and Second International Fuzzy Engineering Symposium*, Yokohama, Japan, Mar. 1995, vol. 3, pp. 1299–1306.

[14] M. Maeda, T. Sato, and S. Murakami, "Design of the self-tuning fuzzy controller," *Proc. International Conference on Fuzzy Logic & Neural Networks*, Iizuka, Japan, July 1990, pp. 393–396.

[15] M. Maeda, M. Tsubone, and M. Murakami, "Fuzzy learning control with multi-unit controller," *Proc. 9th Japan Fuzzy System Symposium*, Sapporo, Japan, May 1993, pp. 285–288.

[16] E. H. Mamdani, "Application of fuzzy algorithms for control of simple dynamic plant," *Proceedings of the Institution of Electrical Engineers*, vol. 121, pp. 1585–1588, 1974.

[17] M. Margaliot and G. Langholz, "Adaptive fuzzy controller design via fuzzy Lyapunov synthesis," *Proc. IEEE Conference on Fuzzy Systems*, Anchorage, AK, May 1998, vol. 1, pp. 354–359.

[18] M. Mizumoto, "Fuzzy controls under new fuzzy reasoning methods," *Proc. Joint Hungarian-Japanese Symposium on Fuzzy Systems and Applications*, Budapest, Hungary, July 1991, pp. 122–126.

[19] M. Mizumoto, "Improvement of fuzzy controls (VII) - Time-variant fuzzy controls by fuzzy singleton-type reasoning method," *Proc. Japan 10th Fuzzy System Symposium*, Osaka, Japan, July 1994, pp. 445–448.

[20] M. Mizumoto, "Products-sum-gravity method = Fuzzy singleton-type reasoning method = Simplified fuzzy reasoning method," *Proc. IEEE Conference on Fuzzy Systems*, New Orleans, LA, Sept. 1996, vol. 3, pp. 2098–2102.

[21] D. H. Nguyen and B. Widrow, "Neural networks for self-learning control systems," *IEEE Control Systems Magazine*, vol. 10, no. 3, pp. 18–23, 1990.

[22] A. Otsubo, K. Hayashi, S. Murakami, and M. Maeda, "Improvement of control performance for tow-dimensional number of fuzzy labellings applying simplified inference method," *Proc. International Joint Conference of CFSA/IFIS/SOFT on Fuzzy Theory and Applications*, Taipei, Taiwan, Dec. 1995, pp. 151–157.

[23] S. Shen, F. Yu, and H. Chung, "Decoupled fuzzy control design with single-input fuzzy logic," *Fuzzy Sets and Systems*, vol. 129, pp. 335–342, 2002.

[24] T. Sugie and M. Okada, "Control of a parallel inverted pendulum system," *Transactions of the Institute of Systems, Control and Information Engineers*, vol. 6, no. 12, pp. 543–551, 1993.

[25] K. Tanaka and M. Sano, "A robust stabilization problem of fuzzy control systems and its application to backing up control of a truck-trailer," *IEEE Transactions on Fuzzy Systems*, vol. 2, no. 2, pp. 119–134, 1994.

[26] T. Taniguchi, K. Tanaka, and H. O. Wang, "Model construction, rule reduction, and robust compensation for generalized form of Takagi-Sugeno fuzzy systems," *IEEE Transactions on Fuzzy Systems*, vol. 9, no. 4, pp. 525–538, 2001.

[27] L. Wang, "Stable and optimal fuzzy control of linear systems," *Proc. IEEE Conference on Fuzzy Systems*, New Orleans, LA, Sept. 1996, vol. 2, pp. 1453–1458.

[28] Q. Wang, M. Mi, G. Ma, and P. Spronck, "Evolving a neural controller for a ball-and-beam system," *Proc. International Conference on Machine Learning and Cybernetics*, Shanghai, China, Aug. 2004, vol. 2, pp. 757–761.

[29] Z. Yeh and H. Chen, "Multi-state inference fuzzy logic control," *Proc. IEEE Conference on Fuzzy Systems*, Barcelona, Spain, July 1997, vol. 2, pp. 1153–1158.

[30] J. Yi, N. Yubazaki, and K. Hirota, "Stabilization fuzzy control of inverted pendulum systems," *Artificial Intelligence in Engineering*, vol. 14, no. 2, pp. 153–163, 2000.

[31] J. Yi, N. Yubazaki, and K. Hirota, "Backing up control of truck-trailer system," *Proc. IEEE Conference on Fuzzy Systems*, Melbourne, Australia, July 2001, vol. 1, pp. 489–492.

[32] J. Yi, N. Yubazaki, and K. Hirota, "Stabilization control of ball and beam systems," *Proc. Joint 9th IFSA World Congress and 20th NAFIPS International Conference*, Vancouver, Canada, July 2001, vol. 4, pp. 2229–2234.

[33] J. Yi, N. Yubazaki, and K. Hirota, "Upswing and stabilization control of inverted pendulum system based on the SIRMs dynamically connected fuzzy inference model," *Fuzzy Sets and Systems*, vol. 122, no. 1, pp. 135–152, 2001.

[34] J. Yi, N. Yubazaki, and K. Hirota, "A proposal of SIRMs dynamically connected fuzzy inference model for plural input fuzzy control," *Fuzzy Sets and Systems*, vol. 125, no. 1, pp. 79–92, 2002.

[35] J. Yi, N. Yubazaki, and K. Hirota, "A new fuzzy controller for stabilization of parallel-type double inverted pendulum system," *Fuzzy Sets and Systems*, vol. 126, no. 1, pp. 105–119, 2002.

[36] H. Ying and L. C. Sheppard, "Tuning parameters of the fuzzy controller based on the golden section search," *Proc. NAFIPS International Conference*, Toronto, Canada, June 1990, pp. 86–89.

[37] N. Yubazaki, M. Otani, T. Ashida, and K. Hirota, "Dynamic fuzzy control method and its application to positioning of induction motor," *Proc. Joint Fourth IEEE Conference on Fuzzy Systems and Second International Fuzzy Engineering Symposium*, Yokohama, Japan, Mar. 1995, vol. 3, pp. 1095–1102.

[38] N. Yubazaki, J. Yi, and K. Hirota, "SIRMs (single input rule modules) connected fuzzy inference model," *Journal of Advanced Computational Intelligence*, vol. 1, no. 1, pp. 23–30, 1997.

[39] N. Yubazaki, J. Yi, M. Otani, and K. Hirota, "SIRM's connected fuzzy inference model and its applications to first-order lag systems and second-order lag systems," *Proc. Asian Fuzzy Systems Symposium*, Kenting, Taiwan, Dec. 1996, pp. 545–550.

[40] H. Zhang, S. Liu, and S. Yang, "A neurodynamics based neuron-PID controller and its application to inverted pendulum," *Proc. International Conference on Machine Learning and Cybernetics*, Shanghai, China, Aug. 2004, vol. 1, pp. 527–532.

[41] J. Zhang, R. Li, and P. Zhang, "Stability analysis and systematic design of fuzzy control system," *Fuzzy Sets and Systems*, vol. 120, pp. 65–72, 2001.

[42] Y. Zheng, S. Luo, Z. Lv, and L. Wu, "Control parallel double inverted pendulum by hierarchical reinforcement learning," *Proc. 7th International Conference on Signal Processing*, Istanbul, Turkey, Dec. 2004, vol. 2, pp. 1614–1617.

Index